Spon's House Improvement Price Book

Spon's House Improvement Price Book

House extensions, storm damage work, Alterations, loft conversions and insulation

Edited by

Bryan Spain

First edition published 1998 by E & FN Spon

This edition published 2003 by Spon Press
11 New Fetter Lane, London EC4P 4EE

Spon Press is an imprint of the Taylor & Francis Group

Printed and bound in Great Britain by MPG Books Ltd, Bodmin

British Library Cataloguing in Publication Data
A catalogue record for this book is available
from the British Library

This book has been prepared from camera-ready copy provided by the author.

Library of Congress Cataloging in Publication Data
Spon's house improvement price book : house extensions, storm damage work, alterations, loft conversions, and insulation / edited by Bryan Spain. – [2nd ed.].
p. cm.
Includes bibliographical references and index.
ISBN 0–415–30938–7
1. Dwellings—Remodelling—Estimates. 2. Dwellings—Remodelling—Costs.
I. Spain, Bryan J. D. II. E & F.N. Spon.

TH4816 .S66 2003
643′.7—dc21

2002036510

Contents

One storey, flat roof (cont'd)

One storey, flat roof (cont'd)

One storey, flat roof (cont'd)

One storey, pitched roof

One storey, pitched roof (cont'd)

One storey, pitched roof (cont'd)

One storey, pitched roof (cont'd)

Two storey, flat roof

Two storey, flat roof (cont'd)

Two storey, flat roof (cont'd)

Two storey, flat roof (cont'd)

Two storey, pitched roof

Two storey, pitched roof (cont'd)

Two storey, pitched roof (cont'd)

Two storey, pitched roof (cont'd)

Loft conversions (cont'd)

Damage Repairs (cont'd)

Preface

This is the second edition of Spon's House Improvement Price Book and is published because of the strong interest shown in the industry, particularly among contractors involved in the domestic construction market, in the first edition. The preface to the first edition forecast that the benefits contractors would receive by using the book, would be both the saving in time in preparing estimates, and the accurate listings of construction costs for a wide range of extension sizes and types. The feedback received by the publisher and the author has proved this forecast to be correct.

In this edition, the costs of house extensions and loft conversions have been updated and altered to include the changes in the Building Regulations governing insulation requirements and the insulation section has been enlarged to cover these changes. In addition, a new section is included to cover the cost of damage caused by fire, floods, gales and theft together with the cost of providing emergency measures such as pumping, tarpaulins and the like.

The costs of the extensions and loft conversions are summarized at the end of their respective sections and presented in a style to show the percentage value of each element together with a cost per square metre for each project. Our research has shown that this data was regarded by users of the first edition as the most helpful in the book.

It will be noted that an index has not been included in the book. I felt that providing an index that repeatedly referred to the same items in the 40 house extensions and 3 loft conversions would not be helpful to the reader. For example, listing 40 page numbers for trench digging would not fulfill the normal function of an index and would probably be counter-productive. Instead a comprehensive contents list is included and this should enable the reader to obtain the maximum benefit from using the book.

I have received great deal of support in the research necessary for this book and I am grateful to those individuals and firms who helped me. I am particularly indebted to John Craggs for his advice on current building regulations. Although every care has been taken in the preparation of the book's contents, neither the publishers nor I can accept any responsibility for the use of the information made by any person or firm. Finally, I would welcome constructive criticism of the book together with suggestions on how it could be improved for the next edition.

Bryan Spain
January 2003

Foreword

I am pleased to have been invited to contribute a foreword to the second edition of Spon's House Improvement Price Book. This type of guide proves both perennially useful and popular with the trade. Having the right information at your fingertips is absolutely crucial, in more ways than one.

There is a growing awareness of the threat to the entire industry and wider public confidence posed by the activities of 'cowboy' builders operating in the domestic repair, maintenance and improvement market. Their poor quality work and irresponsible approach to profiteering from homeowners' need for vital building work preys on the vulnerable elderly and those who can least afford to lose their savings and see work botched in their homes – those most frequently at risk of opting for the cheapest quote offered by these disreputable builders.

The Office of Fair Trading and local council Trading Standards offices all around the country are witnessing a continuous wave of complaint from the public about the damaging activity of the rogue, unsuitable builders. So implementing and operating the Government-backed Quality Mark Scheme to tackle this is of vital importance.

A diverse and widely spread industry such as the building trade, with its huge range of specialities, plethora of small unrelated firms and wide array of trade associations, federations and guilds could not in all fairness be expected to tackle this problem in a totally united and suitably focused way without the direct help and direction of Government.

The Quality Mark Scheme was piloted last year, after detailed consultation and with the invaluable assistance of consumer organisations, representatives of the construction industry and local councils. Lessons have been learned and fine-tuning has been taking place as Quality Mark is rolled out across the regions over the next three to four years, offering the right kind of solution to the problem posed by the rogue builders.

By independently and professionally auditing technical competence with site visits, along with checking business and customer management systems, Quality Mark is able to deliver the protection and peace of mind the public needs to have before hiring a builder to work on their home, safe in the knowledge the job will be done properly, within a specific time frame, represent good value for money and that they will be treated fairly and openly by the workmen they allow through their doors. A standard six-year warranty on all work that comes as part of the Quality Mark package is a pretty impressive selling point too.

Quality Mark also offers general advice to consumers on what to expect from a decent builder, the importance of specifying exactly what work is to be undertaken at the outset, and how to get a clear and straightforward contract in place before work begins. This avoids confusion on both sides and generates better-informed consumers who can turn the market in favour of reputable operators.

By the same token, professional tradesmen who join the scheme are enjoying the benefits of increased turnover and profitability, a broader client base, often improved business procedures and significant cost savings on advertising, improved long-term business prospects, a demonstrable edge over their competitors and the ability to show their professionalism to consumers by gaining Quality Mark approval.

Quality Mark is truly a Scheme that everyone stands to benefit from. This will happen as the public lose their fear of having work done because the growth of Quality Mark reduces the risk to them. At the same time, the industry enhances and improves its reputation by participation in the Scheme and will see business grow. The implications for the economy are obvious and one of the reasons why the government is so committed to the long-term success of Quality Mark. Another important reason was Government recognition that something positive had to be done about the level of consumer complaint and that statutory action on such a diverse industry of small businesses would prove impractical in terms of cost, bureaucracy and enforcement. Quality Mark offers the right degree of sensible self-regulation needed to get the job done properly.

Ultimately, as knowledge of Quality Mark spreads, discerning customers will start to put increased pressure on the building trade to become accredited to the Scheme. Those who lead by example through joining up and advocating the benefits they find from belonging to the Scheme to their colleagues will naturally be the first to benefit from the long-term growth and prosperity implications of Quality Mark.

The wealth of useful information you will find contained within the pages of this book – specific data on the cost of building extensions, loft conversions and insulation work – is yet another example of what Quality Mark is all about: making informed choices with the best available information available to you.

Brian Wilson MP
Minister of State for Energy and Construction

Introduction

The contents of this book cover the cost of domestic construction work in the following areas:

- extensions
- loft conversions
- insulation work
- damage repairs
- alterations.

The house extensions section comprises 40 different sizes and types of extensions ranging from 2 x 3 m to 2 x 6 m, flat roofs and pitched roofs, and one storey and two storey in height. Each extension is presented in a bill of quantities format and individual rates are broken down into labour hours, labour costs, material costs, overheads and profit.

Each extension contains a maximum of thirteen elements:

- preliminaries
- substructure
- external walls
- flat roof*
- pitched roof*
- windows and external doors
- internal partitions and doors*
- wall finishes
- floor finishes
- ceiling finishes
- electrics
- heating work
- alteration work.

* where applicable

Each element has separate totals for labour hours, labour costs, material costs, overheads and profit. These elemental totals are carried to a general summary at the end of each extension to produce final totals.

For example, on page 116 the total cost of a one storey extension size 2 x 3 m with a pitched roof is £6,708.21. This is made up of £895.50 for preliminaries, £1901.67 for labour, £3,152.88 for materials, and £758.18 for overheads and profit. It should be noted that the horizontal and vertical totals do not always coincide due to rounding off.

The total time is shown as 198.39 hours but it should not be assumed that by dividing this figure by 37.5 it would produce the number of weeks needed to construct the work. This is due to many factors including different trades working simultaneously and the inevitable gaps that occur on small projects between the completion of one package of work and the commencement of the next.

The labour rates are based upon current wage awards and are £14.00 for electricians and £12.50 for plumbers. All other work is based upon an all-in rate of £9.50 per hour. This is because the demarcation between trades is less clearly defined on small domestic construction projects.

Further, it is common practice to pay for labour on a daily basis rather an hourly rate and it was felt that using a single all-in hourly labour rate would not only reflect current practice, but would also make it easier for the users of the book to make their own adjustments to the labour rate if necessary.

The purchase of materials at competitive rates can be difficult for small contractors. They cannot obtain large discounts because the size of their orders is generally too small and they often experience a higher than average percentage of waste by not always using all the materials in large packs. Nevertheless, it has been assumed that the contractor has access to facilities to store surplus materials and this should reduce the incidence of high wastage.

The total cost of the one storey extension size 2 x 3 m with a pitched roof on page 116 is transferred to the Summary of Extension Costs on page 355 and entered as £6,709. Each element in the extension is expressed as a percentage of the total cost. The floor area is stated as 4.28 square metres (that is the area inside the external walls) and the cost per square metre is shown as £1,568. This high figure is caused by the small floor area – in larger extensions this figure is reduced to less than £700 per square metre.

Access to square metre prices for a wide range of sizes and types of extensions can be invaluable to a small contractor, particularly in dealing with casual enquiries where the preparation of time-consuming quotations can be expensive. In general terms, the standard of finish is basic but a list of alternative items is included on page 361 so that adjustments can be made to standard extension.

There are eleven elements in the loft conversion section:

preliminaries
preparation
dormer window
roof window

stairs
flooring
partitions and doors
ceilings and soffits
electrics
heating work.

The costs of these elements are transferred to a summary on page 407. Square metre prices are shown in the summary in the same way as in the extensions section. Drawings are included for both the extensions and the loft conversions but they are not drawn to scale and are for illustrative purposes only.

Changes to the building regulations came into effect on 1 April 2002 and three methods of complying with U-values were introduced: Elemental Method, Target U-value Method and the Carbon Index Method.

The Elemental Method is the most relevant to house extension and loft conversion work and the following U-value targets were set:

pitched roofs with insulation between rafters	0.20
pitched roofs with insulation between joists	0.16
flat roofs	0.25
walls	0.35*
	0.30**
floors	0.25

* England and Wales
** Scotland

Materials to meet these targets are included in this book but contractors should seek specialist advice if necessary.

A new section is included in this edition to cover the cost of damage to property caused by fire, flood, gale and theft including some costs of emergency measures. A combination of variations in extreme weather conditions, rising crime levels and an increase in insurance claims has placed demands on small contractors undertaking this type of work. It is hoped that the cost information provided in this section will help those involved in this sector of the industry.

The final section deals with alteration work including the forming and filling in of openings and repair work generally.

Part One

HOUSE EXTENSIONS

	Ref	Qty	Hours	Hours £	Mat'ls £	O & P £	Total £
PART A PRELIMINARIES							
Concrete mixer	A1	wk					45.00
Small tools	A2	wk					35.00
Scaffolding (m2/weeks)	A3	m2/wk					2.25
Skip	A4	wk					100.00
Clean up	A5	hour					8.00
PART B SUBSTRUCTURE TO DPC LEVEL							
Excavate topsoil 150mm thick by hand and deposit on site	B1	m2	0.30	2.85	0.00	0.43	3.28
Excavate to reduce levels by hand and deposit on site	B2	m3	2.50	23.75	0.00	3.56	27.31
Excavate for trench foundations by hand and deposit on site	B3	m3	2.60	24.70	0.00	3.71	28.41
Earthwork support to sides of trenches	B4	m3	0.40	3.80	1.20	0.75	5.75
Backfilling to sides of trenches with excavated material	B5	m2	0.60	5.70	0.00	0.86	6.56
Hardcore filling in bed 225mm thick blinded with sand to receive damp-proof membrane	B6	m2	0.20	1.90	5.08	1.05	8.03
Hardcore filling to sides of trenches	B7	m3	0.50	4.75	5.03	1.47	11.25
Concrete grade 11.5Nmm2 (1:3:6) in foundations	B8	m3	1.35	12.83	67.94	12.11	92.88
Concrete grade 25Nmm2 (1:2:4) in sub-floor 150mm thick	B9	m2	0.30	2.85	10.88	2.06	15.79

	Ref	Qty	Hours	Hours £	Mat'ls £	O & P £	Total £
Concrete grade 25Nmm2 (1:2:4) in filling to hollow wall	B10	m2	0.20	1.90	3.63	0.83	6.36
Polythene damp-proof membrane	B11	m2	0.04	0.38	0.58	0.14	1.10
Steel fabric reinforcement ref A193 in strip foundations	B12	m2	0.12	1.14	1.30	0.37	2.81
Steel fabric reinforcement ref A193 in slab	B13	m2	0.15	1.43	1.30	0.41	3.13
Solid blockwork 140mm thick in skin of cavity wall	B14	m2	1.30	12.35	12.78	3.77	28.90
Common bricks (£180 per 1000) 112.5mm thick in skin of cavity wall	B15	m2	1.70	16.15	13.19	4.40	33.74
Facing bricks (£340 per 1000) 112.5mm thick in skin of cavity wall	B16	m2	1.80	17.10	23.27	6.06	46.43
Form cavity 50mm wide with 5 nr butterfly wall ties per m2 in cavity wall	B17	m2	0.03	0.29	0.60	0.13	1.02
Pitch polymer-based damp-proof course 112mm wide	B18	m2	0.05	0.48	0.81	0.19	1.48
Pitch polymer-based damp-proof course 140mm wide	B19	m2	0.06	0.57	0.97	0.23	1.77
Cut, tooth and bond 140mm thick blockwork to existing faced wall	B20	m2	0.44	4.18	2.18	0.95	7.31
Cut, tooth and bond half brick to existing faced wall	B21	m2	0.35	3.33	2.04	0.80	6.17
50mm thick Polyform Plus insulation board or similar	B22	m2	0.30	2.85	4.48	1.10	8.43
PART C EXTERNAL WALLS							
Solid blockwork 140mm thick in skin of cavity wall	C1	m2	1.30	11.70	12.72	3.66	28.08

	Ref	Qty	Hours	Hours £	Mat'ls £	O & P £	Total £
Facing bricks (£340 per 1000) 112.5mm thick in skin of cavity wall	C2	m2	1.80	17.10	23.06	6.02	46.18
75mm thick Crown Dritherm or similar as cavity filling	C3	m2	0.22	2.09	5.82	1.19	9.10
Standard galvanised steel lintel 256mm high, 2400mm long to hollow wall	C4	nr	0.25	2.38	98.46	15.13	115.96
Standard galvanised steel lintel 256mm high, 1500mm long to hollow wall	C5	nr	0.20	1.90	107.47	16.41	125.78
Standard galvanised steel lintel 256mm high, 1150mm long to hollow wall	C6	nr	0.15	1.43	47.18	7.29	55.90
Close cavity of hollow wall at jambs	C7	m	0.05	0.48	2.04	0.38	2.89
Close cavity of hollow wall at cills	C8	m	0.05	0.48	2.04	0.38	2.89
Close cavity of hollow wall at top	C9	m	0.05	0.48	2.04	0.38	2.89
Pitch polymer-based damp-proof course 112mm wide at jambs	C10	m	0.05	0.48	0.81	0.19	1.48
Pitch polymer-based damp-proof course 112mm wide at cills	C11	m	0.05	0.48	0.81	0.19	1.48
PART D FLAT ROOF							
200 x 50mm sawn softwood joists	D1	m	0.25	2.38	3.10	0.82	6.30
200 x 50mm sawn softwood sprocket pieces 500mm long	D2	nr	0.14	1.33	1.34	0.40	3.07

	Ref	Qty	Hours	Hours £	Mat'ls £	O & P £	Total £
18mm thick WPB grade plywood roof decking fixed to roof joists	D3	m2	0.90	8.55	7.86	2.46	18.87
50mm wide sawn softwood tapered firring pieces average depth 50mm	D4	m	0.18	1.71	2.26	0.60	4.57
High density polyethylene vapour barrier 150mm thick	D5	m2	0.20	1.90	9.47	1.71	13.08
100 x 75mm sawn softwood wall plate bedded in cement mortar	D6	m	0.30	2.85	2.22	0.76	5.83
100 x 75mm sawn softwood tilt fillet	D7	m	0.25	2.38	2.64	0.75	5.77
Build in ends of 150 x 50mm softwood joists to existing faced wall	D8	nr	0.35	3.33	0.15	0.52	4.00
Rake out joint of existing faced brick wall to receive flashing and point up on completion	D9	m	0.35	3.33	0.15	0.52	4.00
6mm thick asbestos-free insulation board soffit 150mm wide	D10	m	0.40	3.80	1.75	0.83	6.38
19mm thick wrought softwood fascia board 200mm high	D11	m	0.50	4.75	1.97	1.01	7.73
Three layer polyester-based mineral-surfaced roofing felt	D12	m2	0.55	5.23	8.41	2.05	15.68
Turn down to edge of roof 100mm girth	D13	m	0.10	0.95	0.84	0.27	2.06
Flashing 150mm girth including dressing over tilt fillet	D14	m	0.10	0.95	1.68	0.39	3.02
PVC-U gutter, 112mm half round with gutter union joints, fixed to softwood fascia board with support brackets at 1m maximum centres	D15	m	0.26	2.47		0.37	2.84

	Ref	Qty	Hours	Hours £	Mat'ls £	O & P £	Total £
Extra for stop end	D16	nr	0.14	1.33	1.11	0.37	2.81
Extra for stop end outlet	D17	nr	0.25	2.38	1.11	0.52	4.01
PVC-U down pipe, 68mm diameter, loose spigot and socket joints, plugged to faced brickwork with pipe clips at 2m maximum centres	D18	m	0.25	2.38	3.39	0.86	6.63
Extra for shoe	D19	nr	0.30	2.85	2.09	0.74	5.68
Apply one coat primer, one oil-based undercoat and one coat gloss paint to fascia and soffit exceeding 300mm girth	D20	m2	0.20	1.90	0.96	0.43	3.29
PART E PITCHED ROOF							
100 x 75mm sawn softwood wall plate bedded in cement mortar	E1	m	0.30	2.85	2.22	0.76	5.83
200 x 50mm sawn softwood pole plate plugged to brick wall	E2	m	0.30	2.85	3.22	0.91	6.98
100 x 50mm sawn softwood rafters	E3	m	0.20	1.90	3.28	0.78	5.96
125 x 50mm sawn softwood purlins	E4	m	0.20	1.90	2.41	0.65	4.96
150 x 50mm sawn softwood joists	E5	m	0.20	1.90	2.28	0.63	4.81
150 x 50mm sawn softwood sprocket pieces 500mm long	E6	nr	0.12	1.14	1.14	0.34	2.62
Crown Wool insulation or similar 100mm thick fixed between joists with chicken wire and 150mm thick layer laid over joists	E7	m2	0.48	4.56	10.91	2.32	17.79
6mm thick asbestos-free insulation board soffit 150mm wide	E8	m	0.40	3.80	1.75	0.83	6.38

	Ref	Qty	Hours	Hours £	Mat'ls £	O & P £	Total £
19mm thick wrought softwood fascia/barge board 200mm high	E9	m	0.50	4.75	1.97	1.01	7.73
Marley Plain roofing tiles size 278 x 165mm, 65mm lap, type 1F reinforced felt and 38 x 19mm softwood battens	E10	m2	1.90	18.05	25.87	6.59	50.51
Extra for double eaves course	E11	m	0.35	3.33	3.22	0.98	7.53
Extra for verge with plain tile undercloak	E12	m	0.25	2.38	1.95	0.65	4.97
Lead flashing, code 5, 200mm girth	E13	m	0.60	5.70	6.68	1.86	14.24
Rake out joint of existing faced brick wall to receive flashing and point up on completion	E14	m	0.35	3.33	0.15	0.52	4.00
PVC-U gutter, 112mm half round with gutter union joints, fixed to softwood fascia board with support brackets at 1m maximum centres	E15	m	0.26	2.47	4.47	1.04	7.98
Extra for stop end	E16	nr	0.14	1.33	1.11	0.37	2.81
Extra for stop end outlet	E17	nr	0.25	2.38	1.11	0.52	4.01
PVC-U down pipe, 68mm diameter, loose spigot and socket joints, plugged to faced brickwork with pipe clips at 2m maximum centres	E18	m	0.25	2.38	3.39	0.86	6.63
Extra for shoe	E19	nr	0.30	2.85	2.09	0.74	5.68
Apply one coat primer, one oil-based undercoat and one coat gloss paint to fascia and soffit exceeding 300mm girth	E20	m2	0.20	1.90	0.96	0.43	3.29

	Ref	Qty	Hours	Hours £	Mat'ls £	O & P £	Total £
PART F WINDOWS AND EXTERNAL DOORS							
PVC-U door size 840 x 1980mm complete (B)	F1	nr	2.50	23.75	209.99	35.06	268.80
PVC-U sliding patio door size 1700 x 2075mm complete (C)	F2	nr	7.00	66.50	293.52	54.00	414.02
PVC-U window size 840 x 1980mm complete (A)	F3	nr	2.00	19.00	157.59	26.49	203.08
25 x 225mm wrought softwood window board with rounded edge	F4	m	0.30	2.85	6.22	1.36	10.43
Apply one coat primer, one oil-based undercoat and one coat gloss paint to window board not exceeding 300mm girth	F5	m	0.14	1.33	0.48	0.27	2.08
PART G INTERNAL PARTITIONS AND DOORS							
50 x 75mm sawn softwood sole plate	G1	m	0.22	2.09	1.12	0.48	3.69
50 x 75mm sawn softwood head	G2	m	0.22	2.09	1.12	0.48	3.69
50 x 75mm sawn softwood studs	G3	m	0.28	2.66	1.12	0.57	4.35
50 x 75mm sawn softwood noggings	G4	m	0.28	2.66	1.12	0.57	4.35
Plasterboard 9.5mm thick fixed to softwood studding, filled joints and taped to receive decoration	G5	m2	0.36	3.42	1.85	0.79	6.06
Flush door 35mm thick, size 762 x 1981mm, internal quality, half-hour fire check, veneered finish both sides	G6	nr	1.25	11.88	55.32	10.08	77.27
38 x 150mm wrought softwood lining	G7	m	0.22	2.09	0.88	0.45	3.42

	Ref	Qty	Hours	Hours £	Mat'ls £	O & P £	Total £
13 x 38mm wrought softwood door stop	G8	m	0.20	1.90	0.49	0.36	2.75
19 x 50mm wrought softwood chamfered architrave	G9	m	0.15	1.43	0.46	0.28	2.17
19 x 100mm wrought softwood chamfered skirting	G10	m	0.17	1.62	1.37	0.45	3.43
100mm rising steel butts	G11	pair	0.30	2.85	3.95	1.02	7.82
Silver anodised aluminium mortice latch with lever furniture	G12	nr	0.80	7.60	14.10	3.26	24.96
Two coats emulsion paint to plasterboard walls	G13	m2	0.26	2.47	0.59	0.46	3.52
Apply one coat primer, one oil-based undercoat and one coat gloss paint to general surfaces exceeding 300mm girth	G14	m2	0.20	1.90	0.96	0.43	3.29
PART H WALL FINISHES							
19 x 100mm wrought softwood chamfered skirting	H1	m	0.17	1.62	1.37	0.45	3.43
12mm plasterboard fixed to walls with dabs	H2	m2	0.39	3.71	1.99	0.85	6.55
12mm plasterboard less than 300mm wide fixed to walls with dabs	H3	m	0.18	1.71	0.92	0.39	3.02
Two coats emulsion paint to walls	H4	m2	0.26	2.47	0.59	0.46	3.52
Apply one coat primer, one oil-based undercoat and one coat gloss paint to general surfaces surfaces not exceeding 300mm	H5	m	0.20	1.90	0.48	0.36	2.74

	Ref	Qty	Hours	Hours £	Mat'ls £	O & P £	Total £
PART J FLOORING/FINISHES							
Cement and sand (1:3) floor screed, 40mm thick, steel trowelled to level floors	J1	m2	0.25	2.38	3.03	0.81	6.22
Vinyl floor tiles, size 300 x 300 x 2mm thick, fixing with adhesive to floor screed	J2	m2	0.17	1.62	6.98	1.29	9.88
25mm thick tongued and grooved softwood flooring	J3	m2	0.74	6.66	9.40	2.41	18.47
150 x 50mm sawn softwood joists	J4	m	0.22	1.98	2.28	0.64	4.90
Cut and pin ends of joists to existing brick wall	J5	nr	0.18	0.10	0.25	0.05	0.40
Build in ends of joists to brickwork	J6	nr	0.10	0.10	0.15	0.04	0.29
PART K CEILING FINISHES							
Plasterboard 9.5mm thick fixed to ceiling joists, joints filled with filler and taped to receive decoration	K1	m2	0.36	3.42	1.85	0.79	6.06
One coat skim plaster to plasterboard ceilings including scrimming joints	K2	m2	0.50	4.75	1.22	0.90	6.87
Two coats emulsion paint to plastered ceilings	K3	m2	0.26	2.47	0.59	0.46	3.52
PART L ELECTRICAL WORK							
13 amp double switched socket outlet with neon	L1	nr	0.40	5.60	8.26	2.08	15.94
Single lighting point	L2	nr	0.35	4.90	6.80	1.76	13.46
Single one way lighting switch	L3	nr	0.35	4.90	3.97	1.33	10.20

	Ref	Qty	Hours	Hours £	Mat'ls £	O & P £	Total £
Lighting wiring	L4	m	0.10	1.40	1.03	0.36	2.79
Power cable	L5	m	0.15	2.10	1.10	0.48	3.68
PART M HEATING WORK							
Copper pipe 15mm diameter, capillary fittings, fixed with pipe clips to softwood	M1	m	0.22	2.75	1.88	0.69	5.32
Extra for elbow	M2	nr	0.28	3.50	1.10	0.69	5.29
Extra for equal tee	M3	nr	0.34	4.25	1.95	0.93	7.13
Radiator, steel panelled double convector, size 1400 x 520mm, plugged and screwed to blockwork with concealed brackets, complete with 3mm chromium-plated air valve, 15mm straight valve with union and 15mm lockshield valve with union	M4	nr	1.30	16.25	115.68	19.79	151.72
Break into 15mm diameter pipe and insert tee	M5	nr	0.75	9.38	3.95	2.00	15.32
PART N ALTERATION WORK							
Take out existing window size 1500 x 1000mm and lintel over, adapt opening to receive 1770 x 2075mm patio door and insert new lintel over (both measured separately) and make good	N1	nr	20.00	190.00	20.00	31.50	241.50
Take out existing window size 1500 x 1000mm and lintel over, adapt opening to receive 840 x 1980mm door (measured separately) and make good	N2	nr	24.00	228.00	45.00	40.95	313.95

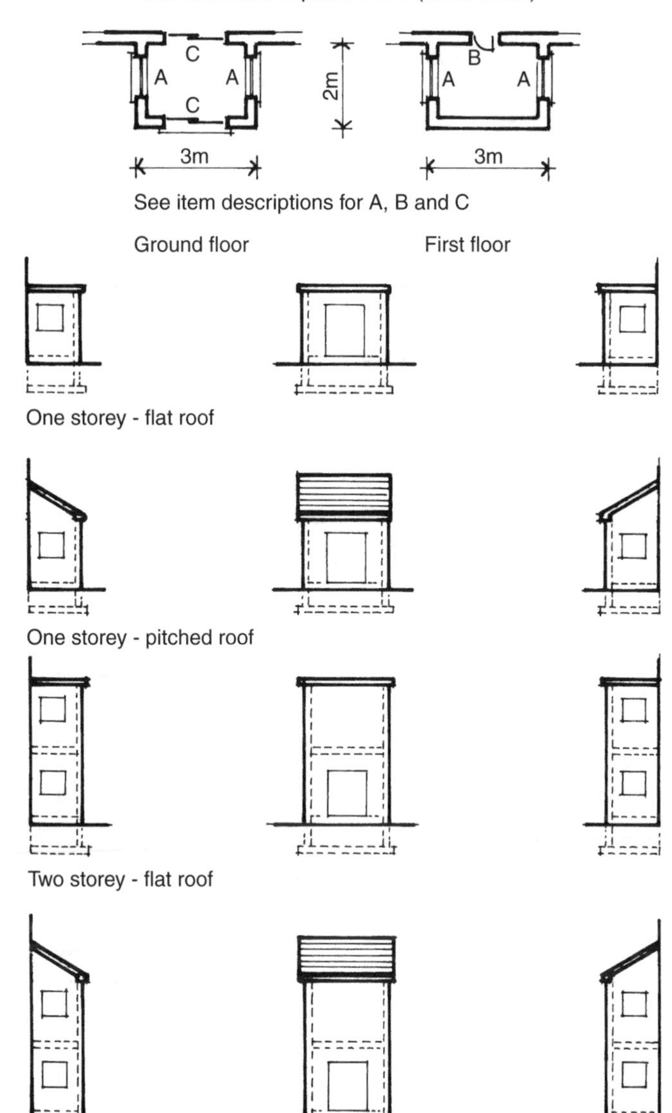
One and two storey extension size 2 × 3m
with either flat or pitched roof (not to scale)
C
A
A
C
2m
B
A
A
3m
3m
See item descriptions for A, B and C
Ground floor
First floor
One storey - flat roof
One storey - pitched roof
Two storey - flat roof
Two storey - pitched roof

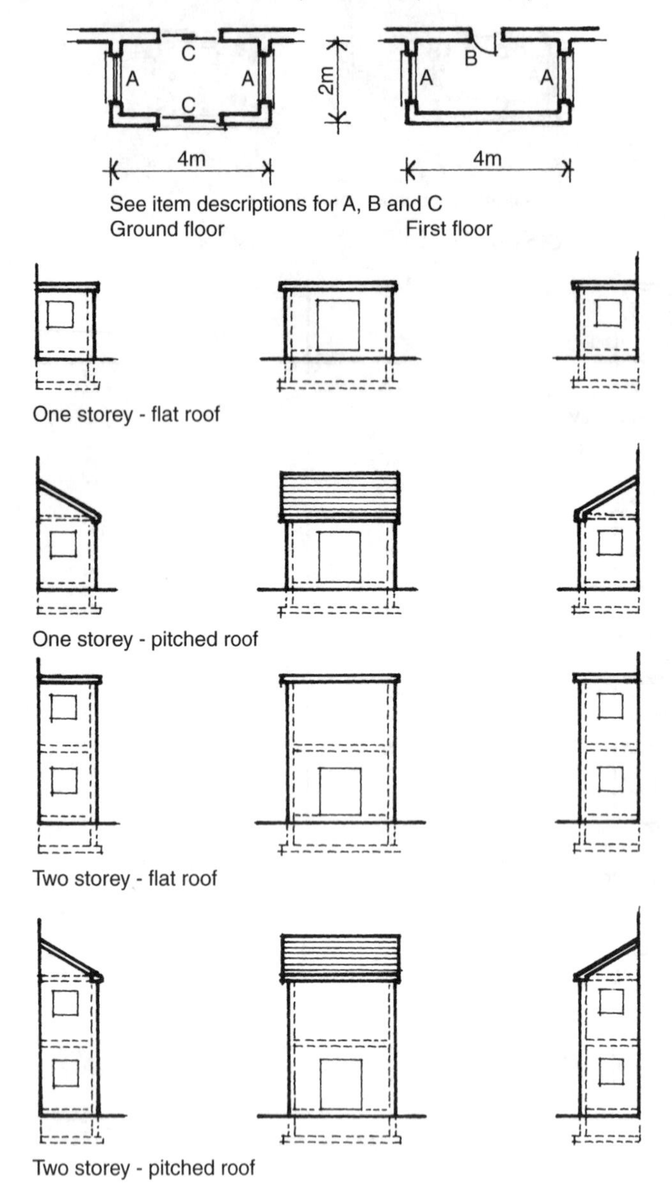
One and two storey extension size 2×4m
with either flat or pitched roof (not to scale)
C
A
A
C
2m
B
A
A
4m
4m
See item descriptions for A, B and C
Ground floor
First floor
One storey - flat roof
One storey - pitched roof
Two storey - flat roof
Two storey - pitched roof

One and two storey extension size 2 × 5m
with either flat or pitched roof (not to scale)

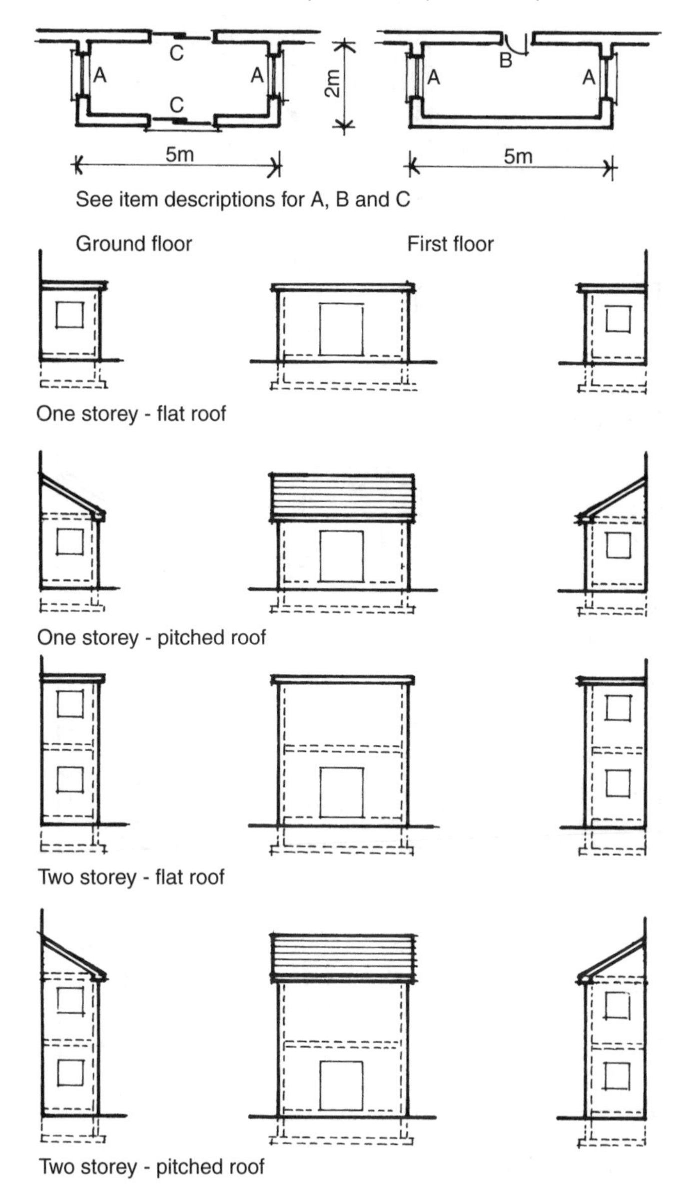

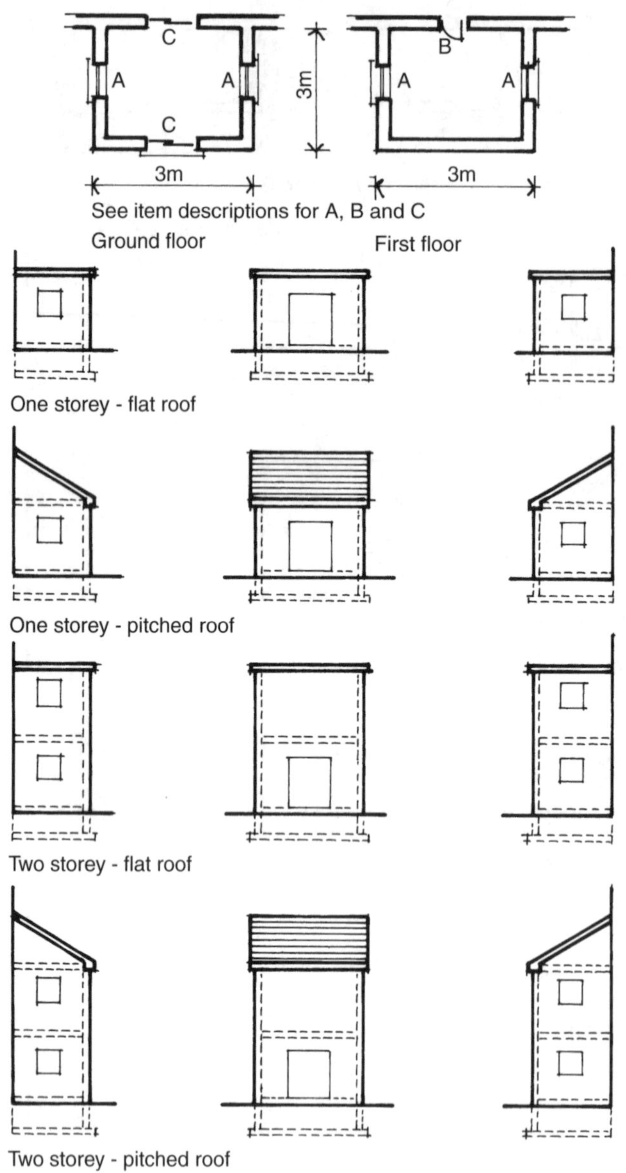
One and two storey extension size 3 × 3m
with either flat or pitched roof (not to scale)
C
A
A
C
3m
B
A
A
3m
3m
See item descriptions for A, B and C
Ground floor
First floor
One storey - flat roof
One storey - pitched roof
Two storey - flat roof
Two storey - pitched roof

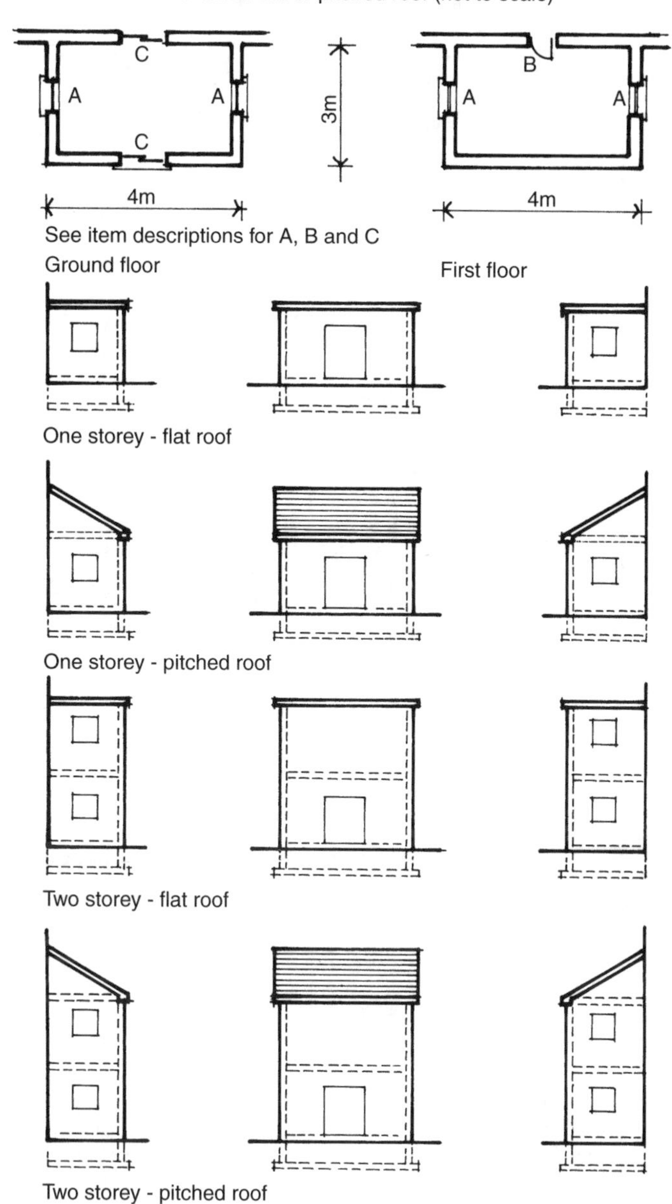
One and two storey extension size 3×4m
with either flat or pitched roof (not to scale)
C
A
A
C
3m
B
A
A
4m
4m
See item descriptions for A, B and C
Ground floor
First floor
One storey - flat roof
One storey - pitched roof
Two storey - flat roof
Two storey - pitched roof

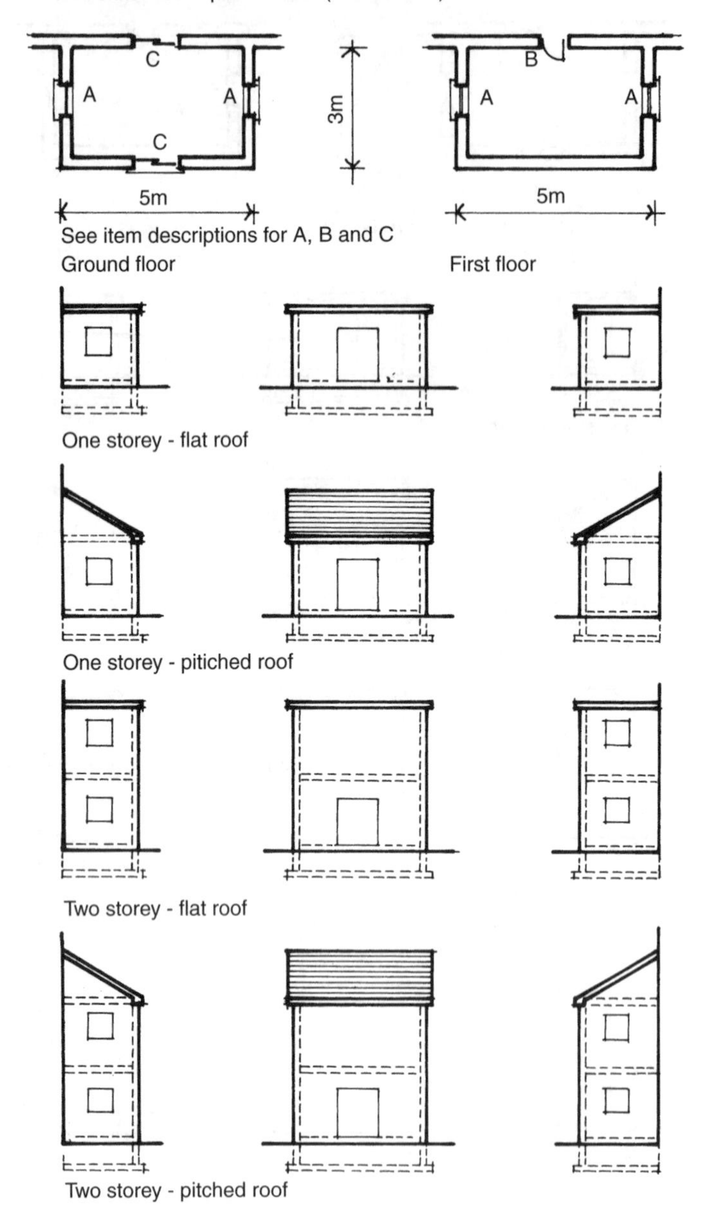
One and two storey extension size 3×5m
with either flat or pitched roof (not to scale)
C
A
A
C
5m
3m
B
A
A
5m
See item descriptions for A, B and C
Ground floor
First floor
One storey - flat roof
One storey - pitched roof
Two storey - flat roof
Two storey - pitched roof

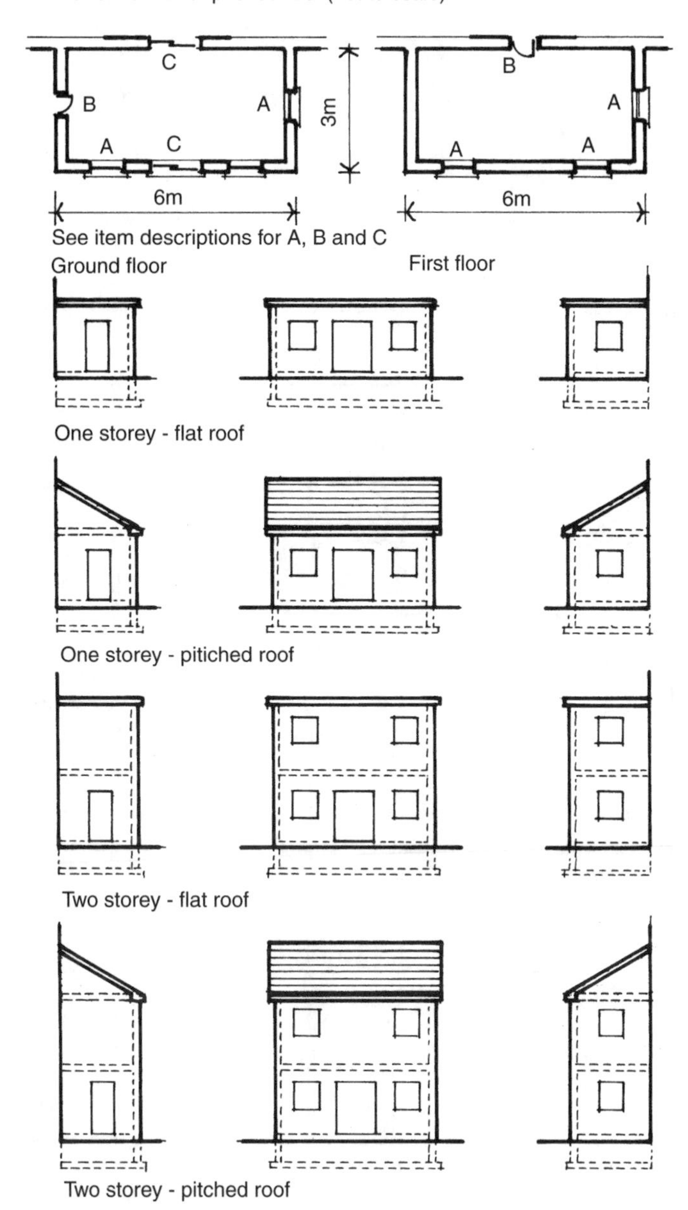
One and two storey extension size 3×6m
with either flat or pitched roof (not to scale)
C
B
A
A
C
3m
B
A
A
A
6m
6m
See item descriptions for A, B and C
Ground floor
First floor
One storey - flat roof
One storey - pitched roof
Two storey - flat roof
Two storey - pitched roof

One and two storey extension size 4 × 4m
with either flat or pitched roof (not to scale)

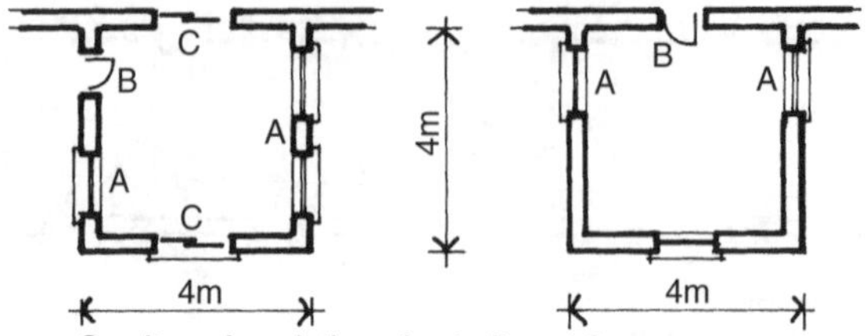

C
B
A
A
C
4m
4m
B
A
A
4m

See item descriptions for A, B and C

Ground floor

First floor

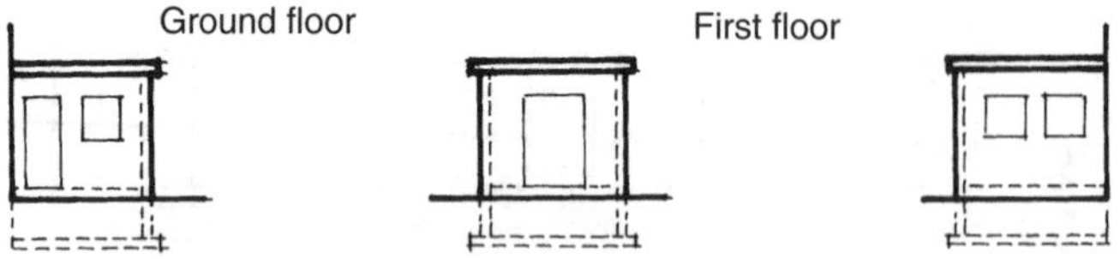

One storey - flat roof

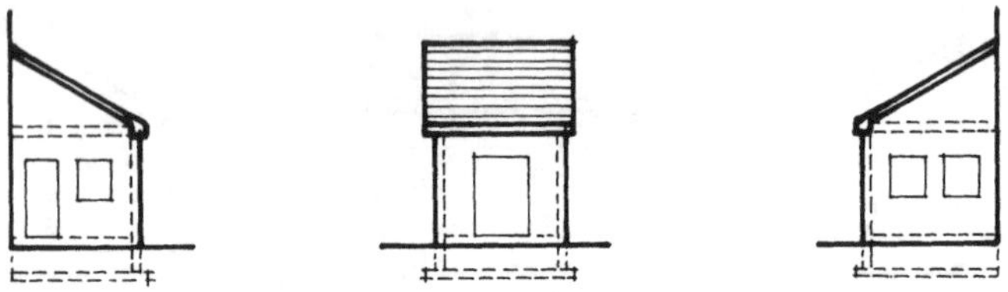

One storey - pitched roof

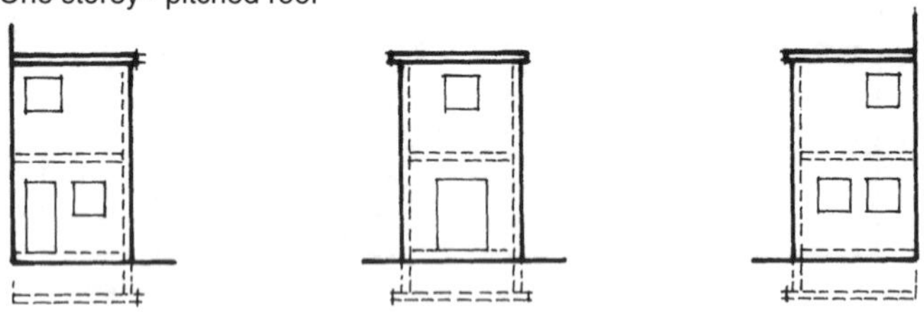

Two storey - flat roof

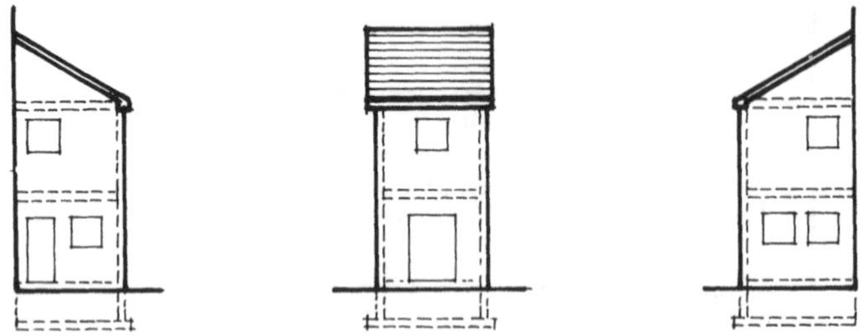

Two storey - pitched roof

One and two storey extension size 4 × 5m
with either flat or pitched roof (not to scale)

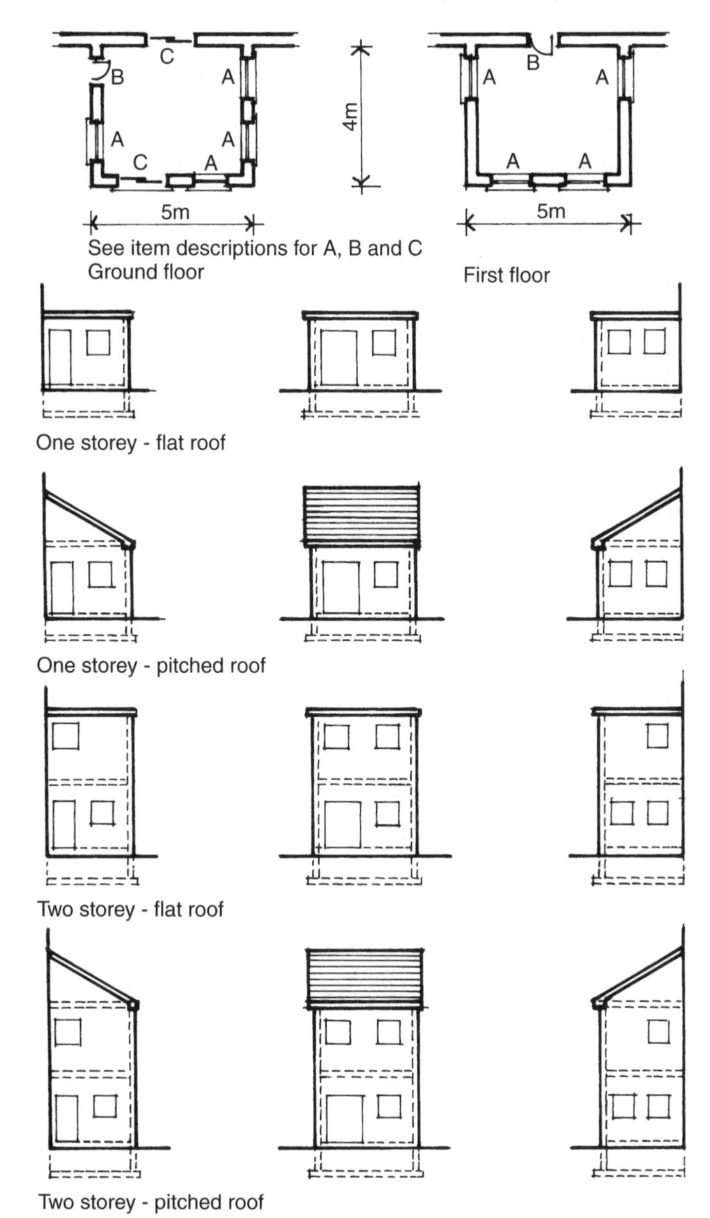

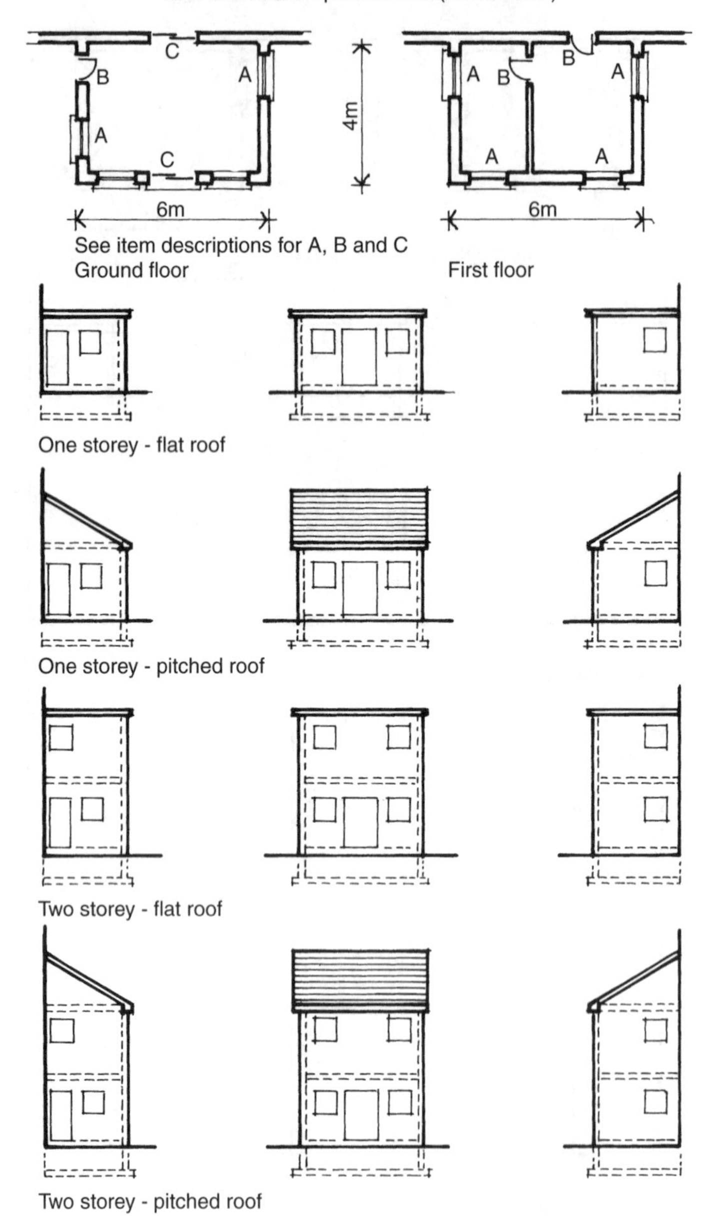
One and two storey extension size 4×6m
with either flat or pitched roof (not to scale)
C
B
A
A
C
4m
6m
B
A
B
A
B
A
A
A
6m
See item descriptions for A, B and C
Ground floor
First floor
One storey - flat roof
One storey - pitched roof
Two storey - flat roof
Two storey - pitched roof

One and two storey extension size 4 × 7m
with either flat or pitched roof (not to scale)

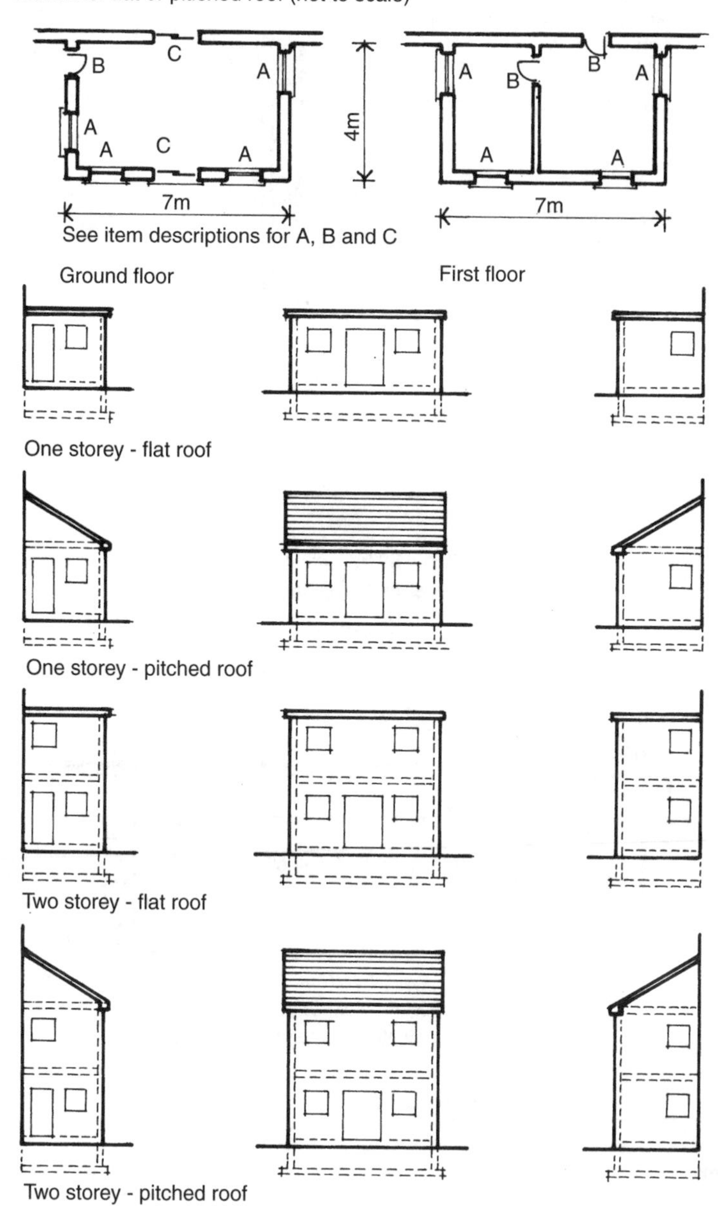

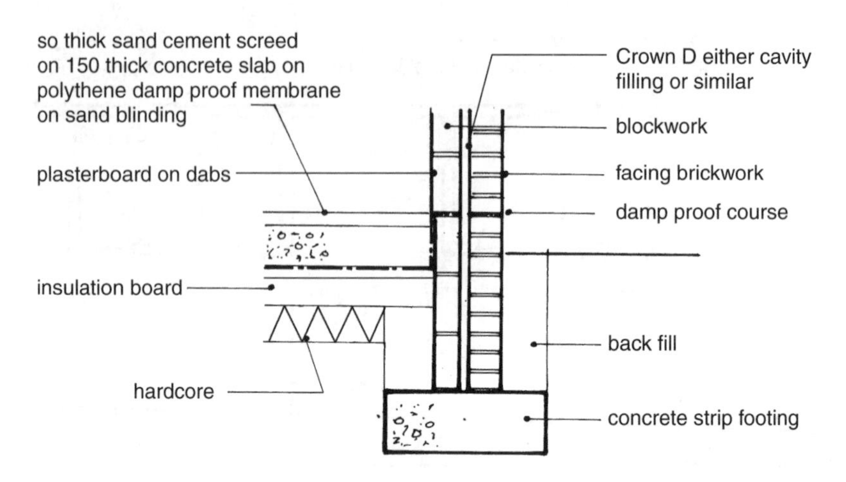
so thick sand cement screed
on 150 thick concrete slab on
polythene damp proof membrane
on sand blinding
plasterboard on dabs
insulation board
hardcore
Crown D either cavity
filling or similar
blockwork
facing brickwork
damp proof course
back fill
concrete strip footing

Foundation detail

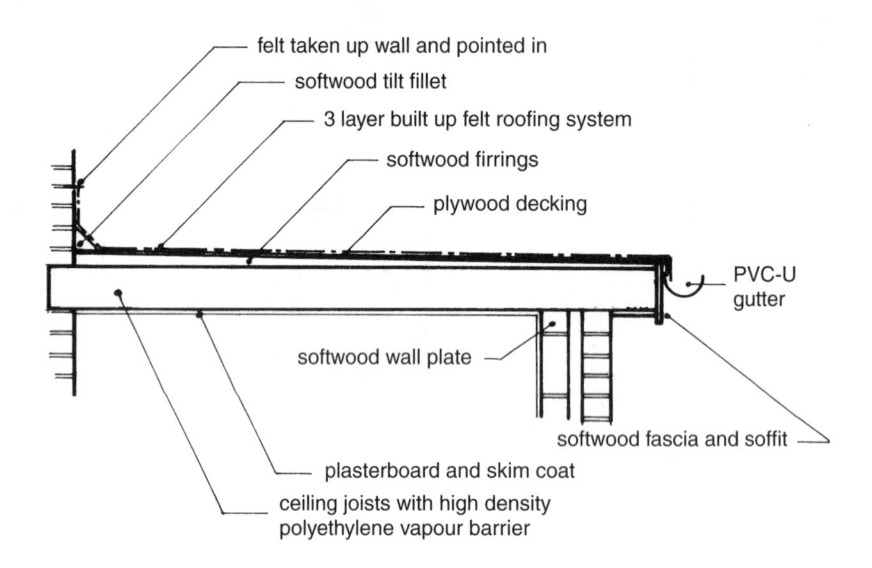
felt taken up wall and pointed in
softwood tilt fillet
3 layer built up felt roofing system
softwood firrings
plywood decking
PVC-U
gutter
softwood wall plate
softwood fascia and soffit
plasterboard and skim coat
ceiling joists with high density
polyethylene vapour barrier

Typical cross section - flat roof

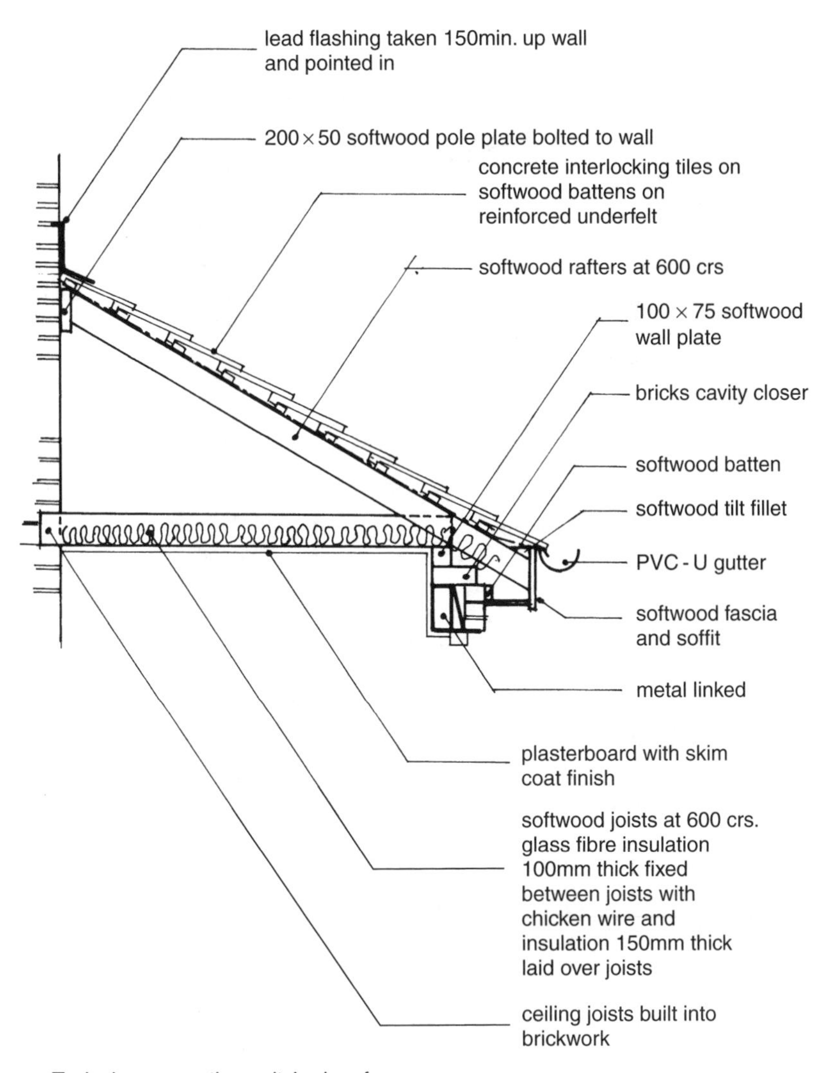
lead flashing taken 150min. up wall and pointed in
200 × 50 softwood pole plate bolted to wall
concrete interlocking tiles on softwood battens on reinforced underfelt
softwood rafters at 600 crs
100 × 75 softwood wall plate
bricks cavity closer
softwood batten
softwood tilt fillet
PVC - U gutter
softwood fascia and soffit
metal linked
plasterboard with skim coat finish
softwood joists at 600 crs. glass fibre insulation 100mm thick fixed between joists with chicken wire and insulation 150mm thick laid over joists
ceiling joists built into brickwork

Typical cross section - pitched roof

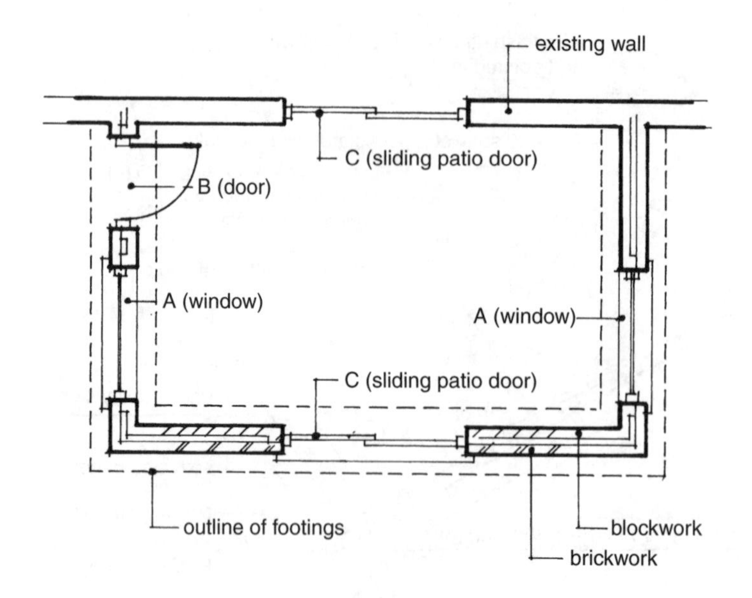

Typical extension plan (ground floor)

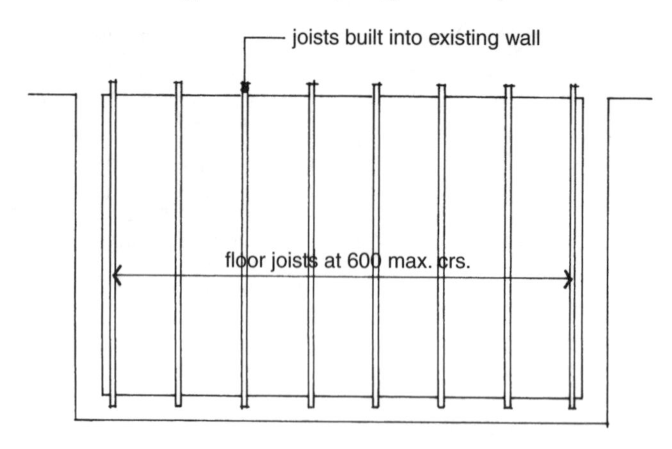

Typical floor joist layout

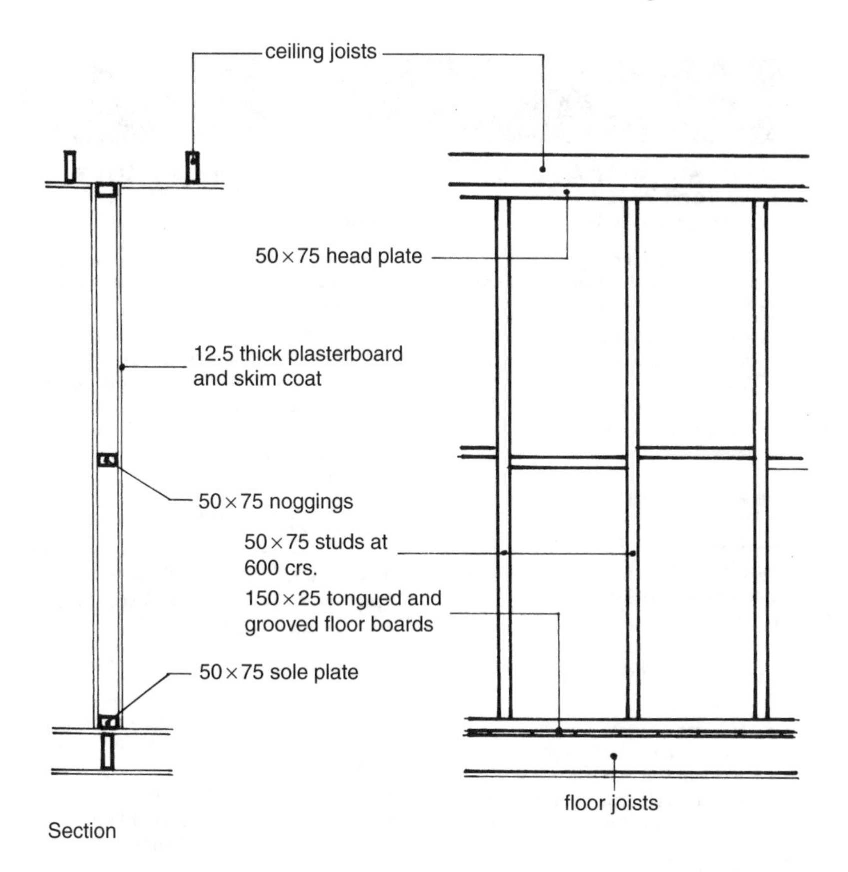
ceiling joists
50 × 75 head plate
12.5 thick plasterboard and skim coat
50 × 75 noggings
50 × 75 studs at 600 crs.
150 × 25 tongued and grooved floor boards
50 × 75 sole plate
floor joists
Section
Elevation

Timber stud wall

Spon Press
Taylor & Francis Group

	Ref	Qty	Hours	Hours £	Mat'ls £	O & P £	Total £
PART A PRELIMINARIES							
Concrete mixer	A1	4 wks					180.00
Small tools	A2	6 wks					210.00
Scaffolding (m2/weeks)	A3	70.00					157.50
Skip	A4	2 wks					300.00
Clean up	A5	4 hrs					48.00
Carried to summary							895.50
PART B SUBSTRUCTURE TO DPC LEVEL							
Excavate topsoil 150mm thick by hand	B1	7.10m2	2.13	20.24	0.00	3.04	23.27
Excavate to reduce levels by hand	B2	2.13m3	5.33	50.64	0.00	7.60	58.23
Excavate for trench foundations by hand	B3	1.06m3	2.76	26.22	0.00	3.93	30.15
Earthwork support to sides of trenches	B4	9.34m2	3.74	35.53	11.21	7.01	53.75
Backfilling with excavated material	B5	0.40m3	0.16	1.52	0.00	0.23	1.75
Hardcore 225mm thick	B6	4.08m2	0.82	7.79	20.73	4.28	32.80
Hardcore filling to trench	B7	0.08m3	0.05	0.48	0.40	0.13	1.01
Concrete grade (1:3:6) in foundations	B8	0.86m3	1.16	11.02	58.43	10.42	79.87
Concrete grade (1:2:4) in bed 150mm thick	B9	4.08m2	1.22	11.59	44.39	8.40	64.38
Concrete (1:2:4) in cavity wall filling	B10	3.20m2	0.64	6.08	11.62	2.66	20.36
Carried forward			18.01	171.10	146.78	47.68	365.56

	Ref	Qty	Hours	Hours £	Mat'ls £	O & P £	Total £
Brought forward			18.01	171.10	146.78	47.68	365.56
Damp-proof membrane	B11	4.38m2	0.17	1.62	2.54	0.62	4.78
Reinforcement ref A193 in foundation	B12	3.22m2	0.38	3.61	4.16	1.17	8.94
Steel fabric reinforcement ref A193 in slab	B13	4.08m2	0.61	5.80	5.30	1.66	12.76
Solid blockwork 140mm thick in cavity wall	B14	4.16m2	5.41	51.40	53.16	15.68	120.24
Common bricks 112.5mm thick in cavity wall	B15	3.20m2	5.41	51.40	42.21	14.04	107.65
Facing bricks in 112.5mm thick in skin of cavity wall	B16	0.96m2	1.73	16.44	22.34	5.82	44.59
Form cavity 50mm wide in cavity wall	B17	4.16m2	0.13	1.24	2.50	0.56	4.30
DPC 112mm wide	B18	6.40m	0.32	3.04	5.18	1.23	9.45
DPC 140mm wide	B19	6.40m	0.38	3.61	6.21	1.47	11.29
Bond in block wall	B20	1.30m	0.58	5.51	2.83	1.25	9.59
Bond in half brick wall	B21	0.30m	0.46	4.37	0.61	0.75	5.73
50mm thick insulation board	B22	4.08m2	1.22	11.59	18.28	4.48	34.35
Carried to summary			34.81	330.70	312.10	96.42	739.21
PART C EXTERNAL WALLS							
Solid blockwork 140mm thick in cavity wall	C1	10.50m2	13.65	129.68	133.56	39.49	302.72
Facing brickwork 112.5mm thick in cavity wall	C2	10.50m2	18.90	179.55	242.13	63.25	484.93
75mm thick insulation in cavity wall	C3	10.50m2	2.31	21.95	61.11	12.46	95.51
Carried forward			34.86	331.17	436.80	115.20	883.17

	Ref	Qty	Hours	Hours £	Mat'ls £	O & P £	Total £
Brought forward			34.86	331.17	436.80	115.20	883.17
Steel lintel 2400mm long	C4	2nr	0.50	4.75	196.92	30.25	231.92
Steel lintel 1500mm long	C5	2nr	0.40	3.80	214.94	32.81	251.55
Close cavity wall at jambs	C7	8.96m	0.45	4.28	18.28	3.38	25.94
Close cavity wall at cills	C8	3.67m	0.18	1.71	7.49	1.38	10.58
Close cavity wall at top	C9	6.40m	0.32	3.04	13.06	2.42	18.52
DPC 112mm wide at jambs	C10	8.96m	0.45	4.28	7.26	1.73	13.27
DPC 112mm wide at cills	C11	3.67m	0.18	1.71	2.97	0.70	5.38
Carried to summary			37.34	354.73	897.72	187.87	1,440.32
PART D FLAT ROOF							
200 x 50mm sawn softwood joists	D1	15.05m	3.76	35.72	46.66	12.36	94.74
200 x 50mm sawn softwood sprocket pieces	D2	8nr	1.12	10.64	10.72	3.20	24.56
18mm thick WPB grade decking	D3	7.10m2	6.39	57.51	55.81	17.00	130.32
50 x 50mm (avg) wide sawn softwood firrings	D4	15.05m	2.71	25.75	34.01	8.96	68.72
High density polyethylene vapour barrier 150mm thick	D5	6.00m2	1.12	10.64	56.82	10.12	77.58
100 x 75mm sawn softwood wall plate	D6	7.00m	2.10	19.95	15.54	5.32	40.81
100 x 75mm sawn softwood tilt fillet	D7	3.30m	0.83	7.89	8.71	2.49	19.08
Build in ends of 200 x 50mm joists	D8	7nr	2.45	23.28	1.05	3.65	27.97
Carried forward			20.48	191.37	229.32	63.10	483.79

	Ref	Qty	Hours	Hours £	Mat'ls £	O & P £	Total £
Brought forward			20.48	191.37	229.32	63.10	483.79
Rake out joint for flashing	D9	3.30m	1.16	11.02	0.50	1.73	13.25
6mm thick soffit 150mm wide	D10	7.30m	2.92	27.74	12.78	6.08	46.60
19mm wrought softwood fascia 200mm high	D11	7.30m	3.65	34.68	14.38	7.36	56.41
Three layer fibre-based roofing felt	D12	7.10m2	3.91	37.15	59.71	14.53	111.38
Felt turn-down 100mm girth	D13	7.30m	0.73	6.94	6.13	1.96	15.02
Felt flashing 150mm girth	D14	3.30m	0.33	3.14	5.54	1.30	9.98
112mm diameter PVC-U gutter	D15	3.30m	0.86	8.17	14.75	3.44	26.36
Stop end	D16	1nr	0.14	1.33	1.11	0.37	2.81
Stop end outlet	D17	1nr	0.25	2.38	1.94	0.65	4.96
68mm diameter PVC-U down pipe	D18	2.50m	0.63	5.99	8.48	2.17	16.63
Shoe	D19	1nr	0.30	2.85	2.09	0.74	5.68
Paint fascia and soffit	D20	2.19m2	1.53	14.54	2.10	2.50	19.13
Carried to summary			36.89	347.26	358.83	105.91	812.00
PART E PITCHED ROOF			N/A	N/A	N/A	N/A	N/A
PART F WINDOWS AND EXTERNAL DOORS							
PVC-U door size 840 x 1980mm complete (B)	F1	0.00	0.00	0.00	0.00	0.00	0.00
PVC-U sliding patio door size 1700 x 2075mm (C)	F2	2nr	14.00	133.00	587.04	108.01	828.05
Carried forward			14.00	133.00	587.04	108.01	828.05

	Ref	Qty	Hours	Hours £	Mat'ls £	O & P £	Total £
Brought forward			14.00	133.00	587.04	108.01	828.05
PVC-U window size 1200 x 1200mm complete (A)	F3	2nr	4.00	38.00	315.18	52.98	406.16
25 x 225mm wrought softwood window board	F4	2.40m	0.72	6.84	12.44	2.89	22.17
Paint window board	F5	2.40m	0.42	3.99	2.30	0.94	7.23
Carried to summary			19.14	181.83	916.96	164.82	1,263.61
PART G INTERNAL PARTITIONS AND DOORS		N/A	N/A	N/A	N/A	N/A	N/A
PART H WALL FINISHES							
19 x 100mm wrought softwood skirting	H1	4.54m	0.91	8.65	6.22	2.23	17.09
12mm plasterboard fixed to walls with dabs	H2	9.00m2	3.51	33.35	17.91	7.69	58.94
12mm plasterboard fixed to walls less than 300mm wide with dabs	H3	12.63m	2.27	21.57	11.62	4.98	38.16
Two coats emulsion paint to walls	H4	10.90m2	2.62	24.89	6.43	4.70	36.02
Paint skirting	H5	4.54m	0.68	6.46	2.18	1.30	9.94
Carried to summary			9.99	94.91	44.36	20.89	160.15
PART J FLOOR FINISHES							
Cement and sand floor screed 40mm thick	J1	4.08m2	1.02	9.69	12.36	3.31	25.36
Vinyl floor tiles, size 300 x 300mm	J2	4.08m2	0.98	9.31	28.47	5.67	43.45
Carried to summary			2.00	19.00	40.83	8.97	68.80

	Ref	Qty	Hours	Hours £	Mat'ls £	O & P £	Total £
PART K CEILING FINISHES							
Plasterboard with taped butt joints fixed to joists	K1	4.08m2	1.47	2.08	7.53	1.44	11.05
5mm skim coat to plasterboard ceilings	K2	4.08m2	2.04	19.38	4.98	3.65	28.01
Two coats emulsion paint to ceilings	K3	4.08m2	1.06	10.07	2.41	1.87	14.35
Carried to summary			4.57	31.53	14.92	6.97	53.42
PART L ELECTRICAL WORK							
13 amp double switched socket outlet with neon	L1	2nr	0.80	11.20	16.52	4.16	31.88
Lighting point	L2	1nr	0.35	4.90	6.80	1.76	13.46
Lighting switch	L3	2nr	0.50	7.00	7.94	2.24	17.18
Lighting wiring	L4	5.00m	0.50	7.00	5.15	1.82	13.97
Power cable	L5	12.00m	1.80	25.20	13.20	5.76	44.16
Carried to summary			3.95	55.30	49.61	15.74	120.65
PART M HEATING WORK							
15mm copper pipe	M1	4.00m	0.88	11.00	7.52	2.78	21.30
Elbow	M2	4nr	1.12	14.00	4.40	2.76	21.16
Tee	M3	1nr	0.22	2.75	1.95	0.71	5.41
Radiator, double convector size 1400 x 520mm	M4	1nr	1.30	16.25	115.68	19.79	151.72
Break into existing pipe and insert tee	M5	1nr	0.75	9.38	3.95	2.00	15.32
Carried to summary			4.27	53.38	133.50	28.03	214.91

	Ref	Qty	Hours	Hours £	Mat'ls £	O & P £	Total £
PART N **ALTERATION WORK**							
Take out existing window size 1500 x 1000mm and lintel over, adapt opening to receive 1770 x 2000mm patio door and insert new lintel over (both measured separately) and make good	N1	1nr	20.00	190.00	20.00	31.50	241.50
Carried to summary			20.00	190.00	20.00	31.50	241.50

SUMMARY

	Hours	Hours £	Mat'ls £	O & P £	Total £
PART A PRELIMINARIES	0.00	0.00	0.00	0.00	895.50
PART B SUBSTRUCTURE TO DPC LEVEL	34.81	330.70	312.10	96.42	739.21
PART C EXTERNAL WALLS	37.34	354.73	897.72	187.87	1,440.32
PART D FLAT ROOF	36.89	347.26	358.83	105.91	812.00
PART E PITCHED ROOF	0.00	0.00	0.00	0.00	0.00
PART F WINDOWS AND EXTERNAL DOORS	19.14	181.83	916.96	164.82	1,263.61
PART G INTERNAL PARTITIONS AND DOORS	0.00	0.00	0.00	0.00	0.00
PART H WALL FINISHES	9.99	94.91	44.36	20.89	160.15
PART J FLOOR FINISHES	2.00	19.00	40.83	8.97	68.80
PART K CEILING FINISHES	4.57	31.53	14.92	6.97	53.42
PART L ELECTRICAL WORK	3.95	55.30	49.61	15.74	120.65
PART M HEATING WORK	4.27	53.38	133.50	28.03	214.91
PART N ALTERATION WORK	20.00	190.00	20.00	31.50	241.50
Final total	172.96	1,658.64	2,788.83	667.12	6,010.07

	Ref	Qty	Hours	Hours £	Mat'ls £	O & P £	Total £
PART A PRELIMINARIES							
Concrete mixer	A1	5 wks					225.00
Small tools	A2	7 wks					245.00
Scaffolding (m2/weeks)	A3	100.00					225.00
Skip	A4	3 wks					300.00
Clean up	A5	6 hrs					48.00
Carried to summary							1,043.00
PART B SUBSTRUCTURE TO DPC LEVEL							
Excavate topsoil 150mm thick by hand	B1	9.25m2	2.78	26.41	0.00	3.96	30.37
Excavate to reduce levels by hand	B2	2.77m3	6.93	65.84	0.00	9.88	75.71
Excavate for trench foundations by hand	B3	1.22m3	3.17	30.12	0.00	4.52	34.63
Earthwork support to sides of trenches	B4	10.80m2	4.32	41.04	12.96	8.10	62.10
Backfilling with excavated material	B5	0.47m3	0.28	2.66	0.00	0.40	3.06
Hardcore 225mm thick	B6	5.78m2	0.61	5.80	29.36	5.27	40.43
Hardcore filling to trench	B7	0.09m3	0.05	0.48	0.45	0.14	1.06
Concrete grade (1:3:6) in foundations	B8	1.00m3	1.35	12.83	67.94	12.11	92.88
Concrete grade (1:2:4) in bed 150mm thick	B9	5.78m2	1.73	16.44	62.87	11.90	91.20
Concrete (1:2:4) in cavity wall filling	B10	3.70m2	0.74	7.03	13.43	3.07	23.53
Carried forward			21.96	208.62	187.01	59.34	454.97

	Ref	Qty	Hours	Hours £	Mat'ls £	O & P £	Total £
Brought forward			21.96	208.62	187.01	59.34	454.97
Damp-proof membrane	B11	6.13m2	0.25	2.38	3.56	0.89	6.83
Reinforcement ref A193 in foundation	B12	3.70m2	0.44	4.18	4.81	1.35	10.34
Steel fabric reinforcement ref A193 in slab	B13	5.78m2	0.87	8.27	7.51	2.37	18.14
Solid blockwork 140mm thick in cavity wall	B14	4.81m2	6.25	59.38	61.47	18.13	138.97
Common bricks 112.5mm thick in cavity wall	B15	3.70m2	6.25	59.38	48.80	16.23	124.40
Facing bricks in 112.5mm thick in skin of cavity wall	B16	1.11m2	2.00	19.00	25.83	6.72	51.55
Form cavity 50mm wide in cavity wall	B17	4.81m2	0.14	1.33	2.89	0.63	4.85
DPC 112mm wide	B18	7.40m	0.37	3.52	5.99	1.43	10.93
DPC 140mm wide	B19	7.40m	0.44	4.18	7.18	1.70	13.06
Bond in block wall	B20	1.30m	0.58	5.51	2.83	1.25	9.59
Bond in half brick wall	B21	0.30m	0.46	4.37	0.61	0.75	5.73
50mm thick insulation board	B22	5.78m2	1.73	16.44	25.89	6.35	48.67
Carried to summary			41.74	396.53	384.38	117.14	898.05
PART C EXTERNAL WALLS							
Solid blockwork 140mm thick in cavity wall	C1	13.00m2	16.90	160.55	165.36	48.89	374.80
Facing brickwork 112.5mm thick in cavity wall	C2	13.00m2	23.40	222.30	299.78	78.31	600.39
75mm thick insulation in cavity wall	C3	13.00m2	2.86	27.17	75.66	15.42	118.25
Carried forward			43.16	410.02	540.80	142.62	1,093.44

	Ref	Qty	Hours	Hours £	Mat'ls £	O & P £	Total £
Brought forward			43.16	410.02	540.80	142.62	1,093.44
Steel lintel 2400mm long	C4	2nr	0.50	4.75	196.92	30.25	231.92
Steel lintel 1500mm long	C5	2nr	0.40	3.80	214.94	32.81	251.55
Close cavity wall at jambs	C7	8.96m	0.45	4.28	18.28	3.38	25.94
Close cavity wall at cills	C8	3.67m	0.18	1.71	7.49	1.38	10.58
Close cavity wall at top	C9	7.40m	0.37	3.52	15.10	2.79	21.41
DPC 112mm wide at jambs	C10	8.96m	0.45	4.28	7.26	1.73	13.27
DPC 112mm wide at cills	C11	3.67m	0.18	1.71	2.97	0.70	5.38
Carried to summary			45.69	434.06	1,003.76	215.67	1,653.49

PART D
FLAT ROOF

	Ref	Qty	Hours	Hours £	Mat'ls £	O & P £	Total £
200 x 50mm sawn softwood joists	D1	19.35m	4.84	45.98	59.98	15.89	121.85
200 x 50mm sawn softwood sprocket pieces	D2	8nr	1.12	10.64	10.72	3.20	24.56
18mm thick WPB grade decking	D3	9.25m2	8.33	74.97	72.71	22.15	169.83
50 x 50mm (avg) wide sawn softwood firrings	D4	19.25m	3.47	32.97	43.51	11.47	87.95
High density polyethylene vapour barrier 150mm thick	D5	6.00m2	1.16	11.02	75.76	13.02	99.80
100 x 75mm sawn softwood wall plate	D6	8.00m	2.40	22.80	17.76	6.08	46.64
100 x 75mm sawn softwood tilt fillet	D7	4.30m	1.08	10.26	11.35	3.24	24.85
Build in ends of 200 x 50mm joists	D8	9nr	3.15	29.93	1.35	4.69	35.97
Carried forward			25.55	238.56	293.14	79.76	611.46

	Ref	Qty	Hours	Hours £	Mat'ls £	O & P £	Total £
Brought forward			25.55	238.56	293.14	79.76	611.46
Rake out joint for flashing	D9	4.30m	1.51	14.35	0.65	2.25	17.24
6mm thick soffit 150mm wide	D10	8.30m	3.32	31.54	14.53	6.91	52.98
19mm wrought softwood fascia 200mm high	D11	8.30m	4.15	39.43	16.35	8.37	64.14
Three layer fibre-based roofing felt	D12	9.25m2	4.63	43.99	77.79	18.27	140.04
Felt turn-down 100mm girth	D13	8.30m	0.83	7.89	6.97	2.23	17.08
Felt flashing 150mm girth	D14	4.30m	0.52	4.94	7.22	1.82	13.98
112mm diameter PVC-U gutter	D15	4.30m	1.12	10.64	19.22	4.48	34.34
Stop end	D16	1nr	0.14	1.33	1.11	0.37	2.81
Stop end outlet	D17	1nr	0.25	2.38	1.94	0.65	4.96
68mm diameter PVC-U down pipe	D18	2.50m	0.63	5.99	8.48	2.17	16.63
Shoe	D19	1nr	0.30	2.85	2.09	0.74	5.68
Paint fascia and soffit	D20	2.49m2	1.74	16.53	2.39	2.84	21.76
Carried to summary			44.69	420.39	451.88	130.84	1,003.11
PART E PITCHED ROOF			N/A	N/A	N/A	N/A	N/A
PART F WINDOWS AND EXTERNAL DOORS							
PVC-U door size 840 x 1980mm complete (B)	F1	0.00	0.00	0.00	0.00	0.00	0.00
PVC-U sliding patio door size 1700 x 2075mm (C)	F2	2nr	14.00	133.00	587.04	108.01	828.05
Carried forward			14.00	133.00	587.04	108.01	828.05

	Ref	Qty	Hours	Hours £	Mat'ls £	O & P £	Total £
Brought forward			14.00	133.00	587.04	108.01	828.05
PVC-U window size 1200 x 1200mm complete (A)	F3	2nr	4.00	38.00	315.18	52.98	406.16
25 x 225mm wrought softwood window board	F4	2.40m	0.72	6.84	12.44	2.89	22.17
Paint window board	F5	2.40m	0.42	3.99	2.30	0.94	7.23
Carried to summary			19.14	181.83	916.96	164.82	1,263.61
PART G INTERNAL PARTITIONS AND DOORS		N/A	N/A	N/A	N/A	N/A	N/A
PART H WALL FINISHES							
19 x 100mm wrought softwood skirting	H1	5.54m	1.11	10.55	7.59	2.72	20.86
12mm plasterboard fixed to walls with dabs	H2	11.50m2	4.49	42.66	22.89	9.83	75.38
12mm plasterboard fixed to walls less than 300mm wide with dabs	H3	12.63m	2.27	21.57	11.62	4.98	38.16
Two coats emulsion paint to walls	H4	13.40m2	3.21	30.50	7.91	5.76	44.17
Paint skirting	H5	5.54m	0.83	7.89	2.66	1.58	12.13
Carried to summary			11.91	113.15	52.67	24.87	190.69
PART J FLOOR FINISHES							
Cement and sand floor screed 40mm thick	J1	5.78m2	1.45	13.78	17.51	4.69	35.98
Vinyl floor tiles, size 300 x 300mm	J2	5.78m2	1.39	13.21	40.03	7.99	61.22
Carried to summary			2.84	26.98	57.54	12.68	97.20

	Ref	Qty	Hours	Hours £	Mat'ls £	O & P £	Total £
PART K CEILING FINISHES							
Plasterboard with taped butt joints fixed to joists	K1	5.78m2	2.08	2.08	10.69	1.92	14.69
5mm skim coat to plasterboard ceilings	K2	5.78m2	2.89	27.46	7.05	5.18	39.68
Two coats emulsion paint to ceilings	K3	5.78m2	1.50	14.25	3.39	2.65	20.29
Carried to summary			6.47	43.79	21.13	9.74	74.65
PART L ELECTRICAL WORK							
13 amp double switched socket outlet with neon	L1	2nr	0.80	11.20	16.52	4.16	31.88
Lighting point	L2	2nr	0.70	9.80	13.60	3.51	26.91
Lighting switch	L3	2nr	0.50	7.00	7.94	2.24	17.18
Lighting wiring	L4	6.00m	0.60	8.40	6.18	2.19	16.77
Power cable	L5	14.00m	2.10	29.40	15.40	6.72	51.52
Carried to summary			4.70	65.80	59.64	18.82	144.26
PART M HEATING WORK							
15mm copper pipe	M1	5.00m	1.10	13.75	9.40	3.47	26.62
Elbow	M2	4nr	1.12	14.00	4.40	2.76	21.16
Tee	M3	1nr	0.22	2.75	1.95	0.71	5.41
Radiator, double convector size 1400 x 520mm	M4	2nr	2.60	32.50	231.36	39.58	303.44
Break into existing pipe and insert tee	M5	1nr	0.75	9.38	3.95	2.00	15.32
Carried to summary			5.79	72.38	251.06	48.52	371.95

	Ref	Qty	Hours	Hours £	Mat'ls £	O & P £	Total £
PART N ALTERATION WORK							
Take out existing window size 1500 x 1000mm and lintel over, adapt opening to receive 1770 x 2000mm patio door and insert new lintel over (both measured separately) and make good	N1	1nr	20.00	190.00	20.00	31.50	241.50
Carried to summary			20.00	190.00	20.00	31.50	241.50

SUMMARY

	Hours	Hours £	Mat'ls £	O & P £	Total £
PART A PRELIMINARIES	0.00	0.00	0.00	0.00	1,043.00
PART B SUBSTRUCTURE TO DPC LEVEL	41.74	396.53	384.38	117.14	898.05
PART C EXTERNAL WALLS	45.69	434.06	1,003.76	215.67	1,653.49
PART D FLAT ROOF	44.69	420.39	451.88	130.84	1,003.11
PART E PITCHED ROOF	0.00	0.00	0.00	0.00	0.00
PART F WINDOWS AND EXTERNAL DOORS	19.14	181.83	916.96	164.82	1,263.61
PART G INTERNAL PARTITIONS AND DOORS	0.00	0.00	0.00	0.00	0.00
PART H WALL FINISHES	11.91	113.15	52.67	24.87	190.69
PART J FLOOR FINISHES	2.84	26.98	57.54	12.68	97.20
PART K CEILING FINISHES	6.47	43.79	21.13	9.74	74.65
PART L ELECTRICAL WORK	4.77	65.80	59.64	18.82	144.26
PART M HEATING WORK	5.79	72.38	251.06	48.52	371.95
PART N ALTERATION WORK	20.00	190.00	20.00	31.50	241.50
Final total	203.04	1,944.91	3,219.02	774.60	6,981.51

	Ref	Qty	Hours	Hours £	Mat'ls £	O & P £	Total £
PART A PRELIMINARIES							
Concrete mixer	A1	6 wks					270.00
Small tools	A2	8 wks					280.00
Scaffolding (m2/weeks)	A3	135.00					303.75
Skip	A4	4 wks					400.00
Clean up	A5	8 hrs					64.00
Carried to summary							1,317.75
PART B SUBSTRUCTURE TO DPC LEVEL							
Excavate topsoil 150mm thick by hand	B1	11.40m2	3.42	32.49	0.00	4.87	37.36
Excavate to reduce levels by hand	B2	3.42m3	8.55	81.23	0.00	12.18	93.41
Excavate for trench foundations by hand	B3	1.39m3	4.90	46.55	0.00	6.98	53.53
Earthwork support to sides of trenches	B4	12.26m2	4.32	41.04	14.71	8.36	64.11
Backfilling with excavated material	B5	0.55m3	0.22	2.09	0.00	0.31	2.40
Hardcore 225mm thick	B6	7.48m2	1.12	10.64	38.00	7.30	55.94
Hardcore filling to trench	B7	0.12m3	0.06	0.57	0.60	0.18	1.35
Concrete grade (1:3:6) in foundations	B8	1.13m3	1.53	14.54	76.77	13.70	105.00
Concrete grade (1:2:4) in bed 150mm thick	B9	7.48m2	2.24	21.28	81.38	15.40	118.06
Concrete (1:2:4) in cavity wall filling	B10	4.20m2	0.84	7.98	15.25	3.48	26.71
Carried forward			27.20	258.40	226.71	72.77	557.88

	Ref	Qty	Hours	Hours £	Mat'ls £	O & P £	Total £
Brought forward			27.20	258.40	226.71	72.77	557.88
Damp-proof membrane	B11	7.88m2	0.32	3.04	4.57	1.14	8.75
Reinforcement ref A193 in foundation	B12	4.20m2	0.50	4.75	5.46	1.53	11.74
Steel fabric reinforcement ref A193 in slab	B13	7.48m2	1.12	10.64	9.72	3.05	23.41
Solid blockwork 140mm thick in cavity wall	B14	5.46m2	7.10	67.45	69.78	20.58	157.81
Common bricks 112.5mm thick in cavity wall	B15	4.20m2	7.10	67.45	55.40	18.43	141.28
Facing bricks in 112.5mm thick in skin of cavity wall	B16	1.26m2	2.27	21.57	29.32	7.63	58.52
Form cavity 50mm wide in cavity wall	B17	5.46m2	0.16	1.52	3.28	0.72	5.52
DPC 112mm wide	B18	8.40m	0.42	3.99	6.80	1.62	12.41
DPC 140mm wide	B19	8.40m	0.50	4.75	8.15	1.94	14.84
Bond in block wall	B20	1.30m	0.58	5.51	2.83	1.25	9.59
Bond in half brick wall	B21	0.30m	0.46	4.37	0.61	0.75	5.73
50mm thick insulation board	B22	7.48m2	2.24	21.28	33.51	8.22	63.01
Carried to summary			49.97	474.72	456.14	139.63	1,070.48
PART C EXTERNAL WALLS							
Solid blockwork 140mm thick in cavity wall	C1	15.50m2	20.15	191.43	197.16	58.29	446.87
Facing brickwork 112.5mm thick in cavity wall	C2	15.50m2	27.90	265.05	357.43	93.37	715.85
75mm thick insulation in cavity wall	C3	15.50m2	3.41	32.40	90.21	18.39	141.00
Carried forward			51.46	488.87	644.80	170.05	1,303.72

	Ref	Qty	Hours	Hours £	Mat'ls £	O & P £	Total £
Brought forward			51.46	488.87	644.80	170.05	1,303.72
Steel lintel 2400mm long	C4	2nr	0.50	4.75	196.92	30.25	231.92
Steel lintel 1500mm long	C5	2nr	0.40	3.80	214.94	32.81	251.55
Close cavity wall at jambs	C7	8.96m	0.45	4.28	18.28	3.38	25.94
Close cavity wall at cills	C8	3.67m	0.18	1.71	7.49	1.38	10.58
Close cavity wall at top	C9	8.40m	0.37	3.52	17.14	3.10	23.75
DPC 112mm wide at jambs	C10	8.96m	0.45	4.28	7.26	1.73	13.27
DPC 112mm wide at cills	C11	3.67m	0.18	1.71	2.97	0.70	5.38
Carried to summary			53.99	512.91	1,109.80	243.41	1,866.11

PART D
FLAT ROOF

	Ref	Qty	Hours	Hours £	Mat'ls £	O & P £	Total £
200 x 50mm sawn softwood joists	D1	23.65m	5.91	56.15	73.32	19.42	148.88
150 x 50mm sawn softwood sprocket pieces	D2	8nr	1.12	10.64	10.72	3.20	24.56
18mm thick WPB grade decking	D3	11.40m2	10.26	92.34	89.60	27.29	209.23
50 x 50mm (avg) wide sawn softwood firrings	D4	23.65m	4.26	40.47	53.45	14.09	108.01
High density polyethylene vapour barrier 150mm thick	D5	10.00m2	2.00	19.00	94.70	17.06	130.76
100 x 75mm sawn softwood wall plate	D6	9.00m	2.70	25.65	19.98	6.84	52.47
100 x 75mm sawn softwood tilt fillet	D7	5.30m	1.33	12.64	13.99	3.99	30.62
Build in ends of 200 x 50mm joists	D8	11nr	3.85	36.58	1.65	5.73	43.96
Carried forward			31.43	293.46	357.41	97.63	748.49

	Ref	Qty	Hours	Hours £	Mat'ls £	O & P £	Total £
Brought forward			31.43	293.46	357.41	97.63	748.49
Rake out joint for flashing	D9	5.30m	1.86	17.67	0.80	2.77	21.24
6mm thick soffit 150mm wide	D10	9.30m	3.72	35.34	16.28	7.74	59.36
19mm wrought softwood fascia 200mm high	D11	9.30m	4.65	44.18	18.32	9.37	71.87
Three layer fibre-based roofing felt	D12	11.40m2	6.27	59.57	95.87	23.32	178.75
Felt turn-down 100mm girth	D13	9.30m	0.93	8.84	7.81	2.50	19.14
Felt flashing 150mm girth	D14	5.30m	0.63	5.99	8.90	2.23	17.12
112mm diameter PVC-U gutter	D15	5.30m	1.12	10.64	23.69	5.15	39.48
Stop end	D16	1nr	0.14	1.33	1.11	0.37	2.81
Stop end outlet	D17	1nr	0.25	2.38	1.94	0.65	4.96
68mm diameter PVC-U down pipe	D18	2.50m	0.63	5.99	8.48	2.17	16.63
Shoe	D19	1nr	0.30	2.85	2.09	0.74	5.68
Paint fascia and soffit	D20	2.79m2	1.74	16.53	2.68	2.88	22.09
Carried to summary			53.67	504.74	545.38	157.52	1,207.63
PART E PITCHED ROOF			N/A	N/A	N/A	N/A	N/A
PART F WINDOWS AND EXTERNAL DOORS							
PVC-U door size 840 x 1980mm complete (B)	F1	1nr	0.00	0.00	0.00	0.00	0.00
PVC-U sliding patio door size 1700 x 2075mm (C)	F2	1nr	14.00	133.00	584.04	107.56	824.60
Carried forward			14.00	133.00	584.04	107.56	824.60

	Ref	Qty	Hours	Hours £	Mat'ls £	O & P £	Total £
Brought forward			14.00	133.00	584.04	107.56	824.60
PVC-U window size 1200 x 1200mm complete (A)	F3	2nr	4.00	38.00	315.18	52.98	406.16
25 x 225mm wrought softwood window board	F4	2.40m	0.72	6.84	12.44	2.89	22.17
Paint window board	F5	2.40m	0.42	3.99	2.30	0.94	7.23
Carried to summary			19.14	181.83	913.96	164.37	1,260.16
PART G INTERNAL PARTITIONS AND DOORS		N/A	N/A	N/A	N/A	N/A	N/A
PART H WALL FINISHES							
19 x 100mm wrought softwood skirting	H1	6.54m	1.31	12.45	8.96	3.21	24.62
12mm plasterboard fixed to walls with dabs	H2	14.00m2	5.46	51.87	27.86	11.96	91.69
12mm plasterboard fixed to walls less than 300mm wide with dabs	H3	12.63m	2.27	21.57	11.62	4.98	38.16
Two coats emulsion paint to walls	H4	15.90m2	3.82	36.29	9.38	6.85	52.52
Paint skirting	H5	6.54m	0.98	9.31	3.14	1.87	14.32
Carried to summary			13.84	131.48	60.96	28.87	221.31
PART J FLOOR FINISHES							
Cement and sand floor screed 40mm thick	J1	7.48m2	1.87	17.77	22.66	6.06	46.49
Vinyl floor tiles, size 300 x 300mm	J2	7.48m2	1.27	12.07	52.22	9.64	73.93
Carried to summary			3.14	29.83	74.88	15.71	120.42

	Ref	Qty	Hours	Hours £	Mat'ls £	O & P £	Total £
PART K **CEILING FINISHES**							
Plasterboard with taped butt joints fixed to joists	K1	7.48m2	2.70	2.08	13.84	2.39	18.31
5mm skim coat to plasterboard ceilings	K2	7.48m2	3.74	35.53	9.13	6.70	51.36
Two coats emulsion paint to ceilings	K3	7.48m2	1.95	18.53	4.41	3.44	26.38
Carried to summary			8.39	56.14	27.38	12.53	96.04
PART L **ELECTRICAL WORK**							
13 amp double switched socket outlet with neon	L1	3nr	1.20	16.80	24.78	6.24	47.82
Lighting point	L2	2nr	0.70	9.80	13.60	3.51	26.91
Lighting switch	L3	2nr	0.50	7.00	7.94	2.24	17.18
Lighting wiring	L4	7.00m	0.70	9.80	7.21	2.55	19.56
Power cable	L5	16.00m	2.40	33.60	17.60	7.68	58.88
Carried to summary			5.50	77.00	71.13	22.22	170.35
PART M **HEATING WORK**							
15mm copper pipe	M1	6.00m	1.32	16.50	11.28	4.17	31.95
Elbow	M2	4nr	1.12	14.00	4.40	2.76	21.16
Tee	M3	1nr	0.22	2.75	1.95	0.71	5.41
Radiator, double convector size 1400 x 520mm	M4	2nr	2.60	32.50	231.36	39.58	303.44
Break into existing pipe and insert tee	M5	1nr	0.75	9.38	3.95	2.00	15.32
Carried to summary			6.01	75.13	252.94	49.21	377.27

	Ref	Qty	Hours	Hours £	Mat'ls £	O & P £	Total £
PART N **ALTERATION WORK**							
Take out existing window size 1500 x 1000mm and lintel over, adapt opening to receive 1770 x 2000mm patio door and insert new lintel over (both measured separately) and make good	N1	1nr	20.00	190.00	20.00	31.50	241.50
Carried to summary			20.00	190.00	20.00	31.50	241.50

SUMMARY

	Hours	Hours £	Mat'ls £	O & P £	Total £
PART A PRELIMINARIES	0.00	0.00	0.00	0.00	1,317.75
PART B SUBSTRUCTURE TO DPC LEVEL	49.97	474.72	456.14	139.63	1,070.48
PART C EXTERNAL WALLS	53.99	512.91	1,109.80	243.41	1,866.11
PART D FLAT ROOF	53.67	504.74	545.38	157.52	1,207.63
PART E PITCHED ROOF	0.00	0.00	0.00	0.00	0.00
PART F WINDOWS AND EXTERNAL DOORS	19.14	181.83	913.96	164.37	1,260.16
PART G INTERNAL PARTITIONS AND DOORS	0.00	0.00	0.00	0.00	0.00
PART H WALL FINISHES	13.84	131.48	60.96	28.87	221.31
PART J FLOOR FINISHES	3.14	29.83	74.88	15.71	120.42
PART K CEILING FINISHES	8.39	56.14	27.38	12.53	96.04
PART L ELECTRICAL WORK	5.50	77.00	71.13	22.22	170.35
PART M HEATING WORK	6.01	75.13	252.94	49.21	377.27
PART N ALTERATION WORK	20.00	190.00	20.00	31.50	241.50
Final total	233.65	2,233.78	3,532.57	864.97	7,949.02

	Ref	Qty	Hours	Hours £	Mat'ls £	O & P £	Total £
PART A PRELIMINARIES							
Concrete mixer	A1	6 wks					270.00
Small tools	A2	7 wks					245.00
Scaffolding (m2/weeks)	A3	90.00					202.50
Skip	A4	3 wks					300.00
Clean up	A5	8 hrs					64.00
Carried to summary							1,081.50
PART B SUBSTRUCTURE TO DPC LEVEL							
Excavate topsoil 150mm thick by hand	B1	10.40m2	3.12	29.64	0.00	4.45	34.09
Excavate to reduce levels by hand	B2	3.12m3	7.80	74.10	0.00	11.12	85.22
Excavate for trench foundations by hand	B3	1.39m3	3.17	30.12	0.00	4.52	34.63
Earthwork support to sides of trenches	B4	12.26m2	4.90	46.55	14.71	9.19	70.45
Backfilling with excavated material	B5	0.47m3	0.19	1.81	0.00	0.27	2.08
Hardcore 225mm thick	B6	6.48m2	0.97	9.22	32.92	6.32	48.46
Hardcore filling to trench	B7	0.09m3	0.05	0.48	0.45	0.14	1.06
Concrete grade (1:3:6) in foundations	B8	1.13m3	1.53	14.54	76.77	13.70	105.00
Concrete grade (1:2:4) in bed 150mm thick	B9	6.48m2	1.94	18.43	70.50	13.34	102.27
Concrete (1:2:4) in cavity wall filling	B10	4.20m2	0.84	7.98	15.25	3.48	26.71
Carried forward			24.51	232.85	210.60	66.52	509.96

	Ref	Qty	Hours	Hours £	Mat'ls £	O & P £	Total £
Brought forward			24.51	232.85	210.60	66.52	509.96
Damp-proof membrane	B11	6.88m2	0.28	2.66	3.99	1.00	7.65
Reinforcement ref A193 in foundation	B12	4.20m2	0.15	1.43	5.46	1.03	7.92
Steel fabric reinforcement ref A193 in slab	B13	6.48m2	0.97	9.22	8.42	2.65	20.28
Solid blockwork 140mm thick in cavity wall	B14	5.46m2	7.10	67.45	69.78	20.58	157.81
Common bricks 112.5mm thick in cavity wall	B15	4.20m2	7.10	67.45	55.40	18.43	141.28
Facing bricks in 112.5mm thick in skin of cavity wall	B16	1.26m2	2.27	21.57	29.32	7.63	58.52
Form cavity 50mm wide in cavity wall	B17	5.46m2	0.16	1.52	3.28	0.72	5.52
DPC 112mm wide	B18	8.40m	0.42	3.99	6.80	1.62	12.41
DPC 140mm wide	B19	8.40m	0.50	4.75	8.15	1.94	14.84
Bond in block wall	B20	1.30m	0.58	5.51	2.83	1.25	9.59
Bond in half brick wall	B21	0.30m	0.46	4.37	0.61	0.75	5.73
50mm thick insulation board	B22	6.48m2	1.94	18.43	29.03	7.12	54.58
Carried to summary			46.44	441.18	433.67	131.23	1,006.08
PART C EXTERNAL WALLS							
Solid blockwork 140mm thick in cavity wall	C1	15.50m2	20.15	191.43	197.16	58.29	446.87
Facing brickwork 112.5mm thick in cavity wall	C2	15.50m2	27.90	265.05	357.43	93.37	715.85
75mm thick insulation in cavity wall	C3	15.50m2	3.41	32.40	90.21	18.39	141.00
Carried forward			51.46	488.87	644.80	170.05	1,303.72

	Ref	Qty	Hours	Hours £	Mat'ls £	O & P £	Total £
Brought forward			51.46	488.87	644.80	170.05	1,303.72
Steel lintel 2400mm long	C4	2nr	0.50	4.75	196.92	30.25	231.92
Steel lintel 1500mm long	C5	2nr	0.40	3.80	214.94	32.81	251.55
Close cavity wall at jambs	C7	8.96m	0.45	4.28	18.28	3.38	25.94
Close cavity wall at cills	C8	3.67m	0.18	1.71	7.49	1.38	10.58
Close cavity wall at top	C9	8.40m	0.37	3.52	17.14	3.10	23.75
DPC 112mm wide at jambs	C10	8.96m	0.45	4.28	7.26	1.73	13.27
DPC 112mm wide at cills	C11	3.67m	0.18	1.71	2.97	0.70	5.38
Carried to summary			53.99	512.91	1,109.80	243.41	1,866.11

PART D
FLAT ROOF

	Ref	Qty	Hours	Hours £	Mat'ls £	O & P £	Total £
150 x 50mm sawn softwood joists	D1	22.05m	5.51	52.35	68.36	18.11	138.81
200 x 50mm sawn softwood sprocket pieces	D2	12nr	1.68	15.96	16.08	4.81	36.85
18mm thick WPB grade decking	D3	10.40m2	9.36	84.24	81.74	24.90	190.88
50 x 50mm (avg) wide sawn softwood firrings	D4	11.05m	3.97	37.72	49.83	13.13	100.68
High density polyethylene vapour barrier 150mm thick	D5	9.00m2	1.80	17.10	85.23	15.35	117.68
100 x 75mm sawn softwood wall plate	D6	9.00m	1.35	12.83	19.98	4.92	37.73
100 x 75mm sawn softwood tilt fillet	D7	3.30m	2.70	25.65	8.71	5.15	39.51
Build in ends of 200 x 50mm joists	D8	7nr	0.83	7.89	1.05	1.34	10.28
Carried forward			27.20	253.72	330.98	87.71	672.41

	Ref	Qty	Hours	Hours £	Mat'ls £	O & P £	Total £
Brought forward			27.20	253.72	330.98	87.71	672.41
Rake out joint for flashing	D9	3.30m	1.16	11.02	0.50	1.73	13.25
6mm thick soffit 150mm wide	D10	9.30m	3.72	35.34	16.28	7.74	59.36
19mm wrought softwood fascia 200mm high	D11	9.30m	4.65	44.18	18.32	9.37	71.87
Three layer fibre-based roofing felt	D12	10.04m2	5.72	54.34	89.99	21.65	165.98
Felt turn-down 100mm girth	D13	9.30m	0.93	8.84	7.81	2.50	19.14
Felt flashing 150mm girth	D14	3.30m	0.40	3.80	5.54	1.40	10.74
112mm diameter PVC-U gutter	D15	3.30m	0.86	8.17	14.75	3.44	26.36
Stop end	D16	1nr	0.14	1.33	1.11	0.37	2.81
Stop end outlet	D17	1nr	0.25	2.38	1.94	0.65	4.96
68mm diameter PVC-U down pipe	D18	2.50m	0.63	5.99	8.48	2.17	16.63
Shoe	D19	1nr	0.30	2.85	2.09	0.74	5.68
Paint fascia and soffit	D20	2.79m2	1.95	18.53	2.68	3.18	24.39
Carried to summary			47.91	450.47	500.47	142.64	1,093.58
PART E PITCHED ROOF			N/A	N/A	N/A	N/A	N/A
PART F WINDOWS AND EXTERNAL DOORS							
PVC-U door size 840 x 1980mm complete (B)	F1	0.00	0.00	0.00	0.00	0.00	0.00
PVC-U sliding patio door size 1700 x 2075mm (C)	F2	2nr	14.00	133.00	587.04	108.01	828.05
Carried forward			14.00	133.00	587.04	108.01	828.05

	Ref	Qty	Hours	Hours £	Mat'ls £	O & P £	Total £
Brought forward			14.00	133.00	587.04	108.01	828.05
PVC-U window size 1200 x 1200mm complete (A)	F3	2nr	4.00	38.00	315.18	52.98	406.16
25 x 225mm wrought softwood window board	F4	2.40m	0.72	6.84	12.44	2.89	22.17
Paint window board	F5	2.40m	0.42	3.99	2.30	0.94	7.23
Carried to summary			19.14	181.83	916.96	164.82	1,263.61
PART G INTERNAL PARTITIONS AND DOORS		N/A	N/A	N/A	N/A	N/A	N/A
PART H WALL FINISHES							
19 x 100mm wrought softwood skirting	H1	6.54m	1.31	12.45	6.22	2.80	21.46
12mm plasterboard fixed to walls with dabs	H2	14.00m2	5.46	51.87	27.86	11.96	91.69
12mm plasterboard fixed to walls less than 300mm wide with dabs	H3	12.63m	2.27	21.57	11.62	4.98	38.16
Two coats emulsion paint to plastered walls	H4	15.40m2	3.82	36.29	9.38	6.85	52.52
Paint skirting	H5	6.54m	0.98	9.31	3.14	1.87	14.32
Carried to summary			13.84	131.48	58.22	28.46	218.16
PART J FLOOR FINISHES							
Cement and sand floor screed 40mm thick	J1	6.48m2	1.62	15.39	19.63	5.25	40.27
Vinyl floor tiles, size 300 x 300mm	J2	6.48m2	1.58	15.01	45.23	9.04	69.28
Carried to summary			3.20	30.40	64.86	14.29	109.55

	Ref	Qty	Hours	Hours £	Mat'ls £	O & P £	Total £
PART K CEILING FINISHES							
Plasterboard with taped butt joints fixed to joists	K1	6.48m2	2.33	2.08	11.99	2.11	16.18
5mm skim coat to plasterboard ceilings	K2	6.48m2	3.24	30.78	7.91	5.80	44.49
Two coats emulsion paint to ceilings	K3	6.48m2	1.69	16.06	3.82	2.98	22.86
Carried to summary			7.26	48.92	23.72	10.90	83.53
PART L ELECTRICAL WORK							
13 amp double switched socket outlet with neon	L1	2nr	0.80	11.20	16.52	4.16	31.88
Lighting point	L2	1nr	0.70	9.80	6.80	2.49	19.09
Lighting switch	L3	2nr	0.50	7.00	7.94	2.24	17.18
Lighting wiring	L4	6.00m	0.60	8.40	6.18	2.19	16.77
Power cable	L5	14.00m	2.10	29.40	15.40	6.72	51.52
Carried to summary			4.70	65.80	52.84	17.80	136.44
PART M HEATING WORK							
15mm copper pipe	M1	5.00m	1.10	13.75	9.40	3.47	26.62
Elbow	M2	4nr	1.12	14.00	4.40	2.76	21.16
Tee	M3	1nr	0.22	2.75	1.95	0.71	5.41
Radiator, double convector size 1400 x 520mm	M4	2nr	2.60	32.50	231.36	39.58	303.44
Break into existing pipe and insert tee	M5	1nr	0.75	9.38	3.95	2.00	15.32
Carried to summary			5.79	72.38	251.06	48.52	371.95

	Ref	Qty	Hours	Hours £	Mat'ls £	O & P £	Total £
PART N **ALTERATION WORK**							
Take out existing window size 1500 x 1000mm and lintel over, adapt opening to receive 1770 x 2000mm patio door and insert new lintel over (both measured separately) and make good	N1	1nr	20.00	190.00	20.00	31.50	241.50
Carried to summary			20.00	190.00	20.00	31.50	241.50

SUMMARY

	Hours	Hours £	Mat'ls £	O & P £	Total £
PART A PRELIMINARIES	0.00	0.00	0.00	0.00	1,081.50
PART B SUBSTRUCTURE TO DPC LEVEL	46.44	441.18	433.67	131.23	1,006.08
PART C EXTERNAL WALLS	53.99	512.91	1,109.80	243.41	1,866.11
PART D FLAT ROOF	47.91	450.47	500.47	142.64	1,093.58
PART E PITCHED ROOF	0.00	0.00	0.00	0.00	0.00
PART F WINDOWS AND EXTERNAL DOORS	19.14	181.83	916.96	164.82	1,263.61
PART G INTERNAL PARTITIONS AND DOORS	0.00	0.00	0.00	0.00	0.00
PART H WALL FINISHES	13.84	131.48	58.22	28.46	218.16
PART J FLOOR FINISHES	3.20	30.40	64.86	14.29	109.55
PART K CEILING FINISHES	7.26	48.92	23.72	10.90	83.53
PART L ELECTRICAL WORK	4.70	65.80	52.84	17.80	136.44
PART M HEATING WORK	5.79	72.38	251.06	48.52	371.95
PART N ALTERATION WORK	20.00	190.00	20.00	31.50	241.50
Final total	222.27	2,125.37	3,431.60	833.57	7,472.01

	Ref	Qty	Hours	Hours £	Mat'ls £	O & P £	Total £
PART A PRELIMINARIES							
Concrete mixer	A1	7 wks					315.00
Small tools	A2	8 wks					280.00
Scaffolding (m2/weeks)	A3	125.00					281.25
Skip	A4	4 wks					400.00
Clean up	A5	6 hrs					48.00
Carried to summary							1,324.25
PART B SUBSTRUCTURE TO DPC LEVEL							
Excavate topsoil 150mm thick by hand	B1	14.19m2	4.26	40.47	0.00	6.07	46.54
Excavate to reduce levels by hand	B2	4.06m3	10.15	96.43	0.00	14.46	110.89
Excavate for trench foundations by hand	B3	1.39m3	3.17	30.12	0.00	4.52	34.63
Earthwork support to sides of trenches	B4	13.72m2	5.49	52.16	16.46	10.29	78.91
Backfilling with excavated material	B5	0.55m3	0.22	2.09	0.00	0.31	2.40
Hardcore 225mm thick	B6	9.18m2	1.38	13.11	46.63	8.96	68.70
Hardcore filling to trench	B7	0.12m3	0.06	0.57	0.60	0.18	1.35
Concrete grade (1:3:6) in foundations	B8	1.27m3	1.71	16.25	86.28	15.38	117.90
Concrete grade (1:2:4) in bed 150mm thick	B9	9.18m2	2.75	26.13	99.88	18.90	144.91
Concrete (1:2:4) in cavity wall filling	B10	4.70m2	0.94	8.93	17.06	3.90	29.89
Carried forward			30.13	286.24	266.91	82.97	636.12

	Ref	Qty	Hours	Hours £	Mat'ls £	O & P £	Total £
Brought forward			30.13	286.24	266.91	82.97	636.12
Damp-proof membrane	B11	9.63m2	0.25	2.38	5.59	1.19	9.16
Reinforcement ref A193 in foundation	B12	4.70m2	0.56	5.32	6.11	1.71	13.14
Steel fabric reinforcement ref A193 in slab	B13	9.18m2	1.38	13.11	11.93	3.76	28.80
Solid blockwork 140mm thick in cavity wall	B14	6.11m2	7.94	75.43	78.08	23.03	176.54
Common bricks 112.5mm thick in cavity wall	B15	4.70m2	7.94	75.43	61.99	20.61	158.03
Facing bricks in 112.5mm thick in skin of cavity wall	B16	1.41m2	2.54	24.13	32.81	8.54	65.48
Form cavity 50mm wide in cavity wall	B17	5.46m2	0.18	1.71	3.67	0.81	6.19
DPC 112mm wide	B18	9.40m	0.00	0.00	7.61	1.14	8.75
DPC 140mm wide	B19	9.40m	0.56	5.32	9.12	2.17	16.61
Bond in block wall	B20	1.30m	0.58	5.51	2.83	1.25	9.59
Bond in half brick wall	B21	0.30m	0.46	4.37	0.61	0.75	5.73
50mm thick insulation board	B22	9.18m2	2.75	26.13	41.13	10.09	77.34
Carried to summary			55.27	525.07	528.39	158.02	1,211.47
PART C EXTERNAL WALLS							
Solid blockwork 140mm thick in cavity wall	C1	18.00m2	23.40	222.30	228.96	67.69	518.95
Facing brickwork 112.5mm thick in cavity wall	C2	18.00m2	32.40	307.80	415.08	108.43	831.31
75mm thick insulation in cavity wall	C3	10.50m2	0.32	3.04	6.30	1.40	10.74
Carried forward			56.12	533.14	650.34	177.52	1,361.00

	Ref	Qty	Hours	Hours £	Mat'ls £	O & P £	Total £
Brought forward			56.12	533.14	650.34	177.52	1,361.00
Steel lintel 2400mm long	C4	2nr	0.50	4.75	196.92	30.25	231.92
Steel lintel 1500mm long	C5	2nr	0.40	3.80	214.94	32.81	251.55
Close cavity wall at jambs	C7	8.96m	0.45	4.28	18.28	3.38	25.94
Close cavity wall at cills	C8	3.67m	0.18	1.71	7.49	1.38	10.58
Close cavity wall at top	C9	9.40m	0.47	4.47	19.18	3.55	27.19
DPC 112mm wide at jambs	C10	8.96m	0.45	4.28	7.26	1.73	13.27
DPC 112mm wide at cills	C11	3.67m	0.18	1.71	2.97	0.70	5.38
Carried to summary			58.75	558.13	1,117.38	251.33	1,926.83

PART D
FLAT ROOF

	Ref	Qty	Hours	Hours £	Mat'ls £	O & P £	Total £
200 x 50mm sawn softwood joists	D1	28.35m	7.09	67.36	87.89	23.29	178.53
200 x 50mm sawn softwood sprocket pieces	D2	12nr	1.68	15.96	16.08	4.81	36.85
18mm thick WPB grade decking	D3	13.55m2	12.20	109.80	106.50	32.45	248.75
50 x 50mm (avg) wide sawn softwood firrings	D4	28.35m	5.10	48.45	64.07	16.88	129.40
High density polyethylene vapour barrier 150mm thick	D5	12.00m2	2.40	22.80	113.64	20.47	156.91
100 x 75mm sawn softwood wall plate	D6	10.00m	3.00	28.50	22.20	7.61	58.31
100 x 75mm sawn softwood tilt fillet	D7	4.30m	1.08	10.26	11.35	3.24	24.85
Build in ends of 200 x 50mm joists	D8	9nr	3.15	29.93	1.35	4.69	35.97
Carried forward			35.70	333.05	423.08	113.42	869.55

	Ref	Qty	Hours	Hours £	Mat'ls £	O & P £	Total £
Brought forward			35.70	333.05	423.08	113.42	869.55
Rake out joint for flashing	D9	4.30m	1.51	14.35	0.65	2.25	17.24
6mm thick soffit 150mm wide	D10	10.30m	4.12	39.14	18.03	8.58	65.75
19mm wrought softwood fascia 200mm high	D11	10.30m	5.15	48.93	22.90	10.77	82.60
Three layer fibre-based roofing felt	D12	13.55m2	7.45	70.78	113.96	27.71	212.45
Felt turn-down 100mm girth	D13	10.30m	1.03	9.79	8.65	2.77	21.20
Felt flashing 150mm girth	D14	4.30m	0.52	4.94	7.22	1.82	13.98
112mm diameter PVC-U gutter	D15	4.30m	1.12	10.64	19.22	4.48	34.34
Stop end	D16	1nr	0.14	1.33	1.11	0.37	2.81
Stop end outlet	D17	1nr	0.25	2.38	1.94	0.65	4.96
68mm diameter PVC-U down pipe	D18	2.50m	0.63	5.99	8.48	2.17	16.63
Shoe	D19	1nr	0.30	2.85	2.09	0.74	5.68
Paint fascia and soffit	D20	3.09m2	2.16	20.52	2.97	3.52	27.01
Carried to summary			60.08	564.66	630.30	179.24	1,374.20
PART E PITCHED ROOF			N/A	N/A	N/A	N/A	N/A
PART F WINDOWS AND EXTERNAL DOORS							
PVC-U door size 840 x 1980mm complete (B)	F1	0.00	0.00	0.00	0.00	0.00	0.00
PVC-U sliding patio door size 1700 x 2075mm (C)	F2	2nr	14.00	133.00	587.04	108.01	828.05
Carried forward			14.00	133.00	587.04	108.01	828.05

	Ref	Qty	Hours	Hours £	Mat'ls £	O & P £	Total £
Brought forward			14.00	133.00	587.04	108.01	828.05
PVC-U window size 1200 x 1200mm complete (A)	F3	2nr	4.00	38.00	315.18	52.98	406.16
25 x 225mm wrought softwood window board	F4	2.40m	0.72	6.84	12.44	2.89	22.17
Paint window board	F5	2.40m	0.42	3.99	2.30	0.94	7.23
Carried to summary			19.14	181.83	916.96	164.82	1,263.61
PART G INTERNAL PARTITIONS AND DOORS		N/A	N/A	N/A	N/A	N/A	N/A
PART H WALL FINISHES							
19 x 100mm wrought softwood skirting	H1	7.54m	1.51	14.35	7.59	3.29	25.23
12mm plasterboard fixed to walls with dabs	H2	16.50m2	6.44	61.18	32.84	14.10	108.12
12mm plasterboard fixed to walls less than 300mm wide with dabs	H3	12.63m	2.27	21.57	11.62	4.98	38.16
Two coats emulsion paint to walls	H4	18.40m2	4.42	41.99	10.86	7.93	60.78
Paint skirting	H5	7.54m	1.13	10.74	3.62	2.15	16.51
Carried to summary			15.77	149.82	66.53	32.45	248.80
PART J FLOOR FINISHES							
Cement and sand floor screed 40mm thick	J1	9.18m2	2.30	21.85	27.82	7.45	57.12
Vinyl floor tiles, size 300 x 300mm	J2	9.18m2	2.20	20.90	64.07	12.75	97.72
Carried to summary			4.50	42.75	91.89	20.20	154.84

	Ref	Qty	Hours	Hours £	Mat'ls £	O & P £	Total £
PART K CEILING FINISHES							
Plasterboard with taped butt joints fixed to joists	K1	9.18m2	3.31	2.08	16.98	2.86	21.92
5mm skim coat to plasterboard ceilings	K2	9.18m2	4.59	43.61	11.20	8.22	63.03
Two coats emulsion paint to ceilings	K3	9.18m2	2.39	22.71	5.42	4.22	32.34
Carried to summary			10.29	68.39	33.60	15.30	117.29
PART L ELECTRICAL WORK							
13 amp double switched socket outlet with neon	L1	3nr	1.20	16.80	24.78	6.24	47.82
Lighting point	L2	2nr	0.70	9.80	13.60	3.51	26.91
Lighting switch	L3	2nr	0.50	7.00	7.94	2.24	17.18
Lighting wiring	L4	6.00m	0.60	8.40	6.18	2.19	16.77
Power cable	L5	16.00m	2.40	33.60	17.60	7.68	58.88
Carried to summary			5.40	75.60	70.10	21.86	167.56
PART M HEATING WORK							
15mm copper pipe	M1	6.00m	1.32	16.50	11.28	4.17	31.95
Elbow	M2	4nr	1.12	14.00	4.40	2.76	21.16
Tee	M3	1nr	0.22	2.75	1.95	0.71	5.41
Radiator, double convector size 1400 x 520mm	M4	2nr	2.60	32.50	231.36	39.58	303.44
Break into existing pipe and insert tee	M5	1nr	0.75	9.38	3.95	2.00	15.32
Carried to summary			6.01	75.13	252.94	49.21	377.27

	Ref	Qty	Hours	Hours £	Mat'ls £	O & P £	Total £
PART N ALTERATION WORK							
Take out existing window size 1500 x 1000mm and lintel over, adapt opening to receive 1770 x 2000mm patio door and insert new lintel over (both measured separately) and make good	N1	1nr	20.00	190.00	20.00	31.50	241.50
Carried to summary			20.00	190.00	20.00	31.50	241.50

SUMMARY

	Hours	Hours £	Mat'ls £	O & P £	Total £
PART A PRELIMINARIES	0.00	0.00	0.00	0.00	1,324.25
PART B SUBSTRUCTURE TO DPC LEVEL	55.27	525.07	528.39	158.02	1,211.47
PART C EXTERNAL WALLS	58.75	558.13	1,117.38	251.33	1,926.83
PART D FLAT ROOF	60.08	564.66	630.30	179.24	1,374.20
PART E PITCHED ROOF	0.00	0.00	0.00	0.00	0.00
PART F WINDOWS AND EXTERNAL DOORS	19.14	181.83	916.96	164.82	1,263.61
PART G INTERNAL PARTITIONS AND DOORS	0.00	0.00	0.00	0.00	0.00
PART H WALL FINISHES	15.77	149.82	66.53	32.45	248.80
PART J FLOOR FINISHES	4.50	42.75	91.89	20.20	154.84
PART K CEILING FINISHES	10.29	68.39	33.60	15.30	117.29
PART L ELECTRICAL WORK	5.40	75.60	70.10	21.86	167.56
PART M HEATING WORK	6.01	75.13	252.94	49.21	377.27
PART N ALTERATION WORK	20.00	190.00	20.00	31.50	241.50
Final total	251.84	2,399.37	2,877.66	791.56	7,392.81

	Ref	Qty	Hours	Hours £	Mat'ls £	O & P £	Total £
PART A PRELIMINARIES							
Concrete mixer	A1	8 wks					360.00
Small tools	A2	9 wks					315.00
Scaffolding (m2/weeks)	A3	165.00					371.25
Skip	A4	5 wks					500.00
Clean up	A5	8 hrs					64.00
Carried to summary							1,610.25
PART B SUBSTRUCTURE TO DPC LEVEL							
Excavate topsoil 150mm thick by hand	B1	17.50m2	5.25	49.88	0.00	7.48	57.36
Excavate to reduce levels by hand	B2	5.00m3	12.50	118.75	0.00	17.81	136.56
Excavate for trench foundations by hand	B3	1.72m3	4.47	42.47	0.00	6.37	48.83
Earthwork support to sides of trenches	B4	15.18m2	6.07	57.67	18.22	11.38	87.27
Backfilling with excavated material	B5	0.63m3	0.37	3.52	0.00	0.53	4.04
Hardcore 225mm thick	B6	1.88m2	2.38	22.61	60.35	12.44	95.40
Hardcore filling to trench	B7	0.14m3	0.07	0.67	0.70	0.20	1.57
Concrete grade (1:3:6) in foundations	B8	1.40m3	1.89	17.96	95.11	16.96	130.02
Concrete grade (1:2:4) in bed 150mm thick	B9	11.88m2	3.56	33.82	129.25	24.46	187.53
Concrete (1:2:4) in cavity wall filling	B10	5.20m2	1.04	9.88	18.88	4.31	33.07
Carried forward			37.60	357.20	322.51	101.96	781.67

	Ref	Qty	Hours	Hours £	Mat'ls £	O & P £	Total £
Brought forward			37.60	357.20	322.51	101.96	781.67
Damp-proof membrane	B11	12.38m2	0.52	4.94	7.18	1.82	13.94
Reinforcement ref A193 in foundation	B12	5.20m2	0.62	5.89	6.76	1.90	14.55
Steel fabric reinforcement ref A193 in slab	B13	11.88m2	1.78	16.91	15.46	4.86	37.23
Solid blockwork 140mm thick in cavity wall	B14	6.76m2	8.79	83.51	95.34	26.83	205.67
Common bricks 112.5mm thick in cavity wall	B15	5.20m2	8.79	83.51	68.69	22.83	175.02
Facing bricks in 112.5mm thick in skin of cavity wall	B16	1.56m2	2.81	26.70	36.30	9.45	72.44
Form cavity 50mm wide in cavity wall	B17	6.76m2	0.20	1.90	4.06	0.89	6.85
DPC 112mm wide	B18	10.40m	0.52	4.94	8.42	2.00	15.36
DPC 140mm wide	B19	10.40m	0.62	5.89	10.09	2.40	18.38
Bond in block wall	B20	1.30m	0.58	5.51	2.83	1.25	9.59
Bond in half brick wall	B21	0.30m	0.46	4.37	0.61	0.75	5.73
50mm thick insulation board	B22	11.88m2	3.56	33.82	53.22	13.06	100.10
Carried to summary			66.85	635.08	631.47	189.98	1,456.53
PART C EXTERNAL WALLS							
Solid blockwork 140mm thick in cavity wall	C1	20.50m2	26.65	253.18	260.76	77.09	591.03
Facing brickwork 112.5mm thick in cavity wall	C2	20.50m2	36.90	350.55	472.73	123.49	946.77
75mm thick insulation in cavity wall	C3	20.50m2	4.51	42.85	119.30	24.32	186.47
Carried forward			68.06	646.57	852.79	224.90	1,724.26

	Ref	Qty	Hours	Hours £	Mat'ls £	O & P £	Total £
Brought forward			68.06	646.57	852.79	224.90	1,724.26
Steel lintel 2400mm long	C4	2nr	0.50	4.75	196.92	30.25	231.92
Steel lintel 1500mm long	C5	2nr	0.40	3.80	214.94	32.81	251.55
Close cavity wall at jambs	C7	8.96m	0.45	4.28	18.28	3.38	25.94
Close cavity wall at cills	C8	3.67m	0.18	1.71	7.49	1.38	10.58
Close cavity wall at top	C9	10.40m	0.52	4.94	21.22	3.92	30.08
DPC 112mm wide at jambs	C10	8.96m	0.45	4.28	7.26	1.73	13.27
DPC 112mm wide at cills	C11	3.67m	0.18	1.71	2.97	0.70	5.38
Carried to summary			70.74	672.03	1,321.87	299.09	2,292.99

PART D
FLAT ROOF

	Ref	Qty	Hours	Hours £	Mat'ls £	O & P £	Total £
200 x 50mm sawn softwood joists	D1	34.65m	8.66	82.27	107.42	28.45	218.14
200 x 50mm sawn softwood sprocket pieces	D2	12nr	1.68	15.96	16.08	4.81	36.85
18mm thick WPB grade decking	D3	16.70m2	15.03	135.27	131.26	39.98	306.51
50 x 50mm (avg) wide sawn softwood firrings	D4	34.65m	6.24	59.28	78.31	20.64	158.23
High density polyethylene vapour barrier 150mm thick	D5	15.00m2	3.00	28.50	142.05	25.58	196.13
100 x 75mm sawn softwood wall plate	D6	11.00m	3.30	31.35	24.42	8.37	64.14
100 x 75mm sawn softwood tilt fillet	D7	5.30m	1.59	15.11	13.99	4.36	33.46
Build in ends of 200 x 50mm joists	D8	11nr	3.85	36.58	1.65	5.73	43.96
Carried forward			43.35	404.31	515.18	137.92	1,057.41

	Ref	Qty	Hours	Hours £	Mat'ls £	O & P £	Total £
Brought forward			43.35	404.31	515.18	137.92	1,057.41
Rake out joint for flashing	D9	5.30m	1.86	17.67	0.80	2.77	21.24
6mm thick soffit 150mm wide	D10	11.30m	4.52	42.94	19.78	9.41	72.13
19mm wrought softwood fascia 200mm high	D11	11.30m	5.65	53.68	22.26	11.39	87.33
Three layer fibre-based roofing felt	D12	16.70m2	9.18	87.21	140.45	34.15	261.81
Felt turn-down 100mm girth	D13	11.30m	1.13	10.74	9.49	3.03	23.26
Felt flashing 150mm girth	D14	5.30m	0.63	5.99	8.90	2.23	17.12
112mm diameter PVC-U gutter	D15	5.30m	2.94	27.93	23.69	7.74	59.36
Stop end	D16	1nr	0.14	1.33	1.11	0.37	2.81
Stop end outlet	D17	1nr	0.25	2.38	1.94	0.65	4.96
68mm diameter PVC-U down pipe	D18	2.50m	0.63	5.99	8.48	2.17	16.63
Shoe	D19	1nr	0.30	2.85	2.09	0.74	5.68
Paint fascia and soffit	D20	3.39m2	2.37	22.52	3.25	3.86	29.63
Carried to summary			72.95	685.51	757.42	216.44	1,659.37
PART E PITCHED ROOF			N/A	N/A	N/A	N/A	N/A
PART F WINDOWS AND EXTERNAL DOORS							
PVC-U door size 840 x 1980mm complete (B)	F1	0.00	0.00	0.00	0.00	0.00	0.00
PVC-U sliding patio door size 1700 x 2075mm (C)	F2	2nr	14.00	133.00	587.04	108.01	828.05
Carried forward			14.00	133.00	587.04	108.01	828.05

	Ref	Qty	Hours	Hours £	Mat'ls £	O & P £	Total £
Brought forward			14.00	133.00	587.04	108.01	828.05
PVC-U window size 1200 x 1200mm complete (A)	F3	2nr	4.00	38.00	315.18	52.98	406.16
25 x 225mm wrought softwood window board	F4	2.40m	0.72	6.84	12.44	2.89	22.17
Paint window board	F5	2.40m	0.42	3.99	2.30	0.94	7.23
Carried to summary			19.14	181.83	916.96	164.82	1,263.61
PART G INTERNAL PARTITIONS AND DOORS		N/A	N/A	N/A	N/A	N/A	N/A
PART H WALL FINISHES							
19 x 100mm wrought softwood skirting	H1	8.54m	1.71	16.25	8.96	3.78	28.99
12mm plasterboard fixed to walls with dabs	H2	19.00m2	7.41	70.40	37.81	16.23	124.44
12mm plasterboard fixed to walls less than 300mm wide with dabs	H3	13.47m	2.43	23.09	12.39	5.32	40.80
Two coats emulsion paint to plastered walls	H4	21.02m2	5.04	47.88	12.40	9.04	69.32
Paint skirting	H5	8.54m	1.28	12.16	4.10	2.44	18.70
Carried to summary			17.87	169.77	75.66	36.81	282.24
PART J FLOOR FINISHES							
Cement and sand floor screed 40mm thick	J1	11.88m2	2.97	28.22	36.00	9.63	73.85
Vinyl floor tiles, size 300 x 300mm	J2	11.88m2	2.85	27.08	83.62	16.60	127.30
Carried to summary			5.82	55.29	119.62	26.24	201.15

	Ref	Qty	Hours	Hours £	Mat'ls £	O & P £	Total £
PART K **CEILING FINISHES**							
Plasterboard with taped butt joints fixed to joists	K1	11.88m2	4.28	2.08	21.98	3.61	27.67
5mm skim coat to plasterboard ceilings	K2	11.88m2	5.94	56.43	14.50	10.64	81.57
Two coats emulsion paint to ceilings	K3	11.88m2	3.09	29.36	7.01	5.45	41.82
Carried to summary			13.31	87.87	43.49	19.70	151.06
PART L **ELECTRICAL WORK**							
13 amp double switched socket outlet with neon	L1	3nr	1.20	16.80	24.78	6.24	47.82
Lighting point	L2	2nr	0.70	9.80	13.60	3.51	26.91
Lighting switch	L3	2nr	0.50	7.00	7.94	2.24	17.18
Lighting wiring	L4	7.00m	0.70	9.80	7.21	2.55	19.56
Power cable	L5	18.00m	2.70	37.80	19.80	8.64	66.24
Carried to summary			5.80	81.20	73.33	23.18	177.71
PART M **HEATING WORK**							
15mm copper pipe	M1	7.00m	1.54	19.25	13.16	4.86	37.27
Elbow	M2	4nr	1.12	14.00	4.40	2.76	21.16
Tee	M3	1nr	0.22	2.75	1.95	0.71	5.41
Radiator, double convector size 1400 x 520mm	M4	2nr	2.60	32.50	231.36	39.58	303.44
Break into existing pipe and insert tee	M5	1nr	0.75	9.38	3.95	2.00	15.32
Carried to summary			6.23	77.88	254.82	49.90	382.60

	Ref	Qty	Hours	Hours £	Mat'ls £	O & P £	Total £
PART N							
ALTERATION WORK							
Take out existing window size 1500 x 1000mm and lintel over, adapt opening to receive 1770 x 2000mm patio door and insert new lintel over (both measured separately) and make good	N1	1nr	20.00	190.00	20.00	31.50	241.50
Carried to summary			20.00	190.00	20.00	31.50	241.50

SUMMARY

	Hours	Hours £	Mat'ls £	O & P £	Total £
PART A PRELIMINARIES	0.00	0.00	0.00	0.00	1,610.25
PART B SUBSTRUCTURE TO DPC LEVEL	66.85	635.08	631.47	189.98	1,456.53
PART C EXTERNAL WALLS	70.74	672.03	1,321.87	299.09	2,292.99
PART D FLAT ROOF	72.95	685.51	757.42	216.44	1,659.37
PART E PITCHED ROOF	0.00	0.00	0.00	0.00	0.00
PART F WINDOWS AND EXTERNAL DOORS	19.14	181.83	916.96	164.82	1,263.61
PART G INTERNAL PARTITIONS AND DOORS	0.00	0.00	0.00	0.00	0.00
PART H WALL FINISHES	17.87	169.77	75.66	36.81	282.24
PART J FLOOR FINISHES	5.82	55.29	119.62	26.24	201.15
PART K CEILING FINISHES	13.31	87.87	43.49	19.70	151.06
PART L ELECTRICAL WORK	5.80	81.20	73.33	23.18	177.71
PART M HEATING WORK	6.23	77.88	254.82	49.90	382.60
PART N ALTERATION WORK	20.00	190.00	20.00	31.50	241.50
Final total	298.71	2,836.46	4,214.64	1,057.66	9,719.01

	Ref	Qty	Hours	Hours £	Mat'ls £	O & P £	Total £
PART A PRELIMINARIES							
Concrete mixer	A1	9 wks					405.00
Small tools	A2	10 wks					350.00
Scaffolding (m2/weeks)	A3	210.00					472.50
Skip	A4	6 wks					600.00
Clean up	A5	10 hrs					80.00
Carried to summary							1,907.50
PART B SUBSTRUCTURE TO DPC LEVEL							
Excavate topsoil 150mm thick by hand	B1	19.85m2	5.96	56.62	0.00	8.49	65.11
Excavate to reduce levels by hand	B2	5.95m3	14.88	141.36	0.00	21.20	162.56
Excavate for trench foundations by hand	B3	1.88m3	3.17	30.12	0.00	4.52	34.63
Earthwork support to sides of trenches	B4	16.64m2	4.90	46.55	19.97	9.98	76.50
Backfilling with excavated material	B5	0.70m3	0.28	2.66	0.00	0.40	3.06
Hardcore 225mm thick	B6	14.58m2	2.19	20.81	74.07	14.23	109.11
Hardcore filling to trench	B7	0.15m3	0.06	0.57	0.76	0.20	1.53
Concrete grade (1:3:6) in foundations	B8	1.54m3	1.53	14.54	104.63	17.87	137.04
Concrete grade (1:2:4) in bed 150mm thick	B9	14.58m2	4.37	41.52	158.63	30.02	230.17
Concrete (1:2:4) in cavity wall filling	B10	5.70m2	0.84	7.98	20.69	4.30	32.97
Carried forward			38.18	362.71	378.75	111.22	852.68

	Ref	Qty	Hours	Hours £	Mat'ls £	O & P £	Total £
Brought forward			38.18	362.71	378.75	111.22	852.68
Damp-proof membrane	B11	15.13m2	0.61	5.80	7.62	2.01	15.43
Reinforcement ref A193 in foundation	B12	5.70m2	0.68	6.46	7.41	2.08	15.95
Steel fabric reinforcement ref A193 in slab	B13	14.58m2	2.19	20.81	18.95	5.96	45.72
Solid blockwork 140mm thick in cavity wall	B14	7.40m2	9.63	91.49	94.70	27.93	214.11
Common bricks 112.5mm thick in cavity wall	B15	5.70m2	9.63	91.49	75.18	25.00	191.66
Facing bricks in 112.5mm thick in skin of cavity wall	B16	1.71m2	2.27	21.57	39.79	9.20	70.56
Form cavity 50mm wide in cavity wall	B17	7.41m2	0.22	2.09	4.45	0.98	7.52
DPC 112mm wide	B18	11.40m	0.57	5.42	9.23	2.20	16.84
DPC 140mm wide	B19	11.40m	0.68	6.46	11.06	2.63	20.15
Bond in block wall	B20	1.30m	0.58	5.51	2.83	1.25	9.59
Bond in half brick wall	B21	0.30m	0.46	4.37	0.61	0.75	5.73
50mm thick insulation board	B22	14.58m2	4.37	41.52	65.32	16.03	122.86
Carried to summary			70.07	665.67	715.90	207.23	1,588.80
PART C EXTERNAL WALLS							
Solid blockwork 140mm thick in cavity wall	C1	19.90m2	33.67	319.87	253.13	85.95	658.94
Facing brickwork 112.5mm thick in cavity wall	C2	19.90m2	46.62	442.89	458.89	135.27	1,037.05
75mm thick insulation in cavity wall	C3	19.90m2	4.38	41.61	115.82	23.61	181.04
Carried forward			84.67	804.37	827.84	244.83	1,877.04

	Ref	Qty	Hours	Hours £	Mat'ls £	O & P £	Total £
Brought forward			84.67	804.37	827.84	244.83	1,877.04
Steel lintel 2400mm long	C4	2nr	0.50	4.75	196.92	30.25	231.92
Steel lintel 1500mm long	C5	3nr	0.60	5.70	322.41	49.22	377.33
Steel lintel 1150mm long	C6	1nr	0.15	1.43	47.28	0.21	1.64
Close cavity wall at jambs	C7	17.32m	0.87	8.27	35.33	6.54	50.13
Close cavity wall at cills	C8	5.71m	0.29	2.76	11.65	2.16	16.57
Close cavity wall at top	C9	11.40m	0.57	5.42	23.26	4.30	32.98
DPC 112mm wide at jambs	C10	17.32m	0.87	8.27	14.03	3.34	25.64
DPC 112mm wide at cills	C11	5.71m	0.29	2.76	4.63	1.11	8.49
Carried to summary			88.66	842.27	1,436.07	341.75	2,620.09
PART D FLAT ROOF							
200 x 50mm sawn softwood joists	D1	40.95m	10.24	97.28	126.95	33.63	257.86
200 x 50mm sawn softwood sprocket pieces	D2	12nr	1.68	15.96	16.08	4.81	36.85
18mm thick WPB grade decking	D3	19.85m2	17.86	160.74	156.02	47.51	364.27
50 x 50mm (avg) wide sawn softwood firrings	D4	40.95m	7.37	70.02	92.55	24.38	186.95
High density polyethylene vapour barrier 150mm thick	D5	18.00m2	3.60	34.20	170.46	30.70	235.36
100 x 75mm sawn softwood wall plate	D6	12.00m	3.69	35.06	26.64	9.25	70.95
100 x 75mm sawn softwood tilt fillet	D7	6.30m	1.58	15.01	16.63	4.75	36.39
Build in ends of 200 x 50mm joists	D8	13nr	4.53	43.04	1.95	6.75	51.73
Carried forward			50.55	471.30	607.28	161.79	1,240.36

	Ref	Qty	Hours	Hours £	Mat'ls £	O & P £	Total £
Brought forward			50.55	471.30	607.28	161.79	1,240.36
Rake out joint for flashing	D9	6.30m	2.21	21.00	0.95	3.29	25.24
6mm thick soffit 150mm wide	D10	12.30m	4.90	46.55	21.30	10.18	78.03
19mm wrought softwood fascia 200mm high	D11	12.30m	6.15	58.43	24.23	12.40	95.05
Three layer fibre-based roofing felt	D12	19.85m2	10.92	103.74	166.93	40.60	311.27
Felt turn-down 100mm girth	D13	12.30m	1.23	11.69	10.33	3.30	25.32
Felt flashing 150mm girth	D14	6.30m	0.76	7.22	10.58	2.67	20.47
112mm diameter PVC-U gutter	D15	6.30m	1.64	15.58	28.16	6.56	50.30
Stop end	D16	1nr	0.14	1.33	1.11	0.37	2.81
Stop end outlet	D17	1nr	0.25	2.38	1.94	0.65	4.96
68mm diameter PVC-U down pipe	D18	2.50m	0.63	5.99	8.48	2.17	16.63
Shoe	D19	1nr	0.30	2.85	2.09	0.74	5.68
Paint fascia and soffit	D20	3.69m2	2.58	24.51	3.54	4.21	32.26
Carried to summary			82.26	772.54	886.92	248.92	1,908.38
PART E PITCHED ROOF			N/A	N/A	N/A	N/A	N/A
PART F WINDOWS AND EXTERNAL DOORS							
PVC-U door size 840 x 1980mm complete (B)	F1	1nr	2.50	23.75	209.99	35.06	268.80
PVC-U sliding patio door size 1700 x 2075mm (C)	F2	2nr	14.00	133.00	587.04	108.01	828.05
Carried forward			16.50	156.75	797.03	143.07	1,096.85

	Ref	Qty	Hours	Hours £	Mat'ls £	O & P £	Total £
Brought forward			16.50	156.75	797.03	143.07	1,096.85
PVC-U window size 1200 x 1200mm complete (A)	F3	3nr	6.00	57.00	472.77	79.47	609.24
25 x 225mm wrought softwood window board	F4	3.60m	1.08	10.26	22.39	4.90	37.55
Paint window board	F5	3.60m	0.57	5.42	3.46	1.33	10.21
Carried to summary			24.15	229.43	1,295.65	228.76	1,753.84
PART G INTERNAL PARTITIONS AND DOORS		N/A	N/A	N/A	N/A	N/A	N/A
PART H WALL FINISHES							
19 x 100mm wrought softwood skirting	H1	8.70m	1.74	16.53	11.92	4.27	32.72
12mm plasterboard fixed to walls with dabs	H2	18.40m2	7.18	68.21	36.62	15.72	120.55
12mm plasterboard fixed to walls less than 300mm wide with dabs	H3	23.03m	4.15	39.43	23.03	9.37	71.82
Two coats emulsion paint to plastered walls	H4	21.85m2	5.22	49.59	12.89	9.37	71.85
Paint skirting	H5	8.70m	1.31	12.45	4.18	2.49	19.12
Carried to summary			19.60	186.20	88.64	41.23	316.07
PART J FLOOR FINISHES							
Cement and sand floor screed 40mm thick	J1	14.58m2	3.65	34.68	44.84	11.93	91.44
Vinyl floor tiles, size 300 x 300mm	J2	14.58m2	3.50	33.25	101.77	20.25	155.27
Carried to summary			7.15	67.93	146.61	32.18	246.72

	Ref	Qty	Hours	Hours £	Mat'ls £	O & P £	Total £
PART K **CEILING FINISHES**							
Plasterboard with taped butt joints fixed to joists	K1	14.58m2	5.25	2.08	26.97	4.36	33.41
5mm skim coat to plasterboard ceilings	K2	14.58m2	7.29	69.26	17.79	13.06	100.10
Two coats emulsion paint to ceilings	K3	14.58m2	3.79	36.01	8.60	6.69	51.30
Carried to summary			16.33	107.34	53.36	24.11	184.81
PART L **ELECTRICAL WORK**							
13 amp double switched socket outlet with neon	L1	4nr	1.60	22.40	33.04	8.32	63.76
Lighting point	L2	2nr	0.70	9.80	13.60	3.51	26.91
Lighting switch	L3	3nr	0.75	10.50	11.91	3.36	25.77
Lighting wiring	L4	8.00m	0.80	11.20	8.24	2.92	22.36
Power cable	L5	20.00m	3.00	42.00	22.00	9.60	73.60
Carried to summary			6.85	95.90	88.79	27.70	212.39
PART M **HEATING WORK**							
15mm copper pipe	M1	8.00m	1.76	22.00	15.04	5.56	42.60
Elbow	M2	4nr	1.12	14.00	4.40	2.76	21.16
Tee	M3	1nr	0.22	2.75	1.95	0.71	5.41
Radiator, double convector size 1400 x 520mm	M4	2nr	2.60	32.50	231.36	39.58	303.44
Break into existing pipe and insert tee	M5	1nr	0.75	9.38	3.95	2.00	15.32
Carried to summary			6.45	80.63	256.70	50.60	387.92

	Ref	Qty	Hours	Hours £	Mat'ls £	O & P £	Total £
PART N **ALTERATION WORK**							
Take out existing window size 1500 x 1000mm and lintel over, adapt opening to receive 1770 x 2000mm patio door and insert new lintel over (both measured separately) and make good	N1	1nr	20.00	190.00	20.00	31.50	241.50
Carried to summary			20.00	190.00	20.00	31.50	241.50

SUMMARY

	Hours	Hours £	Mat'ls £	O & P £	Total £
PART A PRELIMINARIES	0.00	0.00	0.00	0.00	1,907.50
PART B SUBSTRUCTURE TO DPC LEVEL	70.07	665.67	715.90	207.23	1,588.80
PART C EXTERNAL WALLS	88.66	842.27	1,436.07	341.75	2,620.09
PART D FLAT ROOF	82.26	772.54	886. 2	248.92	1,908.38
PART E PITCHED ROOF	0.00	0.00	0.00	0.00	0.00
PART F WINDOWS AND EXTERNAL DOORS	24.15	229.43	1,295.65	228.76	1,753.84
PART G INTERNAL PARTITIONS AND DOORS	0.00	0.00	0.00	0.00	0.00
PART H WALL FINISHES	19.60	186.20	88.64	41.23	316.07
PART J FLOOR FINISHES	7.15	67.93	146.61	32.18	246.72
PART K CEILING FINISHES	16.33	107.34	53.36	24.11	184.81
PART L ELECTRICAL WORK	6.85	95.90	88.79	27.70	212.39
PART M HEATING WORK	6.45	80.63	256.70	50.60	387.92
PART N ALTERATION WORK	20.00	190.00	20.00	31.50	241.50
Final total	280.93	2,583.68	4,237.31	1,023.16	9,751.64

	Ref	Qty	Hours	Hours £	Mat'ls £	O & P £	Total £
PART A PRELIMINARIES							
Concrete mixer	A1	8 wks					360.00
Small tools	A2	9 wks					315.00
Scaffolding (m2/weeks)	A3	270 wks					607.50
Skip	A4	5 wks					500.00
Clean up	A5	10 hrs					80.00
Carried to summary							1,862.50
PART B SUBSTRUCTURE TO DPC LEVEL							
Excavate topsoil 150mm thick by hand	B1	17.85m2	5.36	50.92	0.00	7.64	58.56
Excavate to reduce levels by hand	B2	5.35m3	13.38	127.11	0.00	19.07	146.18
Excavate for trench foundations by hand	B3	1.88m3	4.89	46.46	0.00	6.97	53.42
Earthwork support to sides of trenches	B4	16.64m2	6.66	63.27	19.97	12.49	95.73
Backfilling with excavated material	B5	0.63m3	0.37	3.52	0.00	0.53	4.04
Hardcore 225mm thick	B6	13.69m2	2.74	26.03	69.54	14.34	109.91
Hardcore filling to trench	B7	0.14m3	0.07	0.67	0.70	0.20	1.57
Concrete grade (1:3:6) in foundations	B8	1.54m3	2.08	19.76	104.63	18.66	143.05
Concrete grade (1:2:4) in bed 150mm thick	B9	13.69m2	4.11	39.05	148.94	28.20	216.18
Concrete (1:2:4) in cavity wall filling	B10	5.70m2	1.14	10.83	20.69	4.73	36.25
Carried forward			40.80	387.60	364.47	112.81	864.88

	Ref	Qty	Hours	Hours £	Mat'ls £	O & P £	Total £
Brought forward			40.80	387.60	364.47	112.81	864.88
Damp-proof membrane	B11	13.13m2	0.53	5.04	6.62	1.75	13.40
Reinforcement ref A193 in foundation	B12	5.70m2	0.68	6.46	7.41	2.08	15.95
Steel fabric reinforcement ref A193 in slab	B13	13.69m2	2.05	19.48	17.80	5.59	42.87
Solid blockwork 140mm thick in cavity wall	B14	7.41m2	9.63	91.49	94.70	27.93	214.11
Common bricks 112.5mm thick in cavity wall	B15	5.70m2	9.63	91.49	75.18	25.00	191.66
Facing bricks in 112.5mm thick in skin of cavity wall	B16	1.71m2	2.27	21.57	39.79	9.20	70.56
Form cavity 50mm wide in cavity wall	B17	7.41m2	0.22	2.09	4.45	0.98	7.52
DPC 112mm wide	B18	11.40m	0.57	5.42	9.23	2.20	16.84
DPC 140mm wide	B19	11.40m	0.68	6.46	11.06	2.63	20.15
Bond in block wall	B20	1.30m	0.58	5.51	2.83	1.25	9.59
Bond in half brick wall	B21	0.30m	0.46	4.37	0.61	0.75	5.73
50mm thick insulation board	B22	13.69m2	4.11	39.05	61.33	15.06	115.43
Carried to summary			72.21	686.00	695.48	207.22	1,572.75
PART C EXTERNAL WALLS							
Solid blockwork 140mm thick in cavity wall	C1	19.90m2	25.87	245.77	253.13	74.83	573.73
Facing brickwork 112.5mm thick in cavity wall	C2	19.90m2	35.82	340.29	458.89	119.88	919.06
75mm thick insulation in cavity wall	C3	19.90m2	4.38	41.61	115.82	23.61	181.04
Carried forward			66.07	627.67	827.84	218.33	1,673.83

	Ref	Qty	Hours	Hours £	Mat'ls £	O & P £	Total £
Brought forward			66.07	627.67	827.84	218.33	1,673.83
Steel lintel 2400mm long	C4	2nr	0.50	4.75	196.92	30.25	231.92
Steel lintel 1500mm long	C5	3nr	0.60	5.70	322.41	49.22	377.33
Steel lintel 1150mm long	C6	1nr	0.15	1.43	47.28	7.31	56.01
Close cavity wall at jambs	C7	17.32m	0.87	8.27	35.33	6.54	50.13
Close cavity wall at cills	C8	5.71m	0.29	2.76	11.65	2.16	16.57
Close cavity wall at top	C9	11.40m	0.57	5.42	23.26	4.30	32.98
DPC 112mm wide at jambs	C10	17.32m	0.87	8.27	14.03	3.34	25.64
DPC 112mm wide at cills	C11	5.71m	0.29	2.76	4.63	1.11	8.49
Carried to summary			70.06	665.57	1,436.07	315.25	2,416.89
PART D FLAT ROOF							
200 x 50mm sawn softwood joists	D1	37.35m	9.34	88.73	115.79	30.68	235.20
200 x 50mm sawn softwood sprocket pieces	D2	16nr	2.24	21.28	21.44	6.41	49.13
18mm thick WPB grade decking	D3	17.85m2	16.07	144.63	140.30	42.74	327.67
50 x 50mm (avg) wide sawn softwood firrings	D4	37.35m	6.72	63.84	84.41	22.24	170.49
High density polyethylene vapour barrier 150mm thick	D5	16.00m2	3.20	30.40	151.52	27.29	209.21
100 x 75mm sawn softwood wall plate	D6	12.00m	3.60	34.20	26.64	9.13	69.97
100 x 75mm sawn softwood tilt fillet	D7	4.30m	1.08	10.26	11.36	3.24	24.86
Build in ends of 200 x 50mm joists	D8	9nr	3.15	29.93	1.35	4.69	35.97
Carried forward			45.40	423.27	552.81	146.41	1,122.48

	Ref	Qty	Hours	Hours £	Mat'ls £	O & P £	Total £
Brought forward			45.40	423.27	552.81	146.41	1,122.48
Rake out joint for flashing	D9	4.30m	1.51	14.35	0.65	2.25	17.24
6mm thick soffit 150mm wide	D10	12.30m	4.92	46.74	21.53	10.24	78.51
19mm wrought softwood fascia 200mm high	D11	12.30m	6.15	58.43	24.23	12.40	95.05
Three layer fibre-based roofing felt	D12	17.85m2	9.82	93.29	150.12	36.51	279.92
Felt turn-down 100mm girth	D13	12.30m	1.23	11.69	10.33	3.30	25.32
Felt flashing 150mm girth	D14	4.30m	0.52	4.94	7.22	1.82	13.98
112mm diameter PVC-U gutter	D15	4.30m	1.12	10.64	19.22	4.48	34.34
Stop end	D16	1nr	0.14	1.33	1.11	0.37	2.81
Stop end outlet	D17	1nr	0.25	2.38	1.94	0.65	4.96
68mm diameter PVC-U down pipe	D18	2.50m	0.63	5.99	8.48	2.17	16.63
Shoe	D19	1nr	0.30	2.85	2.09	0.74	5.68
Paint fascia and soffit	D20	3.69m2	1.74	16.53	3.54	3.01	23.08
Carried to summary			73.73	692.40	803.27	224.35	1,720.01
PART E PITCHED ROOF			N/A	N/A	N/A	N/A	N/A
PART F WINDOWS AND EXTERNAL DOORS							
PVC-U door size 840 x 1980mm complete (B)	F1	1nr	2.50	23.75	209.99	35.06	268.80
PVC-U sliding patio door size 1700 x 2075mm (C)	F2	2nr	14.00	133.00	587.04	108.01	828.05
Carried forward			16.50	156.75	797.03	143.07	1,096.85

	Ref	Qty	Hours	Hours £	Mat'ls £	O & P £	Total £
Brought forward			16.50	156.75	797.03	143.07	1,096.85
PVC-U window size 1200 x 1200mm complete (A)	F3	3nr	6.00	57.00	472.77	79.47	609.24
25 x 225mm wrought softwood window board	F4	3.60m	1.08	10.26	22.39	4.90	37.55
Paint window board	F5	3.60m	0.57	5.42	3.46	1.33	10.21
Carried to summary			24.15	229.43	1,295.65	228.76	1,753.84
PART G INTERNAL PARTITIONS AND DOORS		N/A	N/A	N/A	N/A	N/A	N/A
PART H WALL FINISHES							
19 x 100mm wrought softwood skirting	H1	8.70m	1.74	16.53	11.92	4.27	32.72
12mm plasterboard fixed to walls with dabs	H2	18.00m2	7.18	68.21	36.62	15.72	120.55
12mm plasterboard fixed to walls less than 300mm wide with dabs	H3	23.03m	4.15	39.43	23.03	9.37	71.82
Two coats emulsion paint to walls	H4	21.85m2	5.22	49.59	12.89	9.37	71.85
Paint skirting	H5	8.70m	1.31	12.45	4.18	2.49	19.12
Carried to summary			19.60	186.20	88.64	41.23	316.07
PART J FLOOR FINISHES							
Cement and sand floor screed 40mm thick	J1	12.58m2	3.15	29.93	38.12	10.21	78.25
Vinyl floor tiles, size 300 x 300mm	J2	12.58m2	3.02	28.69	87.81	17.48	133.98
Carried to summary			6.17	58.62	125.93	27.68	212.23

	Ref	Qty	Hours	Hours £	Mat'ls £	O & P £	Total £
PART K CEILING FINISHES							
Plasterboard with taped butt joints fixed to joists	K1	12.58m2	4.53	2.08	23.27	3.80	29.15
5mm skim coat to plasterboard ceilings	K2	12.58m2	6.29	59.76	15.35	11.27	86.37
Two coats emulsion paint to ceilings	K3	12.58m2	3.27	31.07	7.42	5.77	44.26
Carried to summary			14.09	92.90	46.04	20.84	159.78
PART L ELECTRICAL WORK							
13 amp double switched socket outlet with neon	L1	3nr	1.20	16.80	24.78	6.24	47.82
Lighting point	L2	2nr	0.70	9.80	13.60	3.51	26.91
Lighting switch	L3	3nr	0.50	7.00	11.91	2.84	21.75
Lighting wiring	L4	8.00m	0.80	11.20	8.24	2.92	22.36
Power cable	L5	18.00m	2.70	37.80	19.80	8.64	66.24
Carried to summary			5.90	82.60	78.33	24.14	185.07
PART M HEATING WORK							
15mm copper pipe	M1	9.00m	1.98	24.75	16.92	6.25	47.92
Elbow	M2	4nr	1.12	14.00	4.40	2.76	21.16
Tee	M3	1nr	0.22	2.75	1.95	0.71	5.41
Radiator, double convector size 1400 x 520mm	M4	2nr	2.60	32.50	231.36	39.58	303.44
Break into existing pipe and insert tee	M5	1nr	0.75	9.38	3.95	2.00	15.32
Carried to summary			6.67	83.38	258.58	51.29	393.25

	Ref	Qty	Hours	Hours £	Mat'ls £	O & P £	Total £
PART N							
ALTERATION WORK							
Take out existing window size 1500 x 1000mm and lintel over, adapt opening to receive 1770 x 2000mm patio door and insert new lintel over (both measured separately) and make good	N1	1nr	20.00	190.00	20.00	31.50	241.50
Carried to summary			20.00	190.00	20.00	31.50	241.50

SUMMARY

	Hours	Hours £	Mat'ls £	O & P £	Total £
PART A PRELIMINARIES	0.00	0.00	0.00	0.00	1,862.50
PART B SUBSTRUCTURE TO DPC LEVEL	72.21	686.00	695.48	207.22	1,572.75
PART C EXTERNAL WALLS	70.06	665.57	1,436.07	315.25	2,416.89
PART D FLAT ROOF	73.73	692.40	803.27	224.35	1,720.01
PART E PITCHED ROOF	0.00	0.00	0.00	0.00	0.00
PART F WINDOWS AND EXTERNAL DOORS	24.15	229.43	1,295.65	228.76	1,753.84
PART G INTERNAL PARTITIONS AND DOORS	0.00	0.00	0.00	0.00	0.00
PART H WALL FINISHES	19.60	186.20	88.64	41.23	316.07
PART J FLOOR FINISHES	6.17	58.62	125.93	27.68	212.23
PART K CEILING FINISHES	14.09	92.90	46.04	20.84	159.78
PART L ELECTRICAL WORK	5.90	82.60	78.33	24.14	185.07
PART M HEATING WORK	6.67	83.38	258.58	51.29	393.25
PART N ALTERATION WORK	20.00	190.00	20.00	31.50	241.50
Final total	312.58	2,967.10	4,847.99	1,172.26	10,833.89

	Ref	Qty	Hours	Hours £	Mat'ls £	O & P £	Total £
PART A PRELIMINARIES							
Concrete mixer	A1	9 wks					405.00
Small tools	A2	10 wks					350.00
Scaffolding (m2/weeks)	A3	325.00					731.25
Skip	A4	6 wks					600.00
Clean up	A5	10 hrs					80.00
Carried to summary							2,166.25
PART B SUBSTRUCTURE TO DPC LEVEL							
Excavate topsoil 150mm thick by hand	B1	22.01m2	6.60	62.70	0.00	9.41	72.11
Excavate to reduce levels by hand	B2	6.60m3	16.60	157.70	0.00	23.66	181.36
Excavate for trench foundations by hand	B3	2.05m3	5.33	50.64	0.00	7.60	58.23
Earthwork support to sides of trenches	B4	18.10m2	7.24	68.78	21.72	13.58	104.08
Backfilling with excavated material	B5	0.63m3	0.42	3.99	0.00	0.60	4.59
Hardcore 225mm thick	B6	16.28m2	2.44	23.18	82.70	15.88	121.76
Hardcore filling to trench	B7	0.15m3	0.08	0.76	0.76	0.23	1.75
Concrete grade (1:3:6) in foundations	B8	1.67	2.25	21.38	113.46	20.23	155.06
Concrete grade (1:2:4) in bed 150mm thick	B9	16.28m2	4.88	46.36	177.13	33.52	257.01
Concrete (1:2:4) in cavity wall filling	B10	6.20m2	1.24	11.78	22.51	5.14	39.43
Carried forward			47.08	447.26	418.28	129.83	995.37

	Ref	Qty	Hours	Hours £	Mat'ls £	O & P £	Total £
Brought forward			47.08	447.26	418.28	129.83	995.37
Damp-proof membrane	B11	16.88m2	0.68	6.46	9.79	2.44	18.69
Reinforcement ref A193 in foundation	B12	6.20m2	0.74	7.03	8.06	2.26	17.35
Steel fabric reinforcement ref A193 in slab	B13	16.28m2	2.44	23.18	21.16	6.65	50.99
Solid blockwork 140mm thick in cavity wall	B14	8.06m2	10.48	99.56	103.00	30.38	232.94
Common bricks 112.5mm thick in cavity wall	B15	6.20m2	10.48	99.56	81.78	27.20	208.54
Facing bricks in 112.5mm thick in skin of cavity wall	B16	1.86m2	3.35	31.83	43.28	11.27	86.37
Form cavity 50mm wide in cavity wall	B17	8.06m2	0.24	2.28	4.84	1.07	8.19
DPC 112mm wide	B18	12.40m	0.62	5.89	10.06	2.39	18.34
DPC 140mm wide	B19	12.40m	0.74	7.03	12.03	2.86	21.92
Bond in block wall	B20	1.30m	0.58	5.51	2.83	1.25	9.59
Bond in half brick wall	B21	0.30m	0.46	4.37	0.61	0.75	5.73
50mm thick insulation board	B22	16.28m2	4.88	46.36	72.93	17.89	137.18
Carried to summary			82.77	786.32	788.65	236.24	1,811.21
PART C EXTERNAL WALLS							
Solid blockwork 140mm thick in cavity wall	C1	20.96m2	27.25	258.88	266.61	78.82	604.31
Facing brickwork 112.5mm thick in cavity wall	C2	20.96m2	37.33	354.64	483.33	125.69	963.66
75mm thick insulation in cavity wall	C3	20.96m2	4.61	43.80	121.99	24.87	190.65
Carried forward			69.19	657.31	871.93	229.39	1,758.62

	Ref	Qty	Hours	Hours £	Mat'ls £	O & P £	Total £
Brought forward			69.19	657.31	871.93	229.39	1,758.62
Steel lintel 2400mm long	C4	2nr	0.50	4.75	196.92	30.25	231.92
Steel lintel 1500mm long	C5	4nr	0.80	7.60	472.77	72.06	552.43
Steel lintel 1150mm long	C6	1nr	0.15	1.43	47.24	7.30	55.96
Close cavity wall at jambs	C7	19.72m	0.99	9.41	40.23	7.45	57.08
Close cavity wall at cills	C8	6.91m	0.35	3.33	14.10	2.61	20.04
Close cavity wall at top	C9	12.40m	0.62	5.89	25.30	4.68	35.87
DPC 112mm wide at jambs	C10	19.72m	0.99	9.41	15.97	3.81	29.18
DPC 112mm wide at cills	C11	6.91m	0.35	3.33	5.60	1.34	10.26
Carried to summary			73.79	701.01	1,642.82	351.57	2,695.40
PART D FLAT ROOF							
200 x 50mm sawn softwood joists	D1	45.65m	11.41	108.40	141.52	37.49	287.40
200 x 50mm sawn softwood sprocket pieces	D2	16nr	2.24	21.28	21.44	6.41	49.13
18mm thick WPB grade decking	D3	22.00m2	19.80	178.20	172.92	52.67	403.79
50 x 50mm (avg) wide sawn softwood firrings	D4	45.65m	8.22	78.09	103.17	27.19	208.45
High density polyethylene vapour barrier 150mm thick	D5	20.00m2	4.00	38.00	189.40	34.11	261.51
100 x 75mm sawn softwood wall plate	D6	13.00m	3.90	37.05	28.86	9.89	75.80
100 x 75mm sawn softwood tilt fillet	D7	5.30m	1.33	12.64	13.99	3.99	30.62
Build in ends of 200 x 50mm joists	D8	11nr	3.85	36.58	1.65	5.73	43.96
Carried forward			54.75	510.23	672.95	177.48	1,360.65

	Ref	Qty	Hours	Hours £	Mat'ls £	O & P £	Total £
Brought forward			54.75	510.23	672.95	177.48	1,360.65
Rake out joint for flashing	D9	5.30m	1.86	17.67	0.80	2.77	21.24
6mm thick soffit 150mm wide	D10	13.30m	5.32	50.54	23.28	11.07	84.89
19mm wrought softwwod fascia 200mm high	D11	13.30m	6.65	63.18	26.20	13.41	102.78
Three layer fibre-based roofing felt	D12	22.00m2	12.10	114.95	185.02	45.00	344.97
Felt turn-down 100mm girth	D13	13.30m	1.33	12.64	11.17	3.57	27.38
Felt flashing 150mm girth	D14	5.30m	0.63	5.99	8.90	2.23	17.12
112mm diameter PVC-U gutter	D15	5.30m	1.38	13.11	23.69	5.52	42.32
Stop end	D16	1nr	0.14	1.33	1.11	0.37	2.81
Stop end outlet	D17	1nr	0.25	2.38	1.94	0.65	4.96
68mm diameter PVC-U down pipe	D18	2.50m	0.63	5.99	8.48	2.17	16.63
Shoe	D19	1nr	0.30	2.85	2.09	0.74	5.68
Paint fascia and soffit	D20	3.99m2	2.79	26.51	3.83	4.55	34.89
Carried to summary			88.13	827.34	969.46	269.52	2,066.31
PART E PITCHED ROOF			N/A	N/A	N/A	N/A	N/A
PART F WINDOWS AND EXTERNAL DOORS							
PVC-U door size 840 x 1980mm complete (B)	F1	1nr	2.50	23.75	209.99	35.06	268.80
PVC-U sliding patio door size 1700 x 2075mm (C)	F2	2nr	14.00	133.00	587.04	108.01	828.05
Carried forward			16.50	156.75	797.03	143.07	1,096.85

	Ref	Qty	Hours	Hours £	Mat'ls £	O & P £	Total £
Brought forward			16.50	156.75	797.03	143.07	1,096.85
PVC-U window size 1200 x 1200mm complete (A)	F3	4nr	8.00	76.00	630.36	105.95	812.31
25 x 225mm wrought softwood window board	F4	4.80m	1.44	13.68	29.86	6.53	50.07
Paint window board	F5	4.80m	0.76	7.22	4.61	1.77	13.60
Carried to summary			26.70	253.65	1,461.86	257.33	1,972.84
PART G INTERNAL PARTITIONS AND DOORS		N/A	N/A	N/A	N/A	N/A	N/A
PART H WALL FINISHES							
19 x 100mm wrought softwood skirting	H1	9.70m	1.94	18.43	13.29	4.76	36.48
12mm plasterboard fixed to walls with dabs	H2	19.60m2	7.64	72.58	39.00	16.74	128.32
12mm plasterboard fixed to walls less than 300mm wide with dabs	H3	26.63m	4.79	45.51	24.50	10.50	80.51
Two coats emulsion paint to plastered walls	H4	23.46m2	5.63	53.49	13.86	10.10	77.45
Paint skirting	H5	9.70m	1.46	13.87	4.66	2.78	21.31
Carried to summary			21.46	203.87	95.31	44.88	344.06
PART J FLOOR FINISHES							
Cement and sand floor screed 40mm thick	J1	16.28m2	4.07	38.67	49.33	13.20	101.19
Vinyl floor tiles, size 300 x 300mm	J2	16.28m2	3.90	37.05	113.63	22.60	173.28
Carried to summary			7.97	75.72	162.96	35.80	274.48

	Ref	Qty	Hours	Hours £	Mat'ls £	O & P £	Total £
PART K **CEILING FINISHES**							
Plasterboard with taped butt joints fixed to joists	K1	16.28m2	5.96	2.08	30.12	4.83	37.03
5mm skim coat to plasterboard ceilings	K2	16.28m2	8.14	77.33	19.86	14.58	111.77
Two coats emulsion paint to ceilings	K3	16.28m2	4.23	40.19	9.60	7.47	57.25
Carried to summary			18.33	119.60	59.58	26.88	206.05
PART L **ELECTRICAL WORK**							
13 amp double switched socket outlet with neon	L1	4.00	1.60	22.40	33.04	8.32	63.76
Lighting point	L2	3nr	1.05	14.70	20.40	5.27	40.37
Lighting switch	L3	3nr	0.75	10.50	11.91	3.36	25.77
Lighting wiring	L4	9.00m	0.70	9.80	27.00	5.52	42.32
Power cable	L5	20.00m	3.00	42.00	22.00	9.60	73.60
Carried to summary			7.10	99.40	114.35	32.06	245.81
PART M **HEATING WORK**							
15mm copper pipe	M1	10.00m	2.20	27.50	18.80	6.95	53.25
Elbow	M2	4nr	1.12	14.00	4.40	2.76	21.16
Tee	M3	1nr	0.22	2.75	1.95	0.71	5.41
Radiator, double convector size 1400 x 520mm	M4	3nr	3.90	48.75	347.04	59.37	455.16
Break into existing pipe and insert tee	M5	1nr	0.75	9.38	3.95	2.00	15.32
Carried to summary			8.19	102.38	376.14	71.78	550.29

	Ref	Qty	Hours	Hours £	Mat'ls £	O & P £	Total £
PART N ALTERATION WORK							
Take out existing window size 1500 x 1000mm and lintel over, adapt opening to receive 1770 x 2000mm patio door and insert new lintel over (both measured separately) and make good	N1	1nr	20.00	190.00	20.00	31.50	241.50
Carried to summary			20.00	190.00	20.00	31.50	241.50

SUMMARY

	Hours	Hours £	Mat'ls £	O & P £	Total £
PART A PRELIMINARIES	0.00	0.00	0.00	0.00	2,166.25
PART B SUBSTRUCTURE TO DPC LEVEL	82.77	786.32	788.65	236.24	1,811.21
PART C EXTERNAL WALLS	73.79	701.01	1,642.82	351.57	2,695.40
PART D FLAT ROOF	88.13	827.34	969.46	269.52	2,066.31
PART E PITCHED ROOF	0.00	0.00	0.00	0.00	0.00
PART F WINDOWS AND EXTERNAL DOORS	26.70	253.65	1,461.86	257.33	1,972.84
PART G INTERNAL PARTITIONS AND DOORS	0.00	0.00	0.00	0.00	0.00
PART H WALL FINISHES	21.46	203.87	95.31	44.88	344.06
PART J FLOOR FINISHES	7.97	75.72	162.96	35.80	274.48
PART K CEILING FINISHES	18.33	119.60	59.58	26.88	206.05
PART L ELECTRICAL WORK	7.10	99.40	114.35	32.06	245.81
PART M HEATING WORK	8.19	102.38	376.14	71.78	550.29
PART N ALTERATION WORK	20.00	190.00	20.00	31.50	241.50
Final total	354.44	3,359.29	5,691.13	1,357.56	12,574.20

	Ref	Qty	Hours	Hours £	Mat'ls £	O & P £	Total £
PART A PRELIMINARIES							
Concrete mixer	A1	10 wks					450.00
Small tools	A2	11 wks					385.00
Scaffolding (m2/weeks)	A3	385					866.25
Skip	A4	7 wks					700.00
Clean up	A5	10 hrs					80.00
Carried to summary							2,481.25
PART B SUBSTRUCTURE TO DPC LEVEL							
Excavate topsoil 150mm thick by hand	B1	26.15m2	7.85	74.58	0.00	11.19	85.76
Excavate to reduce levels by hand	B2	7.85m3	19.63	186.49	0.00	27.97	214.46
Excavate for trench foundations by hand	B3	2.21m3	5.75	54.63	0.00	8.19	62.82
Earthwork support to sides of trenches	B4	19.56m2	7.81	74.20	23.47	14.65	112.31
Backfilling with excavated material	B5	0.77m3	0.31	2.95	0.00	0.44	3.39
Hardcore 225mm thick	B6	19.98m2	3.00	28.50	101.50	19.50	149.50
Hardcore filling to trench	B7	0.17m3	0.09	0.86	0.86	0.26	1.97
Concrete grade (1:3:6) in foundations	B8	1.81m3	2.44	23.18	122.97	21.92	168.07
Concrete grade (1:2:4) in bed 150mm thick	B9	19.98m2	5.99	56.91	211.94	40.33	309.17
Concrete (1:2:4) in cavity wall filling	B10	6.70m2	1.34	12.73	24.32	5.56	42.61
Carried forward			54.21	515.00	485.06	150.01	1,150.06

	Ref	Qty	Hours	Hours £	Mat'ls £	O & P £	Total £
Brought forward			54.21	515.00	485.06	150.01	1,150.06
Damp-proof membrane	B11	20.63m2	0.83	7.89	11.97	2.98	22.83
Reinforcement ref A193 in foundation	B12	6.70m2	0.80	7.60	8.71	2.45	18.76
Steel fabric reinforcement ref A193 in slab	B13	19.98m2	3.00	28.50	25.97	8.17	62.64
Solid blockwork 140mm thick in cavity wall	B14	8.71m2	11.32	107.54	111.31	32.83	251.68
Common bricks 112.5mm thick in cavity wall	B15	6.70m2	11.32	107.54	88.37	29.39	225.30
Facing bricks in 112.5mm thick in skin of cavity wall	B16	2.01m2	3.62	34.39	46.77	12.17	93.33
Form cavity 50mm wide in cavity wall	B17	8.71m2	0.26	2.47	5.23	1.16	8.86
DPC 112mm wide	B18	13.40m	0.67	6.37	10.85	2.58	19.80
DPC 140mm wide	B19	13.40m	0.80	7.60	13.00	3.09	23.69
Bond in block wall	B20	1.30m	0.58	5.51	2.83	1.25	9.59
Bond in half brick wall	B21	0.30m	0.46	4.37	0.61	0.75	5.73
50mm thick insulation board	B22	19.98m2	5.99	56.91	89.51	21.96	168.38
Carried to summary			93.86	891.67	900.19	268.78	2,060.64

PART C
EXTERNAL WALLS

	Ref	Qty	Hours	Hours £	Mat'ls £	O & P £	Total £
Solid blockwork 140mm thick in cavity wall	C1	22.46m2	29.20	277.40	285.69	84.46	647.55
Facing brickwork 112.5mm thick in cavity wall	C2	22.46m2	40.42	383.99	517.93	135.29	1,037.21
75mm thick insulation in cavity wall	C3	22.46m2	4.96	47.12	130.72	26.68	204.52
Carried forward			74.58	708.51	934.34	246.43	1,889.28

	Ref	Qty	Hours	Hours £	Mat'ls £	O & P £	Total £
Brought forward			74.58	708.51	934.34	246.43	1,889.28
Steel lintel 2400mm long	C4	2nr	0.50	4.75	196.92	30.25	231.92
Steel lintel 1500mm long	C5	4nr	0.80	7.60	429.48	65.56	502.64
Steel lintel 1150mm long	C6	1nr	0.15	1.43	47.28	7.31	56.01
Close cavity wall at jambs	C7	19.72m	0.99	9.41	40.23	7.45	57.08
Close cavity wall at cills	C8	6.91m	0.35	3.33	14.10	2.61	20.04
Close cavity wall at top	C9	13.40m	0.67	6.37	27.34	5.06	38.76
DPC 112mm wide at jambs	C10	19.72m	0.99	9.41	15.97	3.81	29.18
DPC 112mm wide at cills	C11	6.91m	0.35	3.33	5.60	1.34	10.26
Carried to summary			79.23	752.69	1,663.98	362.50	2,779.16
PART D FLAT ROOF							
200 x 50mm sawn softwood joists	D1	53.95m	13.49	128.16	167.25	44.31	339.72
200 x 50mm sawn softwood sprocket pieces	D2	16nr	2.24	21.28	21.44	6.41	49.13
18mm thick WPB grade decking	D3	26.15m2	23.54	211.86	205.54	62.61	480.01
50 x 50mm (avg) wide sawn softwood firrings	D4	53.95m	9.71	92.25	121.93	32.13	246.30
High density polyethylene vapour barrier 150mm thick	D5	24.00m2	4.80	45.60	227.28	40.93	313.81
100 x 75mm sawn softwood wall plate	D6	14.00m	5.72	54.34	31.08	12.81	98.23
100 x 75mm sawn softwood tilt fillet	D7	6.30m	1.58	15.01	16.63	4.75	36.39
Build in ends of 200 x 50mm joists	D8	13nr	4.55	43.23	1.95	6.78	51.95
Carried forward			65.63	611.72	793.10	210.72	1,615.54

	Ref	Qty	Hours	Hours £	Mat'ls £	O & P £	Total £
Brought forward			65.63	611.72	793.10	210.72	1,615.54
Rake out joint for flashing	D9	6.30m	2.21	21.00	0.95	3.29	25.24
6mm thick soffit 150mm wide	D10	14.30m	5.72	54.34	25.03	11.91	91.28
19mm wrought softwood fascia 200mm high	D11	14.30m	7.15	67.93	28.17	14.41	110.51
Three layer fibre-based roofing felt	D12	26.15m2	14.38	136.61	219.92	53.48	410.01
Felt turn-down 100mm girth	D13	14.30m	1.43	13.59	12.01	3.84	29.43
Felt flashing 150mm girth	D14	6.30m	0.76	7.22	10.58	2.67	20.47
112mm diameter PVC-U gutter	D15	6.30m	1.64	15.58	28.16	6.56	50.30
Stop end	D16	1nr	0.14	1.33	1.11	0.37	2.81
Stop end outlet	D17	1nr	0.25	2.38	1.94	0.65	4.96
68mm diameter PVC-U down pipe	D18	2.50m	0.63	5.99	8.48	2.17	16.63
Shoe	D19	1nr	0.30	2.85	2.09	0.74	5.68
Paint fascia and soffit	D20	4.29m2	3.00	28.50	4.12	4.89	37.51
Carried to summary			103.24	969.01	1,135.66	315.70	2,420.37
PART E **PITCHED ROOF**			N/A	N/A	N/A	N/A	N/A
PART F **WINDOWS AND EXTERNAL DOORS**							
PVC-U door size 840 x 1980mm complete (B)	F1	1nr	2.50	23.75	209.99	35.06	268.80
PVC-U sliding patio door size 1700 x 2075mm (C)	F2	2nr	14.00	133.00	587.04	108.01	828.05
Carried forward			16.50	156.75	797.03	143.07	1,096.85

	Ref	Qty	Hours	Hours £	Mat'ls £	O & P £	Total £
Brought forward			16.50	156.75	797.03	143.07	1,096.85
PVC-U window size 1200 x 1200mm complete (A)	F3	4nr	8.00	76.00	630.36	105.95	812.31
25 x 225mm wrought softwood window board	F4	4.80m	1.44	13.68	29.86	6.53	50.07
Paint window board	F5	4.80m	0.76	7.22	4.61	1.77	13.60
Carried to summary			26.70	253.65	1,461.86	257.33	1,972.84
PART G INTERNAL PARTITIONS AND DOORS		N/A	N/A	N/A	N/A	N/A	N/A
PART H WALL FINISHES							
19 x 100mm wrought softwood skirting	H1	10.07m	2.14	20.33	14.66	5.25	40.24
12mm plasterboard fixed to walls with dabs	H2	21.96m2	8.56	81.32	43.70	18.75	143.77
12mm plasterboard fixed to walls less than 300mm wide with dabs	H3	26.63m	4.79	45.51	24.50	10.50	80.51
Two coats emulsion paint to walls	H4	25.96m2	6.23	59.19	15.32	11.18	85.68
Paint skirting	H5	10.70m	1.61	15.30	5.14	3.07	23.50
Carried to summary			23.33	221.64	103.32	48.74	373.70
PART J FLOOR FINISHES							
Cement and sand floor screed 40mm thick	J1	19.88m2	5.00	47.50	60.54	16.21	124.25
Vinyl floor tiles, size 300 x 300mm	J2	19.88m2	4.79	45.51	139.46	27.74	212.71
Carried to summary			9.79	93.01	200.00	43.95	336.96

	Ref	Qty	Hours	Hours £	Mat'ls £	O & P £	Total £
PART K **CEILING FINISHES**							
Plasterboard with taped butt joints fixed to joists	K1	19.88m2	7.19	2.08	36.96	5.86	44.90
5mm skim coat to plasterboard ceilings	K2	19.88m2	9.99	94.91	24.00	17.84	136.74
Two coats emulsion paint to ceilings	K3	19.88m2	5.19	49.31	11.79	9.16	70.26
Carried to summary			22.37	146.29	72.75	32.86	251.90
PART L **ELECTRICAL WORK**							
13 amp double switched socket outlet with neon	L1	5nr	2.00	28.00	41.30	10.40	79.70
Lighting point	L2	3nr	1.05	14.70	20.40	5.27	40.37
Lighting switch	L3	3nr	0.75	10.50	11.91	3.36	25.77
Lighting wiring	L4	11.00m	1.10	15.40	11.33	4.01	30.74
Power cable	L5	22.00m	3.30	46.20	24.20	10.56	80.96
Carried to summary			8.20	114.80	109.14	33.59	257.53
PART M **HEATING WORK**							
15mm copper pipe	M1	11.00m	2.42	30.25	26.68	8.54	65.47
Elbow	M2	4nr	1.12	14.00	4.40	2.76	21.16
Tee	M3	1nr	0.22	2.75	1.95	0.71	5.41
Radiator, double convector size 1400 x 520mm	M4	3nr	3.90	48.75	347.04	59.37	455.16
Break into existing pipe and insert tee	M5	1nr	0.75	9.38	3.95	2.00	15.32
Carried to summary			8.41	105.13	384.02	73.37	562.52

	Ref	Qty	Hours	Hours £	Mat'ls £	O & P £	Total £
PART N **ALTERATION WORK**							
Take out existing window size 1500 x 1000mm and lintel over, adapt opening to receive 1770 x 2000mm patio door and insert new lintel over (both measured separately) and make good	N1	1nr	20.00	190.00	20.00	31.50	241.50
Carried to summary			20.00	190.00	20.00	31.50	241.50

SUMMARY

	Hours	Hours £	Mat'ls £	O & P £	Total £
PART A PRELIMINARIES	0.00	0.00	0.00	0.00	2,481.25
PART B SUBSTRUCTURE TO DPC LEVEL	93.86	891.67	900.19	268.78	2,060.64
PART C EXTERNAL WALLS	79.23	752.69	1,663.98	362.50	2,779.16
PART D FLAT ROOF	103.24	969.01	1,135.66	315.70	2,420.37
PART E PITCHED ROOF	0.00	0.00	0.00	0.00	0.00
PART F WINDOWS AND EXTERNAL DOORS	26.70	253.65	1,461.86	257.83	1,972.84
PART G INTERNAL PARTITIONS AND DOORS	0.00	0.00	0.00	0.00	0.00
PART H WALL FINISHES	23.33	221.64	103.32	48.74	373.70
PART J FLOOR FINISHES	9.79	93.01	200.00	43.95	336.96
PART K CEILING FINISHES	22.37	146.29	72.75	32.86	251.90
PART L ELECTRICAL WORK	8.20	114.80	109.14	33.59	257.53
PART M HEATING WORK	8.41	105.13	384.02	73.37	562.52
PART N ALTERATION WORK	20.00	190.00	20.00	31.50	241.50
Final total	395.13	3,737.89	6,050.92	1,468.82	13,738.37

	Ref	Qty	Hours	Hours £	Mat'ls £	O & P £	Total £
PART A PRELIMINARIES							
Concrete mixer	A1	4 wks					180.00
Small tools	A2	6 wks					210.00
Scaffolding (m2/weeks)	A3	70.00					157.50
Skip	A4	2 wks					300.00
Clean up	A5	4 hrs					48.00
Carried to summary							895.50
PART B SUBSTRUCTURE TO DPC LEVEL							
Excavate topsoil 150mm thick by hand	B1	7.10m2	2.13	20.24	0.00	3.04	23.27
Excavate to reduce levels by hand	B2	2.13m3	5.33	50.64	0.00	7.60	58.23
Excavate for trench foundations by hand	B3	1.06m3	2.76	26.22	0.00	3.93	30.15
Earthwork support to sides of trenches	B4	9.34m2	3.74	35.53	11.21	7.01	53.75
Backfilling with excavated material	B5	0.40m3	0.16	1.52	0.00	0.23	1.75
Hardcore 225mm thick	B6	4.08m2	0.82	7.79	20.73	4.28	32.80
Hardcore filling to trench	B7	0.08m3	0.05	0.48	0.40	0.13	1.01
Concrete grade (1:3:6) in foundations	B8	0.86m3	1.16	11.02	58.43	10.42	79.87
Concrete grade (1:2:4) in bed 150mm thick	B9	4.08m2	1.22	11.59	44.39	8.40	64.38
Concrete (1:2:4) in cavity wall filling	B10	3.20m2	0.64	6.08	11.62	2.66	20.36
Carried forward			18.01	171.10	146.78	47.68	365.56
Carried forward			18.01	171.10	146.78	47.68	365.56

	Ref	Qty	Hours	Hours £	Mat'ls £	O & P £	Total £
Brought forward			18.01	171.10	146.78	47.68	365.56
Damp-proof membrane	B11	4.38m2	0.17	1.62	2.54	0.62	4.78
Reinforcement ref A193 in foundation	B12	3.22m2	0.38	3.61	4.16	1.17	8.94
Steel fabric reinforcement ref A193 in slab	B13	4.08m2	0.61	5.80	5.30	1.66	12.76
Solid blockwork 140mm thick in cavity wall	B14	4.16m2	5.41	51.40	53.16	15.68	120.24
Common bricks 112.5mm thick in cavity wall	B15	3.20m2	5.41	51.40	42.21	14.04	107.65
Facing bricks in 112.5mm thick in skin of cavity wall	B16	0.96m2	1.73	16.44	22.34	5.82	44.59
Form cavity 50mm wide in cavity wall	B17	4.16m2	0.13	1.24	2.50	0.56	4.30
DPC 112mm wide	B18	6.40m	0.32	3.04	5.18	1.23	9.45
DPC 140mm wide	B19	6.40m	0.38	3.61	6.21	1.47	11.29
Bond in block wall	B20	1.30m	0.58	5.51	2.83	1.25	9.59
Bond in half brick wall	B21	0.30m	0.46	4.37	0.61	0.75	5.73
50mm thick insulation board	B22	4.08m2	1.22	11.59	18.28	4.48	34.35
Carried to summary			34.81	330.70	312.10	96.42	739.21
PART C EXTERNAL WALLS							
Solid blockwork 140mm thick in cavity wall	C1	13.50m2	17.55	166.73	171.72	50.77	389.21
Facing brickwork 112.5mm thick in cavity wall	C2	13.50m2	24.30	230.85	311.31	81.32	623.48
75mm thick insulation in cavity wall	C3	13.50m2	2.97	28.22	78.57	16.02	122.80
Carried forward			44.82	425.79	561.60	148.11	1,135.50

	Ref	Qty	Hours	Hours £	Mat'ls £	O & P £	Total £
Brought forward			44.82	425.79	561.60	148.11	1,135.50
Steel lintel 2400mm long	C4	2nr	0.50	4.75	196.92	30.25	231.92
Steel lintel 1500mm long	C5	2nr	0.40	3.80	214.94	32.81	251.55
Close cavity wall at jambs	C7	8.96m	0.45	4.28	18.28	3.38	25.94
Close cavity wall at cills	C8	3.67m	0.18	1.71	7.49	1.38	10.58
Close cavity wall at top	C9	6.40m	0.32	3.04	13.06	2.42	18.52
DPC 112mm wide at jambs	C10	8.96m	0.45	4.28	7.26	1.73	13.27
DPC 112mm wide at cills	C11	3.67m	0.18	1.71	2.97	0.70	5.38
Carried to summary			47.30	449.35	1,022.52	220.78	1,692.65
PART D FLAT ROOF			N/A	N/A	N/A	N/A	N/A
PART E PITCHED ROOF							
100 x 75mm sawn softwood wall plate	E1	3.00m	1.05	9.98	6.66	2.50	19.13
150 x 50mm sawn softwood pole plate plugged to brickwork	E2	3.30m	0.99	9.41	10.63	3.01	23.04
100 x 50mm sawn softwood rafters	E3	17.50m	3.50	31.50	57.40	13.34	102.24
125 x 25mm sawn softwood purlin	E4	3.30m	0.66	6.27	7.91	2.13	16.31
150 x 50mm sawn softwood joists	E5	12.50m	2.75	26.13	28.50	8.19	62.82
150 x 50mm sawn softwood sprockets	E6	14nr	1.68	15.96	15.96	4.79	36.71
100mm layer of insulation quilt between joists fixed with chicken wire and 150mm layer over joists	E7	6.00m2	2.88	27.36	65.46	13.92	106.74
6mm softwood soffit 150mm wide	E8	8.30m	3.32	31.54	14.53	6.91	52.98
Carried forward			16.83	158.14	207.05	54.78	419.96

	Ref	Qty	Hours	Hours £	Mat'ls £	O & P £	Total £
Brought forward			16.83	158.14	207.05	54.78	419.96
19mm wrought softwood fascia/ barge board 200mm high	E9	3.30m	4.15	39.43	6.50	6.89	52.81
Marley Plain roof tiles on felt and battens	E10	11.55m2	21.95	208.53	298.79	76.10	583.41
Double eaves course	E11	3.30m	0.83	7.89	10.63	2.78	21.29
Verge with plain tile undercloak	E12	7.00m	1.75	16.63	22.54	5.87	45.04
Lead flashing code 5, 200mm girth	E13	3.30m	1.98	18.81	22.04	6.13	46.98
Rake out joint for flashing	E14	3.30m	1.16	11.02	0.50	1.73	13.25
112mm diameter PVC-U gutter	E15	3.30m	0.86	8.17	14.75	3.44	26.36
Stop end	E16	1nr	0.14	1.33	1.11	0.37	2.81
Stop end outlet	E17	1nr	0.25	2.38	1.11	0.52	4.01
68mm diameter PVC-U down pipe	E18	2.50m	0.63	5.99	8.48	2.17	16.63
Shoe	E19	1nr	0.30	2.85	2.09	0.74	5.68
Paint fascia and soffit	E20	2.59m2	1.53	14.54	2.49	2.55	19.58
Carried to summary			52.36	495.67	598.08	164.06	1,257.81
PART F WINDOWS AND EXTERNAL DOORS							
PVC-U door size 840 x 1980mm complete (B)	F1	0.00	0.00	0.00	0.00	0.00	0.00
PVC-U sliding patio door size 1700 x 2075mm (C)	F2	2nr	14.00	133.00	587.04	108.01	828.05
Carried forward			14.00	133.00	587.04	108.01	828.05

	Ref	Qty	Hours	Hours £	Mat'ls £	O & P £	Total £
Brought forward			14.00	133.00	587.04	108.01	828.05
PVC-U window size 1200 x 1200mm complete (A)	F3	2nr	4.00	38.00	315.18	52.98	406.16
25 x 225mm wrought softwood window board	F4	2.40m	0.72	6.84	12.44	2.89	22.17
Paint window board	F5	2.40m	0.42	3.99	2.30	0.94	7.23
Carried to summary			19.14	181.83	916.96	164.82	1,263.61
PART G INTERNAL PARTITIONS AND DOORS		N/A	N/A	N/A	N/A	N/A	N/A
PART H WALL FINISHES							
19 x 100mm wrought softwood skirting	H1	4.54m	0.91	8.65	6.22	2.23	17.09
12mm plasterboard fixed to walls with dabs	H2	9.00m2	3.51	33.35	17.91	7.69	58.94
12mm plasterboard fixed to walls less than 300mm wide with dabs	H3	12.63m	2.27	21.57	11.62	4.98	38.16
Two coats emulsion paint to walls	H4	10.90m2	2.62	24.89	6.43	4.70	36.02
Paint skirting	H5	4.54m	0.68	6.46	2.18	1.30	9.94
Carried to summary			9.99	94.91	44.36	20.89	160.15
PART J FLOOR FINISHES							
Cement and sand floor screed 40mm thick	J1	4.08m2	1.02	9.69	12.36	3.31	25.36
Vinyl floor tiles, size 300 x 300mm	J2	4.08m2	0.98	9.31	28.47	5.67	43.45
Carried to summary			2.00	19.00	40.83	8.97	68.80

	Ref	Qty	Hours	Hours £	Mat'ls £	O & P £	Total £
PART K CEILING FINISHES							
Plasterboard with taped butt joints fixed to joists	K1	4.08m2	1.47	2.08	7.53	1.44	11.05
5mm skim coat to plasterboard ceilings	K2	4.08m2	2.04	19.38	4.98	3.65	28.01
Two coats emulsion paint to ceilings	K3	4.08m2	1.06	10.07	2.41	1.87	14.35
Carried to summary			4.57	31.53	14.92	6.97	53.42
PART L ELECTRICAL WORK							
13 amp double switched socket outlet with neon	L1	2nr	0.80	11.20	16.52	4.16	31.88
Lighting point	L2	1nr	0.35	4.90	6.80	1.76	13.46
Lighting switch	L3	2nr	0.50	7.00	7.94	2.24	17.18
Lighting wiring	L4	5.00m	0.50	7.00	5.15	1.82	13.97
Power cable	L5	12.00m	1.80	25.20	13.20	5.76	44.16
Carried to summary			3.95	55.30	49.61	15.74	120.65
PART M HEATING WORK							
15mm copper pipe	M1	4.00m	0.88	11.00	7.52	2.78	21.30
Elbow	M2	4nr	1.12	14.00	4.40	2.76	21.16
Tee	M3	1nr	0.22	2.75	1.95	0.71	5.41
Radiator, double convector size 1400 x 520mm	M4	1nr	1.30	16.25	115.68	19.79	151.72
Break into existing pipe and insert tee	M5	1nr	0.75	9.38	3.95	2.00	15.32
Carried to summary			4.27	53.38	133.50	28.03	214.91

	Ref	Qty	Hours	Hours £	Mat'ls £	O & P £	Total £
PART N **ALTERATION WORK**							
Take out existing window size 1500 x 1000mm and lintel over, adapt opening to receive 1770 x 2000mm patio door and insert new lintel over (both measured separately) and make good	N1	1nr	20.00	190.00	20.00	31.50	241.50
Carried to summary			20.00	190.00	20.00	31.50	241.50

SUMMARY

	Hours	Hours £	Mat'ls £	O & P £	Total £
PART A PRELIMINARIES	0.00	0.00	0.00	0.00	895.50
PART B SUBSTRUCTURE TO DPC LEVEL	34.81	330.70	312.10	96.42	739.21
PART C EXTERNAL WALLS	47.30	449.35	1,022.52	220.78	1,692.65
PART D FLAT ROOF	0.00	0.00	0.00	0.00	0.00
PART E PITCHED ROOF	52.36	495.67	598.08	164.06	1,257.81
PART F WINDOWS AND EXTERNAL DOORS	19.14	181.83	916.96	164.82	1,263.61
PART G INTERNAL PARTITIONS AND DOORS	0.00	0.00	0.00	0.00	0.00
PART H WALL FINISHES	9.99	94.91	44.36	20.89	160.15
PART J FLOOR FINISHES	2.00	19.00	40.83	8.97	68.80
PART K CEILING FINISHES	4.57	31.53	14.92	6.97	53.42
PART L ELECTRICAL WORK	3.95	55.30	49.61	15.74	120.65
PART M HEATING WORK	4.27	53.38	133.50	28.03	214.91
PART N ALTERATION WORK	20.00	190.00	20.00	31.50	241.50
Final total	198.39	1,901.67	3,152.88	758.18	6,708.21

	Ref	Qty	Hours	Hours £	Mat'ls £	O & P £	Total £
PART A PRELIMINARIES							
Concrete mixer	A1	5 wks					225.00
Small tools	A2	7 wks					245.00
Scaffolding (m2/weeks)	A3	100.00					225.00
Skip	A4	3 wks					300.00
Clean up	A5	6 hrs					48.00
Carried to summary							1,043.00
PART B SUBSTRUCTURE TO DPC LEVEL							
Excavate topsoil 150mm thick by hand	B1	9.25m2	2.78	26.41	0.00	3.96	30.37
Excavate to reduce levels by hand	B2	2.77m3	6.93	65.84	0.00	9.88	75.71
Excavate for trench foundations by hand	B3	1.22m3	3.17	30.12	0.00	4.52	34.63
Earthwork support to sides of trenches	B4	10.80m2	4.32	41.04	12.96	8.10	62.10
Backfilling with excavated material	B5	0.47m3	0.28	2.66	0.00	0.40	3.06
Hardcore 225mm thick	B6	5.78m2	0.61	5.80	29.36	5.27	40.43
Hardcore filling to trench	B7	0.09m3	0.05	0.48	0.45	0.14	1.06
Concrete grade (1:3:6) in foundations	B8	1.00m3	1.35	12.83	67.94	12.11	92.88
Concrete grade (1:2:4) in bed 150mm thick	B9	5.78m2	1.73	16.44	62.87	11.90	91.20
Concrete (1:2:4) in cavity wall filling	B10	3.70m2	0.74	7.03	13.43	3.07	23.53
Carried forward			21.96	208.62	187.01	59.34	454.97

	Ref	Qty	Hours	Hours £	Mat'ls £	O & P £	Total £
Brought forward			21.96	208.62	187.01	59.34	454.97
Damp-proof membrane	B11	6.13m2	0.25	2.38	3.56	0.89	6.83
Reinforcement ref A193 in foundation	B12	3.70m2	0.44	4.18	4.81	1.35	10.34
Steel fabric reinforcement ref A193 in slab	B13	5.78m2	0.87	8.27	7.51	2.37	18.14
Solid blockwork 140mm thick in cavity wall	B14	4.81m2	6.25	59.38	61.47	18.13	138.97
Common bricks 112.5mm thick in cavity wall	B15	3.70m2	6.25	59.38	48.80	16.23	124.40
Facing bricks in 112.5mm thick in skin of cavity wall	B16	1.11m2	2.00	19.00	25.83	6.72	51.55
Form cavity 50mm wide in cavity wall	B17	4.81m2	0.14	1.33	2.89	0.63	4.85
DPC 112mm wide	B18	7.40m	0.37	3.52	5.99	1.43	10.93
DPC 140mm wide	B19	7.40m	0.44	4.18	7.18	1.70	13.06
Bond in block wall	B20	1.30m	0.58	5.51	2.83	1.25	9.59
Bond in half brick wall	B21	0.30m	0.46	4.37	0.61	0.75	5.73
50mm thick insulation board	B22	5.78m2	1.73	16.44	25.89	6.35	48.67
Carried to summary			41.74	396.53	384.38	117.14	898.05
PART C EXTERNAL WALLS							
Solid blockwork 140mm thick in cavity wall	C1	16.00m2	20.80	197.60	203.52	60.17	461.29
Facing brickwork 112.5mm thick in cavity wall	C2	16.00m2	28.80	273.60	368.96	96.38	738.94
75mm thick insulation in cavity wall	C3	16.00m2	3.52	33.44	93.12	18.98	145.54
Carried forward			53.12	504.64	665.60	175.54	1,345.78

	Ref	Qty	Hours	Hours £	Mat'ls £	O & P £	Total £
Brought forward			53.12	504.64	665.60	175.54	1,345.78
Steel lintel 2400mm long	C4	2nr	0.50	4.75	196.92	30.25	231.92
Steel lintel 1500mm long	C5	2nr	0.40	3.80	214.94	32.81	251.55
Close cavity wall at jambs	C7	8.96m	0.45	4.28	18.28	3.38	25.94
Close cavity wall at cills	C8	3.67m	0.18	1.71	7.49	1.38	10.58
Close cavity wall at top	C9	7.40m	0.37	3.52	15.10	2.79	21.41
DPC 112mm wide at jambs	C10	8.96m	0.45	4.28	7.26	1.73	13.27
DPC 112mm wide at cills	C11	3.67m	0.18	1.71	2.97	0.70	5.38
Carried to summary			55.65	528.68	1,128.56	248.59	1,905.82
PART D FLAT ROOF			N/A	N/A	N/A	N/A	N/A
PART E PITCHED ROOF							
100 x 75mm sawn softwood wall plate	E1	4.00m	1.40	13.30	8.88	3.33	25.51
150 x 50mm sawn softwood pole plate plugged to brickwork	E2	4.30m	1.29	12.26	13.85	3.92	30.02
100 x 50mm sawn softwood rafters	E3	17.50m	3.50	31.50	57.40	13.34	102.24
125 x 25mm sawn softwood purlin	E4	4.30m	0.86	8.17	10.36	2.78	21.31
150 x 50mm sawn softwood joists	E5	12.50m	2.75	26.13	28.50	8.19	62.82
150 x 50mm sawn softwood sprockets	E6	14nr	1.68	15.96	15.96	4.79	36.71
100mm layer of insulation quilt between joists fixed with chicken wire and 150mm layer over joists	E7	8.00m2	3.84	36.48	87.28	18.56	142.32
6mm softwood soffit 150mm wide	E8	9.30m	3.72	35.34	16.28	7.74	59.36
Carried forward			18.18	170.96	228.15	59.87	458.98

	Ref	Qty	Hours	Hours £	Mat'ls £	O & P £	Total £
Brought forward			18.18	170.96	228.15	59.87	458.98
19mm wrought softwood fascia/ barge board 200mm high	E9	4.30m	4.65	44.18	8.47	7.90	60.54
Marley Plain roof tiles on felt and battens	E10	15.05m2	28.60	271.70	389.34	99.16	760.20
Double eaves course	E11	4.30m	1.51	14.35	13.85	4.23	32.42
Verge with plain tile undercloak	E12	7.00m	1.08	10.26	22.54	4.92	37.72
Lead flashing code 5, 200mm girth	E13	4.30m	2.58	24.51	27.82	7.85	60.18
Rake out joint for flashing	E14	4.30m	1.51	14.35	0.65	2.25	17.24
112mm diameter PVC-U gutter	E15	4.30m	1.12	10.64	19.22	4.48	34.34
Stop end	E16	1nr	0.14	1.33	1.11	0.37	2.81
Stop end outlet	E17	1nr	0.25	2.38	1.11	0.52	4.01
68mm diameter PVC-U down pipe	E18	2.50m	0.63	5.99	8.48	2.17	16.63
Shoe	E19	1nr	0.30	2.85	2.09	0.74	5.68
Paint fascia and soffit	E20	2.79m2	1.81	17.20	2.68	2.98	22.86
Carried to summary			62.36	590.67	725.51	197.43	1,513.61

PART F
WINDOWS AND EXTERNAL DOORS

	Ref	Qty	Hours	Hours £	Mat'ls £	O & P £	Total £
PVC-U door size 840 x 1980mm complete (B)	F1	0.00	0.00	0.00	0.00	0.00	0.00
PVC-U sliding patio door size 1700 x 2075mm (C)	F2	2nr	14.00	133.00	587.04	108.01	828.05
Carried forward			14.00	133.00	587.04	108.01	828.05

	Ref	Qty	Hours	Hours £	Mat'ls £	O & P £	Total £
Brought forward			14.00	133.00	587.04	108.01	828.05
PVC-U window size 1200 x 1200mm complete (A)	F3	2nr	4.00	38.00	315.18	52.98	406.16
25 x 225mm wrought softwood window board	F4	2.40m	0.72	6.84	12.44	2.89	22.17
Paint window board	F5	2.40m	0.42	3.99	2.30	0.94	7.23
Carried to summary			19.14	181.83	916.96	164.82	1,263.61
PART G INTERNAL PARTITIONS AND DOORS		N/A	N/A	N/A	N/A	N/A	N/A
PART H WALL FINISHES							
19 x 100mm wrought softwood skirting	H1	5.54m	1.11	10.55	7.59	2.72	20.86
12mm plasterboard fixed to walls with dabs	H2	11.50m2	4.49	42.66	22.89	9.83	75.38
12mm plasterboard fixed to walls less than 300mm wide with dabs	H3	12.63m	2.27	21.57	11.62	4.98	38.16
Two coats emulsion paint to walls	H4	13.40m2	3.21	30.50	7.91	5.76	44.17
Paint skirting	H5	5.54m	0.83	7.89	2.66	1.58	12.13
Carried to summary			11.91	113.15	52.67	24.87	190.69
PART J FLOOR FINISHES							
Cement and sand floor screed 40mm thick	J1	5.78m2	1.45	13.78	17.51	4.69	35.98
Vinyl floor tiles, size 300 x 300mm	J2	5.78m2	1.39	13.21	40.03	7.99	61.22
Carried to summary			2.84	26.98	57.54	12.68	97.20

	Ref	Qty	Hours	Hours £	Mat'ls £	O & P £	Total £
PART K **CEILING FINISHES**							
Plasterboard with taped butt joints fixed to joists	K1	5.78m2	2.08	2.08	10.69	1.92	14.69
5mm skim coat to plasterboard ceilings	K2	5.78m2	2.89	27.46	7.05	5.18	39.68
Two coats emulsion paint to ceilings	K3	5.78m2	1.50	14.25	3.39	2.65	20.29
Carried to summary			6.47	43.79	21.13	9.74	74.65
PART L **ELECTRICAL WORK**							
13 amp double switched socket outlet with neon	L1	2nr	0.80	11.20	16.52	4.16	31.88
Lighting point	L2	2nr	0.70	9.80	13.60	3.51	26.91
Lighting switch	L3	2nr	0.50	7.00	7.94	2.24	17.18
Lighting wiring	L4	6.00m	0.60	8.40	6.18	2.19	16.77
Power cable	L5	14.00m	2.10	29.40	15.40	6.72	51.52
Carried to summary			4.70	65.80	59.64	18.82	144.26
PART M **HEATING WORK**							
15mm copper pipe	M1	5.00m	1.10	13.75	9.40	3.47	26.62
Elbow	M2	4nr	1.12	14.00	4.40	2.76	21.16
Tee	M3	1nr	0.22	2.75	1.95	0.71	5.41
Radiator, double convector size 1400 x 520mm	M4	2nr	2.60	32.50	231.36	39.58	303.44
Break into existing pipe and insert tee	M5	1nr	0.75	9.38	3.95	2.00	15.32
Carried to summary			5.79	72.38	251.06	48.52	371.95

	Ref	Qty	Hours	Hours £	Mat'ls £	O & P £	Total £
PART N **ALTERATION WORK**							
Take out existing window size 1500 x 1000mm and lintel over, adapt opening to receive 1770 x 2000mm patio door and insert new lintel over (both measured separately) and make good	N1	1nr	20.00	190.00	20.00	31.50	241.50
Carried to summary			20.00	190.00	20.00	31.50	241.50

SUMMARY

	Hours	Hours £	Mat'ls £	O & P £	Total £
PART A PRELIMINARIES	0.00	0.00	0.00	0.00	1,043.00
PART B SUBSTRUCTURE TO DPC LEVEL	41.74	396.53	384.38	117.14	898.05
PART C EXTERNAL WALLS	55.65	528.68	1,128.56	248.59	1,905.82
PART D FLAT ROOF	0.00	0.00	0.00	0.00	0.00
PART E PITCHED ROOF	62.36	590.67	725.51	197.43	1,513.61
PART F WINDOWS AND EXTERNAL DOORS	19.14	181.83	916.96	164.82	1,263.61
PART G INTERNAL PARTITIONS AND DOORS	0.00	0.00	0.00	0.00	0.00
PART H WALL FINISHES	11.91	113.15	52.67	24.87	190.69
PART J FLOOR FINISHES	2.84	26.98	57.54	12.68	97.20
PART K CEILING FINISHES	6.47	43.79	21.13	9.74	74.65
PART L ELECTRICAL WORK	4.70	65.80	59.64	18.82	144.26
PART M HEATING WORK	5.79	72.38	251.06	48.52	371.95
PART N ALTERATION WORK	20.00	190.00	20.00	31.50	241.50
Final total	230.60	2,209.81	3,617.45	874.11	7,744.34

	Ref	Qty	Hours	Hours £	Mat'ls £	O & P £	Total £
PART A PRELIMINARIES							
Concrete mixer	A1	6 wks					270.00
Small tools	A2	8 wks					280.00
Scaffolding (m2/weeks)	A3	135.00					303.75
Skip	A4	4 wks					400.00
Clean up	A5	8 hrs					64.00
Carried to summary							1,317.75
PART B SUBSTRUCTURE TO DPC LEVEL							
Excavate topsoil 150mm thick by hand	B1	11.40m2	3.42	32.49	0.00	4.87	37.36
Excavate to reduce levels by hand	B2	3.42m3	8.55	81.23	0.00	12.18	93.41
Excavate for trench foundations by hand	B3	1.39m3	4.90	46.55	0.00	6.98	53.53
Earthwork support to sides of trenches	B4	12.26m2	4.32	41.04	14.71	8.36	64.11
Backfilling with excavated material	B5	0.55m3	0.22	2.09	0.00	0.31	2.40
Hardcore 225mm thick	B6	7.48m2	1.12	10.64	38.00	7.30	55.94
Hardcore filling to trench	B7	0.12m3	0.06	0.57	0.60	0.18	1.35
Concrete grade (1:3:6) in foundations	B8	1.13m3	1.53	14.54	76.77	13.70	105.00
Concrete grade (1:2:4) in bed 150mm thick	B9	7.48m2	2.24	21.28	81.38	15.40	118.06
Concrete (1:2:4) in cavity wall filling	B10	4.20m2	0.84	7.98	15.25	3.48	26.71
Carried forward			27.20	258.40	226.71	72.77	557.88

	Ref	Qty	Hours	Hours £	Mat'ls £	O & P £	Total £
Brought forward			27.20	258.40	226.71	72.77	557.88
Damp-proof membrane	B11	7.88m2	0.32	3.04	4.57	1.14	8.75
Reinforcement ref A193 in foundation	B12	4.20m2	0.50	4.75	5.46	1.53	11.74
Steel fabric reinforcement ref A193 in slab	B13	7.48m2	1.12	10.64	9.72	3.05	23.41
Solid blockwork 140mm thick in cavity wall	B14	5.46m2	7.10	67.45	69.78	20.58	157.81
Common bricks 112.5mm thick in cavity wall	B15	4.20m2	7.10	67.45	55.40	18.43	141.28
Facing bricks in 112.5mm thick in skin of cavity wall	B16	1.26m2	2.27	21.57	29.32	7.63	58.52
Form cavity 50mm wide in cavity wall	B17	5.46m2	0.16	1.52	3.28	0.72	5.52
DPC 112mm wide	B18	8.40m	0.42	3.99	6.80	1.62	12.41
DPC 140mm wide	B19	8.40m	0.50	4.75	8.15	1.94	14.84
Bond in block wall	B20	1.30m	0.58	5.51	2.83	1.25	9.59
Bond in half brick wall	B21	0.30m	0.46	4.37	0.61	0.75	5.73
50mm thick insulation board	B22	7.48m2	2.24	21.28	33.51	8.22	63.01
Carried to summary			49.97	474.72	456.14	139.63	1,070.48
PART C EXTERNAL WALLS							
Solid blockwork 140mm thick in cavity wall	C1	18.50m2	24.05	228.48	235.32	69.57	533.36
Facing brickwork 112.5mm thick in cavity wall	C2	18.50m2	33.30	316.35	426.61	111.44	854.40
75mm thick insulation in cavity wall	C3	18.50m2	4.07	38.67	107.67	21.95	168.29
Carried forward			61.42	583.49	769.60	202.96	1,556.05

	Ref	Qty	Hours	Hours £	Mat'ls £	O & P £	Total £
Brought forward			61.42	583.49	769.60	202.96	1,556.05
Steel lintel 2400mm long	C4	2nr	0.50	4.75	196.92	30.25	231.92
Steel lintel 1500mm long	C5	2nr	0.40	3.80	214.94	32.81	251.55
Close cavity wall at jambs	C7	8.96m	0.45	4.28	18.28	3.38	25.94
Close cavity wall at cills	C8	3.67m	0.18	1.71	7.49	1.38	10.58
Close cavity wall at top	C9	8.40m	0.37	3.52	17.14	3.10	23.75
DPC 112mm wide at jambs	C10	8.96m	0.45	4.28	7.26	1.73	13.27
DPC 112mm wide at cills	C11	3.67m	0.18	1.71	2.97	0.70	5.38
Carried to summary			63.95	607.53	1,234.60	276.32	2,118.44
PART D FLAT ROOF			N/A	N/A	N/A	N/A	N/A
PART E PITCHED ROOF							
100 x 75mm sawn softwood wall plate	E1	5.00m	1.75	16.63	11.10	4.16	31.88
150 x 50mm sawn softwood pole plate plugged to brickwork	E2	5.30m	1.89	17.96	17.07	5.25	40.28
100 x 50mm sawn softwood rafters	E3	17.50m	3.50	31.50	57.40	13.34	102.24
125 x 25mm sawn softwood purlin	E4	5.30m	1.06	10.07	12.77	3.43	26.27
150 x 50mm sawn softwood joists	E5	12.50m	2.75	26.13	28.80	8.24	63.16
150 x 50mm sawn softwood sprockets	E6	14nr	1.68	15.96	15.96	4.79	36.71
100mm layer of insulation quilt between joists fixed with chicken wire and 150mm layer over joists	E7	10.00m2	4.80	45.60	109.10	23.21	177.91
6mm softwood soffit 150mm wide	E8	10.30m	4.12	39.14	18.03	8.58	65.75
Carried forward			19.87	141.42	145.17	42.99	329.57

	Ref	Qty	Hours	Hours £	Mat'ls £	O & P £	Total £
Brought forward			19.87	141.42	145.17	42.99	329.57
19mm wrought softwood fascia/ barge board 200mm high	E9	5.30m	5.15	48.93	10.44	8.90	68.27
Marley Plain roof tiles on felt and battens	E10	18.55m2	35.25	334.88	479.89	122.21	936.98
Double eaves course	E11	5.30m	1.86	17.67	17.07	5.21	39.95
Verge with plain tile undercloak	E12	7.00m	1.08	10.26	22.54	4.92	37.72
Lead flashing code 5, 200mm girth	E13	5.30m	3.18	30.21	35.40	9.84	75.45
Rake out joint for flashing	E14	5.30m	1.86	17.67	0.80	2.77	21.24
112mm diameter PVC-U gutter	E15	5.30m	1.38	13.11	23.69	5.52	42.32
Stop end	E16	1nr	0.14	1.33	1.11	0.37	2.81
Stop end outlet	E17	1nr	0.25	2.38	1.11	0.52	4.01
68mm diameter PVC-U down pipe	E18	2.50m	0.63	5.99	8.48	2.17	16.63
Shoe	E19	1nr	0.30	2.85	2.09	0.74	5.68
Paint fascia and soffit	E20	3.09m2	2.16	20.52	2.97	3.52	27.01
Carried to summary			73.11	647.20	750.76	209.69	1,607.65

PART F
WINDOWS AND
EXTERNAL DOORS

	Ref	Qty	Hours	Hours £	Mat'ls £	O & P £	Total £
PVC-U door size 840 x 1980mm complete (B)	F1	0.00	0.00	0.00	0.00	0.00	0.00
PVC-U sliding patio door size 1700 x 2075mm (C)	F2	2nr	14.00	133.00	584.04	107.56	824.60
Carried forward			14.00	133.00	584.04	107.56	824.60

	Ref	Qty	Hours	Hours £	Mat'ls £	O & P £	Total £
Brought forward			14.00	133.00	584.04	107.56	824.60
PVC-U window size 1200 x 1200mm complete (A)	F3	2nr	4.00	38.00	315.18	52.98	406.16
25 x 225mm wrought softwood window board	F4	2.40m	0.72	6.84	12.44	2.89	22.17
Paint window board	F5	2.40m	0.42	3.99	2.30	0.94	7.23
Carried to summary			19.14	181.83	913.96	164.37	1,260.16
PART G INTERNAL PARTITIONS AND DOORS		N/A	N/A	N/A	N/A	N/A	N/A
PART H WALL FINISHES							
19 x 100mm wrought softwood skirting	H1	6.54m	1.31	12.45	8.96	3.21	24.62
12mm plasterboard fixed to walls with dabs	H2	14.00m2	5.46	51.87	27.86	11.96	91.69
12mm plasterboard fixed to walls less than 300mm wide with dabs	H3	12.63m	2.27	21.57	11.62	4.98	38.16
Two coats emulsion paint to walls	H4	15.90m2	3.82	36.29	9.38	6.85	52.52
Paint skirting	H5	6.54m	0.98	9.31	3.14	1.87	14.32
Carried to summary			13.84	131.48	60.96	28.87	221.31
PART J FLOOR FINISHES							
Cement and sand floor screed 40mm thick	J1	7.48m2	1.87	17.77	22.66	6.06	46.49
Vinyl floor tiles, size 300 x 300mm	J2	7.48m2	1.27	12.07	52.22	9.64	73.93
Carried to summary			3.14	29.83	74.88	15.71	120.42

	Ref	Qty	Hours	Hours £	Mat'ls £	O & P £	Total £
PART K **CEILING FINISHES**							
Plasterboard with taped butt joints fixed to joists	K1	7.48m2	2.70	2.08	13.84	2.39	18.31
5mm skim coat to plasterboard ceilings	K2	7.48m2	3.74	35.53	9.13	6.70	51.36
Two coats emulsion paint to ceilings	K3	7.48m2	1.95	18.53	4.41	3.44	26.38
Carried to summary			8.39	56.14	27.38	12.53	96.04
PART L **ELECTRICAL WORK**							
13 amp double switched socket outlet with neon	L1	3nr	1.20	16.80	24.78	6.24	47.82
Lighting point	L2	2nr	0.70	9.80	13.60	3.51	26.91
Lighting switch	L3	2nr	0.50	7.00	7.94	2.24	17.18
Lighting wiring	L4	7.00m	0.70	9.80	7.21	2.55	19.56
Power cable	L5	16.00m	2.40	33.60	17.60	7.68	58.88
Carried to summary			5.50	77.00	71.13	22.22	170.35
PART M **HEATING WORK**							
15mm copper pipe	M1	6.00m	1.32	16.50	11.28	4.17	31.95
Elbow	M2	4nr	1.12	14.00	4.40	2.76	21.16
Tee	M3	1nr	0.22	2.75	1.95	0.71	5.41
Radiator, double convector size 1400 x 520mm	M4	2nr	2.60	32.50	231.36	39.58	303.44
Break into existing pipe and insert tee	M5	1nr	0.75	9.38	3.95	2.00	15.32
Carried to summary			6.01	75.13	252.94	49.21	377.27

	Ref	Qty	Hours	Hours £	Mat'ls £	O & P £	Total £
PART N ALTERATION WORK							
Take out existing window size 1500 x 1000mm and lintel over, adapt opening to receive 1770 x 2000mm patio door and insert new lintel over (both measured separately) and make good	N1	1nr	20.00	190.00	20.00	31.50	241.50
Carried to summary			20.00	190.00	20.00	31.50	241.50

SUMMARY

	Hours	Hours £	Mat'ls £	O & P £	Total £
PART A PRELIMINARIES	0.00	0.00	0.00	0.00	1,317.75
PART B SUBSTRUCTURE TO DPC LEVEL	49.97	474.72	456.14	139.63	1,070.48
PART C EXTERNAL WALLS	63.95	607.53	1,234.60	276.32	2,118.44
PART D FLAT ROOF	0.00	0.00	0.00	0.00	0.00
PART E PITCHED ROOF	73.11	647.20	750.76	209.69	1,607.65
PART F WINDOWS AND EXTERNAL DOORS	19.14	181.83	913.96	164.37	1,260.16
PART G INTERNAL PARTITIONS AND DOORS	0.00	0.00	0.00	0.00	0.00
PART H WALL FINISHES	13.84	131.48	60.96	28.87	221.31
PART J FLOOR FINISHES	3.14	29.83	74.88	15.71	120.42
PART K CEILING FINISHES	8.39	56.14	27.38	12.53	96.04
PART L ELECTRICAL WORK	5.50	77.00	71.13	22.22	170.35
PART M HEATING WORK	6.01	75.13	252.94	49.21	377.27
PART N ALTERATION WORK	20.00	190.00	20.00	31.50	241.50
Final total	263.05	2,470.86	3,862.75	950.05	8,601.37

	Ref	Qty	Hours	Hours £	Mat'ls £	O & P £	Total £
PART A PRELIMINARIES							
Concrete mixer	A1	6 wks					270.00
Small tools	A2	7 wks					245.00
Scaffolding (m2/weeks)	A3	90.00					202.50
Skip	A4	3 wks					300.00
Clean up	A5	8 hrs					64.00
Carried to summary							1,081.50
PART B SUBSTRUCTURE TO DPC LEVEL							
Excavate topsoil 150mm thick by hand	B1	10.40m2	3.12	29.64	0.00	4.45	34.09
Excavate to reduce levels by hand	B2	3.12m3	7.80	74.10	0.00	11.12	85.22
Excavate for trench foundations by hand	B3	1.39m3	3.17	30.12	0.00	4.52	34.63
Earthwork support to sides of trenches	B4	12.26m2	4.90	46.55	14.71	9.19	70.45
Backfilling with excavated material	B5	0.47m3	0.19	1.81	0.00	0.27	2.08
Hardcore 225mm thick	B6	6.48m2	0.97	9.22	32.92	6.32	48.46
Hardcore filling to trench	B7	0.09m3	0.05	0.48	0.45	0.14	1.06
Concrete grade (1:3:6) in foundations	B8	1.13m3	1.53	14.54	76.77	13.70	105.00
Concrete grade (1:2:4) in bed 150mm thick	B9	6.48m2	1.94	18.43	70.50	13.34	102.27
Concrete (1:2:4) in cavity wall filling	B10	4.20m2	0.84	7.98	15.25	3.48	26.71
Carried forward			24.51	232.85	210.60	66.52	509.96

	Ref	Qty	Hours	Hours £	Mat'ls £	O & P £	Total £
Brought forward			24.51	232.85	210.60	66.52	509.96
Damp-proof membrane	B11	6.88m2	0.28	2.66	3.99	1.00	7.65
Reinforcement ref A193 in foundation	B12	4.20m2	0.15	1.43	5.46	1.03	7.92
Steel fabric reinforcement ref A193 in slab	B13	6.48m2	0.97	9.22	8.42	2.65	20.28
Solid blockwork 140mm thick in cavity wall	B14	5.46m2	7.10	67.45	69.78	20.58	157.81
Common bricks 112.5mm thick in cavity wall	B15	4.20m2	7.10	67.45	55.40	18.43	141.28
Facing bricks in 112.5mm thick in skin of cavity wall	B16	1.26m2	2.27	21.57	29.32	7.63	58.52
Form cavity 50mm wide in cavity wall	B17	5.46m2	0.16	1.52	3.28	0.72	5.52
DPC 112mm wide	B18	8.40m	0.42	3.99	6.80	1.62	12.41
DPC 140mm wide	B19	8.40m	0.50	4.75	8.15	1.94	14.84
Bond in block wall	B20	1.30m	0.58	5.51	2.83	1.25	9.59
Bond in half brick wall	B21	0.30m	0.46	4.37	0.61	0.75	5.73
50mm thick insulation board	B22	6.48m2	1.94	18.43	29.03	7.12	54.58
Carried to summary			46.44	441.18	433.67	131.23	1,006.08
PART C EXTERNAL WALLS							
Solid blockwork 140mm thick in cavity wall	C1	21.50m2	27.95	265.53	273.48	80.85	619.86
Facing brickwork 112.5mm thick in cavity wall	C2	21.50m2	38.70	367.65	495.79	129.52	992.96
75mm thick insulation in cavity wall	C3	21.50m2	4.73	44.94	125.13	25.51	195.57
Carried forward			71.38	678.11	894.40	235.88	1,808.39

	Ref	Qty	Hours	Hours £	Mat'ls £	O & P £	Total £
Brought forward			71.38	678.11	894.40	235.88	1,808.39
Steel lintel 2400mm long	C4	2nr	0.50	4.75	196.92	30.25	231.92
Steel lintel 1500mm long	C5	2nr	0.40	3.80	214.94	32.81	251.55
Close cavity wall at jambs	C7	8.96m	0.45	4.28	18.28	3.38	25.94
Close cavity wall at cills	C8	3.67m	0.18	1.71	7.49	1.38	10.58
Close cavity wall at top	C9	8.40m	0.37	3.52	17.14	3.10	23.75
DPC 112mm wide at jambs	C10	8.96m	0.45	4.28	7.26	1.73	13.27
DPC 112mm wide at cills	C11	3.67m	0.18	1.71	2.97	0.70	5.38
Carried to summary			73.91	702.15	1,359.40	309.23	2,370.78
PART D FLAT ROOF			N/A	N/A	N/A	N/A	N/A
PART E PITCHED ROOF							
100 x 75mm sawn softwood wall plate	E1	3.00m	1.05	9.98	6.66	2.50	19.13
150 x 50mm sawn softwood pole plate plugged to brickwork	E2	3.30m	0.99	9.41	10.63	3.01	23.04
100 x 50mm sawn softwood rafters	E3	31.50	3.50	31.50	103.32	20.22	155.04
125 x 25mm sawn softwood purlin	E4	3.30m	0.66	6.27	7.91	2.13	16.31
150 x 50mm sawn softwood joists	E5	24.50m	2.75	26.13	55.86	12.30	94.28
150 x 50mm sawn softwood sprockets	E6	18nr	1.68	15.96	20.52	5.47	41.95
100mm layer of insulation quilt between joists fixed with chicken wire and 150mm layer over joists	E7	9.00m2	4.32	41.04	98.19	20.88	160.11
6mm softwood soffit 150mm wide	E8	10.30m	3.32	31.54	18.03	7.44	57.01
Carried forward			18.27	171.82	321.12	73.94	566.88

	Ref	Qty	Hours	Hours £	Mat'ls £	O & P £	Total £
Brought forward			18.27	171.82	321.12	73.94	566.88
19mm wrought softwood fascia/ barge board 200mm high	E9	3.30m	4.15	39.43	6.50	6.89	52.81
Marley Plain roof tiles on felt and battens	E10	14.85m2	21.95	208.53	384.17	88.90	681.60
Double eaves course	E11	3.30m	0.83	7.89	10.63	2.78	21.29
Verge with plain tile undercloak	E12	9.00m	1.75	16.63	28.98	6.84	52.45
Lead flashing code 5, 200mm girth	E13	3.30m	1.98	18.81	22.04	6.13	46.98
Rake out joint for flashing	E14	3.30m	1.16	11.02	0.50	1.73	13.25
112mm diameter PVC-U gutter	E15	3.30m	0.86	8.17	14.75	3.44	26.36
Stop end	E16	1nr	0.14	1.33	1.11	0.37	2.81
Stop end outlet	E17	1nr	0.25	2.38	1.11	0.52	4.01
68mm diameter PVC-U down pipe	E18	2.50m	0.63	5.99	8.48	2.17	16.63
Shoe	E19	1nr	0.30	2.85	2.09	0.74	5.68
Paint fascia and soffit	E20	3.09m2	1.53	14.54	2.97	2.63	20.13
Carried to summary			53.80	509.35	804.45	197.07	1,510.87
PART F WINDOWS AND EXTERNAL DOORS							
PVC-U door size 840 x 1980mm complete (B)	F1	0.00	0.00	0.00	0.00	0.00	0.00
PVC-U sliding patio door size 1700 x 2075mm (C)	F2	2nr	14.00	133.00	587.04	108.01	828.05
Carried forward			14.00	133.00	587.04	108.01	828.05

	Ref	Qty	Hours	Hours £	Mat'ls £	O & P £	Total £
Brought forward			14.00	133.00	587.04	108.01	828.05
PVC-U window size 1200 x 1200mm complete (A)	F3	2nr	4.00	38.00	315.18	52.98	406.16
25 x 225mm wrought softwood window board	F4	2.40m	0.72	6.84	12.44	2.89	22.17
Paint window board	F5	2.40m	0.42	3.99	2.30	0.94	7.23
Carried to summary			19.14	181.83	916.96	164.82	1,263.61
PART G INTERNAL PARTITIONS AND DOORS		N/A	N/A	N/A	N/A	N/A	N/A
PART H WALL FINISHES							
19 x 100mm wrought softwood skirting	H1	6.54m	1.31	12.45	6.22	2.80	21.46
12mm plasterboard fixed to walls with dabs	H2	14.00m2	5.46	51.87	27.86	11.96	91.69
12mm plasterboard fixed to walls less than 300mm wide with dabs	H3	12.63m	2.27	21.57	11.62	4.98	38.16
Two coats emulsion paint to walls	H4	15.40m2	3.82	36.29	9.38	6.85	52.52
Paint skirting	H5	6.54m	0.98	9.31	3.14	1.87	14.32
Carried to summary			13.84	131.48	58.22	28.46	218.16
PART J FLOOR FINISHES							
Cement and sand floor screed 40mm thick	J1	6.48m2	1.62	15.39	19.63	5.25	40.27
Vinyl floor tiles, size 300 x 300mm	J2	6.48m2	1.58	15.01	45.23	9.04	69.28
Carried to summary			3.20	30.40	64.86	14.29	109.55

	Ref	Qty	Hours	Hours £	Mat'ls £	O & P £	Total £
PART K							
CEILING FINISHES							
Plasterboard with taped butt joints fixed to joists	K1	6.48m2	2.33	2.08	11.99	2.11	16.18
5mm skim coat to plasterboard ceilings	K2	6.48m2	3.24	30.78	7.91	5.80	44.49
Two coats emulsion paint to ceilings	K3	6.48m2	1.69	16.06	3.82	2.98	22.86
Carried to summary			7.26	48.92	23.72	10.90	83.53
PART L							
ELECTRICAL WORK							
13 amp double switched socket outlet with neon	L1	2nr	0.80	11.20	16.52	4.16	31.88
Lighting point	L2	1nr	0.70	9.80	6.80	2.49	19.09
Lighting switch	L3	2nr	0.50	7.00	7.94	2.24	17.18
Lighting wiring	L4	6.00m	0.60	8.40	6.18	2.19	16.77
Power cable	L5	14.00m	2.10	29.40	15.40	6.72	51.52
Carried to summary			4.70	65.80	52.84	17.80	136.44
PART M							
HEATING WORK							
15mm copper pipe	M1	5.00m	1.10	13.75	9.40	3.47	26.62
Elbow	M2	4nr	1.12	14.00	4.40	2.76	21.16
Tee	M3	1nr	0.22	2.75	1.95	0.71	5.41
Radiator, double convector size 1400 x 520mm	M4	2nr	2.60	32.50	231.36	39.58	303.44
Break into existing pipe and insert tee	M5	1nr	0.75	9.38	3.95	2.00	15.32
Carried to summary			5.79	72.38	251.06	48.52	371.95

	Ref	Qty	Hours	Hours £	Mat'ls £	O & P £	Total £
PART N **ALTERATION WORK**							
Take out existing window size 1500 x 1000mm and lintel over, adapt opening to receive 1770 x 2000mm patio door and insert new lintel over (both measured separately) and make good	N1	1nr	20.00	190.00	20.00	31.50	241.50
Carried to summary			20.00	190.00	20.00	31.50	241.50

SUMMARY

	Hours	Hours £	Mat'ls £	O & P £	Total £
PART A PRELIMINARIES	0.00	0.00	0.00	0.00	1,081.50
PART B SUBSTRUCTURE TO DPC LEVEL	46.44	441.18	433.67	131.23	1,006.08
PART C EXTERNAL WALLS	73.91	702.15	1,359.40	309.23	2,370.78
PART D FLAT ROOF	0.00	0.00	0.00	0.00	0.00
PART E PITCHED ROOF	53.80	509.35	804.45	197.07	1,510.87
PART F WINDOWS AND EXTERNAL DOORS	19.14	181.83	916.96	164.82	1,263.61
PART G INTERNAL PARTITIONS AND DOORS	0.00	0.00	0.00	0.00	0.00
PART H WALL FINISHES	13.84	131.48	58.22	28.46	218.16
PART J FLOOR FINISHES	3.20	30.40	64.86	14.29	109.55
PART K CEILING FINISHES	7.26	48.92	23.72	10.90	83.53
PART L ELECTRICAL WORK	4.70	65.80	52.84	17.80	136.44
PART M HEATING WORK	5.79	72.38	251.06	48.52	371.95
PART N ALTERATION WORK	20.00	190.00	20.00	31.50	241.50
Final total	248.08	2,373.49	3,985.18	953.82	8,393.97

	Ref	Qty	Hours	Hours £	Mat'ls £	O & P £	Total £
PART A PRELIMINARIES							
Concrete mixer	A1	7 wks					315.00
Small tools	A2	8 wks					280.00
Scaffolding (m2/weeks)	A3	125.00					281.25
Skip	A4	4 wks					400.00
Clean up	A5	6 hrs					48.00
Carried to summary							1,324.25
PART B SUBSTRUCTURE TO DPC LEVEL							
Excavate topsoil 150mm thick by hand	B1	14.19m2	4.26	40.47	0.00	6.07	46.54
Excavate to reduce levels by hand	B2	4.06m3	10.15	96.43	0.00	14.46	110.89
Excavate for trench foundations by hand	B3	1.39m3	3.17	30.12	0.00	4.52	34.63
Earthwork support to sides of trenches	B4	13.72m2	5.49	52.16	16.46	10.29	78.91
Backfilling with excavated material	B5	0.55m3	0.22	2.09	0.00	0.31	2.40
Hardcore 225mm thick	B6	9.18m2	1.38	13.11	46.63	8.96	68.70
Hardcore filling to trench	B7	0.12m3	0.06	0.57	0.60	0.18	1.35
Concrete grade (1:3:6) in foundations	B8	1.27m3	1.71	16.25	86.28	15.38	117.90
Concrete grade (1:2:4) in bed 150mm thick	B9	9.18m2	2.75	26.13	99.88	18.90	144.91
Concrete (1:2:4) in cavity wall filling	B10	4.70m2	0.94	8.93	17.06	3.90	29.89
Carried forward			30.13	286.24	266.91	82.97	636.12

	Ref	Qty	Hours	Hours £	Mat'ls £	O & P £	Total £
Brought forward			30.13	286.24	266.91	82.97	636.12
Damp-proof membrane	B11	9.63m2	0.25	2.38	5.59	1.19	9.16
Reinforcement ref A193 in foundation	B12	4.70m2	0.56	5.32	6.11	1.71	13.14
Steel fabric reinforcement ref A193 in slab	B13	9.18m2	1.38	13.11	11.93	3.76	28.80
Solid blockwork 140mm thick in cavity wall	B14	6.11m2	7.94	75.43	78.08	23.03	176.54
Common bricks 112.5mm thick in cavity wall	B15	4.70m2	7.94	75.43	61.99	20.61	158.03
Facing bricks in 112.5mm thick in skin of cavity wall	B16	1.41m2	2.54	24.13	32.81	8.54	65.48
Form cavity 50mm wide in cavity wall	B17	5.46m2	0.18	1.71	3.67	0.81	6.19
DPC 112mm wide	B18	9.40m	0.00	0.00	7.61	1.14	8.75
DPC 140mm wide	B19	9.40m	0.56	5.32	9.12	2.17	16.61
Bond in block wall	B20	1.30m	0.58	5.51	2.83	1.25	9.59
Bond in half brick wall	B21	0.30m	0.46	4.37	0.61	0.75	5.73
50mm thick insulation board	B22	9.18m2	2.75	26.13	41.13	10.09	77.34
Carried to summary			55.27	525.07	528.39	158.02	1,211.47
PART C EXTERNAL WALLS							
Solid blockwork 140mm thick in cavity wall	C1	24.00m2	31.20	296.40	305.28	90.25	691.93
Facing brickwork 112.5mm thick in cavity wall	C2	24.00m2	43.20	410.40	553.44	144.58	1,108.42
75mm thick insulation in cavity wall	C3	24.00m2	5.28	50.16	139.68	28.48	218.32
Carried forward			79.68	756.96	998.40	263.30	2,018.66

	Ref	Qty	Hours	Hours £	Mat'ls £	O & P £	Total £
Brought forward			79.68	756.96	998.40	263.30	2,018.66
Steel lintel 2400mm long	C4	2nr	0.50	4.75	196.92	30.25	231.92
Steel lintel 1500mm long	C5	2nr	0.80	7.60	214.94	33.38	255.92
Close cavity wall at jambs	C7	8.96m	0.69	6.56	18.28	3.73	28.56
Close cavity wall at cills	C8	3.67m	0.25	2.38	7.49	1.48	11.34
Close cavity wall at top	C9	9.40m	0.47	4.47	19.18	3.55	27.19
DPC 112mm wide at jambs	C10	8.96m	0.69	6.56	7.26	2.07	15.89
DPC 112mm wide at cills	C11	3.67m	0.25	2.38	2.97	0.80	6.15
Carried to summary			83.33	791.64	1,465.44	338.56	2,595.64
PART D FLAT ROOF			N/A	N/A	N/A	N/A	N/A
PART E PITCHED ROOF							
100 x 75mm sawn softwood wall plate	E1	4.00m	1.40	13.30	8.85	3.32	25.47
150 x 50mm sawn softwood pole plate plugged to brickwork	E2	4.30m	1.29	12.26	13.85	3.92	30.02
100 x 50mm sawn softwood rafters	E3	31.50m	4.90	44.10	103.32	22.11	169.53
125 x 25mm sawn softwood purlin	E4	4.30m	0.86	8.17	10.30	2.77	21.24
150 x 50mm sawn softwood joists	E5	24.50m	5.39	51.21	55.80	16.05	123.06
150 x 50mm sawn softwood sprockets	E6	18nr	2.16	20.52	20.52	6.16	47.20
100mm layer of insulation quilt between joists fixed with chicken wire and 150mm layer over joists	E7	12.00m2	5.76	54.72	130.92	27.85	213.49
6mm softwood soffit 150mm wide	E8	11.30m	4.92	46.74	19.78	9.98	76.50
Carried forward			26.68	251.01	363.34	92.15	706.50

	Ref	Qty	Hours	Hours £	Mat'ls £	O & P £	Total £
Brought forward			26.68	251.01	363.34	92.15	706.50
19mm wrought softwood fascia/ barge board 200mm high	E9	4.30m	6.15	58.43	8.47	10.03	76.93
Marley Plain roof tiles on felt and battens	E10	19.55m2	36.77	349.32	500.59	127.49	977.39
Double eaves course	E11	4.30m	1.51	14.35	13.85	4.23	32.42
Verge with plain tile undercloak	E12	9.00m	2.25	21.38	28.98	7.55	57.91
Lead flashing code 5, 200mm girth	E13	4.30m	2.58	24.51	28.72	7.98	61.21
Rake out joint for flashing	E14	4.30m	1.51	14.35	0.65	2.25	17.24
112mm diameter PVC-U gutter	E15	4.30m	1.12	10.64	19.22	4.48	34.34
Stop end	E16	1nr	0.14	1.33	1.11	0.37	2.81
Stop end outlet	E17	1nr	0.25	2.38	1.11	0.52	4.01
68mm diameter PVC-U down pipe	E18	2.50m	0.63	5.99	8.48	2.17	16.63
Shoe	E19	1nr	0.30	2.85	2.09	0.74	5.68
Paint fascia and soffit	E20	3.39m2	2.37	22.52	3.25	3.86	29.63
Carried to summary			82.26	779.02	979.86	263.83	2,022.71

PART F
WINDOWS AND
EXTERNAL DOORS

	Ref	Qty	Hours	Hours £	Mat'ls £	O & P £	Total £
PVC-U door size 840 x 1980mm complete (B)	F1	0.00	0.00	0.00	0.00	0.00	0.00
PVC-U sliding patio door size 1700 x 2075mm (C)	F2	2nr	14.00	133.00	587.04	108.01	828.05
Carried forward			14.00	133.00	587.04	108.01	828.05

	Ref	Qty	Hours	Hours £	Mat'ls £	O & P £	Total £
Brought forward			14.00	133.00	587.04	108.01	828.05
PVC-U window size 1200 x 1200mm complete (A)	F3	2nr	4.00	38.00	315.18	52.98	406.16
25 x 225mm wrought softwood window board	F4	2.40m	0.72	6.84	12.44	2.89	22.17
Paint window board	F5	2.40m	0.42	3.99	2.30	0.94	7.23
Carried to summary			19.14	181.83	916.96	164.82	1,263.61
PART G INTERNAL PARTITIONS AND DOORS		N/A	N/A	N/A	N/A	N/A	N/A
PART H WALL FINISHES							
19 x 100mm wrought softwood skirting	H1	7.54m	1.51	14.35	7.59	3.29	25.23
12mm plasterboard fixed to walls with dabs	H2	16.50m2	6.44	61.18	32.84	14.10	108.12
12mm plasterboard fixed to walls less than 300mm wide with dabs	H3	12.63m	2.27	21.57	11.62	4.98	38.16
Two coats emulsion paint to walls	H4	18.40m2	4.42	41.99	10.86	7.93	60.78
Paint skirting	H5	7.54m	1.13	10.74	3.62	2.15	16.51
Carried to summary			15.77	149.82	66.53	32.45	248.80
PART J FLOOR FINISHES							
Cement and sand floor screed 40mm thick	J1	9.18m2	2.30	21.85	27.82	7.45	57.12
Vinyl floor tiles, size 300 x 300mm	J2	9.18m2	2.20	20.90	64.07	12.75	97.72
Carried to summary			4.50	42.75	91.89	20.20	154.84

	Ref	Qty	Hours	Hours £	Mat'ls £	O & P £	Total £
PART K **CEILING FINISHES**							
Plasterboard with taped butt joints fixed to joists	K1	9.18m2	3.31	2.08	16.98	2.86	21.92
5mm skim coat to plasterboard ceilings	K2	9.18m2	4.59	43.61	11.20	8.22	63.03
Two coats emulsion paint to ceilings	K3	9.18m2	2.39	22.71	5.42	4.22	32.34
Carried to summary			10.29	68.39	33.60	15.30	117.29
PART L **ELECTRICAL WORK**							
13 amp double switched socket outlet with neon	L1	3nr	1.20	16.80	24.78	6.24	47.82
Lighting point	L2	2nr	0.70	9.80	13.60	3.51	26.91
Lighting switch	L3	2nr	0.50	7.00	7.94	2.24	17.18
Lighting wiring	L4	6.00m	0.60	8.40	6.18	2.19	16.77
Power cable	L5	16.00m	2.40	33.60	17.60	7.68	58.88
Carried to summary			5.40	75.60	70.10	21.86	167.56
PART M **HEATING WORK**							
15mm copper pipe	M1	6.00m	1.32	16.50	11.28	4.17	31.95
Elbow	M2	4nr	1.12	14.00	4.40	2.76	21.16
Tee	M3	1nr	0.22	2.75	1.95	0.71	5.41
Radiator, double convector size 1400 x 520mm	M4	2nr	2.60	32.50	231.36	39.58	303.44
Break into existing pipe and insert tee	M5	1nr	0.75	9.38	3.95	2.00	15.32
Carried to summary			6.01	75.13	252.94	49.21	377.27

	Ref	Qty	Hours	Hours £	Mat'ls £	O & P £	Total £
PART N **ALTERATION WORK**							
Take out existing window size 1500 x 1000mm and lintel over, adapt opening to receive 1770 x 2000mm patio door and insert new lintel over (both measured separately) and make good	N1	1nr	20.00	190.00	20.00	31.50	241.50
Carried to summary			20.00	190.00	20.00	31.50	241.50

SUMMARY

	Hours	Hours £	Mat'ls £	O & P £	Total £
PART A PRELIMINARIES	0.00	0.00	0.00	0.00	1,324.25
PART B SUBSTRUCTURE TO DPC LEVEL	55.27	525.07	528.39	158.02	1,211.47
PART C EXTERNAL WALLS	83.33	791.64	1,465.44	338.56	2,595.64
PART D FLAT ROOF	0.00	0.00	0.00	0.00	0.00
PART E PITCHED ROOF	82.26	779.02	979.86	263.83	2,022.71
PART F WINDOWS AND EXTERNAL DOORS	19.14	181.83	916.96	164.82	1,263.61
PART G INTERNAL PARTITIONS AND DOORS	0.00	0.00	0.00	0.00	0.00
PART H WALL FINISHES	15.77	149.82	66.53	32.45	248.80
PART J FLOOR FINISHES	4.50	42.75	91.89	20.20	154.84
PART K CEILING FINISHES	10.29	68.39	33.60	15.30	117.29
PART L ELECTRICAL WORK	5.40	75.60	70.10	21.86	167.56
PART M HEATING WORK	6.01	75.13	252.94	49.21	377.27
PART N ALTERATION WORK	20.00	190.00	20.00	31.50	241.50
Final total	301.97	2,879.25	4,425.71	1,095.75	9,724.94

	Ref	Qty	Hours	Hours £	Mat'ls £	O & P £	Total £
PART A PRELIMINARIES							
Concrete mixer	A1	8 wks					360.00
Small tools	A2	9 wks					315.00
Scaffolding (m2/weeks)	A3	165.00					371.25
Skip	A4	5 wks					500.00
Clean up	A5	8 hrs					64.00
Carried to summary							1,610.25
PART B SUBSTRUCTURE TO DPC LEVEL							
Excavate topsoil 150mm thick by hand	B1	17.50m2	5.25	49.88	0.00	7.48	57.36
Excavate to reduce levels by hand	B2	5.00m3	12.50	118.75	0.00	17.81	136.56
Excavate for trench foundations by hand	B3	1.72m3	4.47	42.47	0.00	6.37	48.83
Earthwork support to sides of trenches	B4	15.18m2	6.07	57.67	18.22	11.38	87.27
Backfilling with excavated material	B5	0.63m3	0.37	3.52	0.00	0.53	4.04
Hardcore 225mm thick	B6	1.88m2	2.38	22.61	60.35	12.44	95.40
Hardcore filling to trench	B7	0.14m3	0.07	0.67	0.70	0.20	1.57
Concrete grade (1:3:6) in foundations	B8	1.40m3	1.89	17.96	95.11	16.96	130.02
Concrete grade (1:2:4) in bed 150mm thick	B9	11.88m2	3.56	33.82	129.25	24.46	187.53
Concrete (1:2:4) in cavity wall filling	B10	5.20m2	1.04	9.88	18.88	4.31	33.07
Carried forward			37.60	357.20	322.51	101.96	781.67

	Ref	Qty	Hours	Hours £	Mat'ls £	O & P £	Total £
Brought forward			37.60	357.20	322.51	101.96	781.67
Damp-proof membrane	B11	12.38m2	0.52	4.94	7.18	1.82	13.94
Reinforcement ref A193 in foundation	B12	5.20m2	0.62	5.89	6.76	1.90	14.55
Steel fabric reinforcement ref A193 in slab	B13	11.88m2	1.78	16.91	15.46	4.86	37.23
Solid blockwork 140mm thick in cavity wall	B14	6.76m2	8.79	83.51	95.34	26.83	205.67
Common bricks 112.5mm thick in cavity wall	B15	5.20m2	8.79	83.51	68.69	22.83	175.02
Facing bricks in 112.5mm thick in skin of cavity wall	B16	1.56m2	2.81	26.70	36.30	9.45	72.44
Form cavity 50mm wide in cavity wall	B17	6.76m2	0.20	1.90	4.06	0.89	6.85
DPC 112mm wide	B18	10.40m	0.52	4.94	8.42	2.00	15.36
DPC 140mm wide	B19	10.40m	0.62	5.89	10.09	2.40	18.38
Bond in block wall	B20	1.30m	0.58	5.51	2.83	1.25	9.59
Bond in half brick wall	B21	0.30m	0.46	4.37	0.61	0.75	5.73
50mm thick insulation board	B22	11.88m2	3.56	33.82	53.22	13.06	100.10
Carried to summary			66.85	689.26	666.25	264.93	1,444.43
PART C EXTERNAL WALLS							
Solid blockwork 140mm thick in cavity wall	C1	26.50m2	34.45	327.28	337.08	99.65	764.01
Facing brickwork 112.5mm thick in cavity wall	C2	26.50m2	47.70	453.15	611.09	159.64	1,223.88
75mm thick insulation in cavity wall	C3	26.50m2	5.83	55.39	154.23	31.44	241.06
Carried forward			87.98	835.81	1,102.40	290.73	2,228.94

	Ref	Qty	Hours	Hours £	Mat'ls £	O & P £	Total £
Brought forward			87.98	835.81	1,102.40	290.73	2,228.94
Steel lintel 2400mm long	C4	2nr	0.50	4.75	196.92	30.25	231.92
Steel lintel 1500mm long	C5	2nr	0.40	3.80	214.94	32.81	251.55
Close cavity wall at jambs	C7	8.96m	0.45	4.28	18.28	3.38	25.94
Close cavity wall at cills	C8	3.67m	0.18	1.71	7.49	1.38	10.58
Close cavity wall at top	C9	10.40m	0.52	4.94	21.22	3.92	30.08
DPC 112mm wide at jambs	C10	8.96m	0.45	4.28	7.26	1.73	13.27
DPC 112mm wide at cills	C11	3.87m	0.18	1.71	7.49	1.38	10.58
Carried to summary			90.66	861.27	1,576.00	365.59	2,802.86
PART D FLAT ROOF			N/A	N/A	N/A	N/A	N/A
PART E PITCHED ROOF							
100 x 75mm sawn softwood wall plate	E1	5.00m	1.05	9.98	11.10	3.16	24.24
150 x 50mm sawn softwood pole plate plugged to brickwork	E2	5.30m	0.99	9.41	17.07	3.97	30.45
100 x 50mm sawn softwood rafters	E3	31.50m	3.50	31.50	103.32	20.22	155.04
125 x 25mm sawn softwood purlin	E4	5.30m	0.66	6.27	12.77	2.86	21.90
150 x 50mm sawn softwood joists	E5	24.50m	2.75	26.13	55.88	12.30	94.31
150 x 50mm sawn softwood sprockets	E6	18nr	1.68	15.96	20.52	5.47	41.95
100mm layer of insulation quilt between joists fixed with chicken wire and 150mm layer over joists	E7	15.00m2	7.20	68.40	163.65	34.81	266.86
6mm softwood soffit 150mm wide	E8	12.30m	3.32	31.54	21.53	7.96	61.03
Carried forward			21.15	199.18	405.84	90.75	695.77

	Ref	Qty	Hours	Hours £	Mat'ls £	O & P £	Total £
Brought forward			21.15	199.18	405.84	90.75	695.77
19mm wrought softwood fascia/ barge board 200mm high	E9	5.30m	4.15	39.43	10.44	7.48	57.34
Marley Plain roof tiles on felt and battens	E10	23.85m2	21.95	208.53	617.00	123.83	949.35
Double eaves course	E11	5.30m	0.83	7.89	17.07	3.74	28.70
Verge with plain tile undercloak	E12	9.00m	1.75	16.63	28.98	6.84	52.45
Lead flashing code 5, 200mm girth	E13	5.30m	1.98	18.81	35.40	8.13	62.34
Rake out joint for flashing	E14	5.30m	1.16	11.02	0.80	1.77	13.59
112mm diameter PVC-U gutter	E15	5.30m	0.86	8.17	23.69	4.78	36.64
Stop end	E16	1nr	0.14	1.33	1.11	0.37	2.81
Stop end outlet	E17	1nr	0.25	2.38	1.11	0.52	4.01
68mm diameter PVC-U down pipe	E18	2.50m	0.63	5.99	8.48	2.17	16.63
Shoe	E19	1nr	0.30	2.85	2.09	0.74	5.68
Paint fascia and soffit	E20	3.69m2	1.53	14.54	3.54	2.71	20.79
Carried to summary			56.68	536.71	1,155.55	253.84	1,946.10
PART F WINDOWS AND EXTERNAL DOORS							
PVC-U door size 840 x 1980mm complete (B)	F1	0.00	0.00	0.00	0.00	0.00	0.00
PVC-U sliding patio door size 1700 x 2075mm (C)	F2	2nr	14.00	133.00	587.04	108.01	828.05
Carried forward			14.00	133.00	587.04	108.01	828.05

	Ref	Qty	Hours	Hours £	Mat'ls £	O & P £	Total £
Brought forward			14.00	133.00	587.04	108.01	828.05
PVC-U window size 1200 x 1200mm complete (A)	F3	2nr	4.00	38.00	315.18	52.98	406.16
25 x 225mm wrought softwood window board	F4	2.40m	0.72	6.84	12.44	2.89	22.17
Paint window board	F5	2.40m	0.42	3.99	2.30	0.94	7.23
Carried to summary			19.14	181.83	916.96	164.82	1,263.61
PART G INTERNAL PARTITIONS AND DOORS		N/A	N/A	N/A	N/A	N/A	N/A
PART H WALL FINISHES							
19 x 100mm wrought softwood skirting	H1	8.54m	1.71	16.25	8.96	3.78	28.99
12mm plasterboard fixed to walls with dabs	H2	19.00m2	7.41	70.40	37.81	16.23	124.44
12mm plasterboard fixed to walls less than 300mm wide with dabs	H3	13.47m	2.43	23.09	12.39	5.32	40.80
Two coats emulsion paint to walls	H4	21.02m2	5.04	47.88	12.40	9.04	69.32
Paint skirting	H5	8.54m	1.28	12.16	4.10	2.44	18.70
Carried to summary			17.87	169.77	75.66	36.81	282.24
PART J FLOOR FINISHES							
Cement and sand floor screed 40mm thick	J1	11.88m2	2.97	28.22	36.00	9.63	73.85
Vinyl floor tiles, size 300 x 300mm	J2	11.88m2	2.85	27.08	83.62	16.60	127.30
Carried to summary			5.82	55.29	119.62	26.24	201.15

	Ref	Qty	Hours	Hours £	Mat'ls £	O & P £	Total £
PART K **CEILING FINISHES**							
Plasterboard with taped butt joints fixed to joists	K1	11.88m2	4.28	2.08	21.98	3.61	27.67
5mm skim coat to plasterboard ceilings	K2	11.88m2	5.94	56.43	14.50	10.64	81.57
Two coats emulsion paint to ceilings	K3	11.88m2	3.09	29.36	7.01	5.45	41.82
Carried to summary			13.31	87.87	43.49	19.70	151.06
PART L **ELECTRICAL WORK**							
13 amp double switched socket outlet with neon	L1	3nr	1.20	16.80	24.78	6.24	47.82
Lighting point	L2	2nr	0.70	9.80	13.60	3.51	26.91
Lighting switch	L3	2nr	0.50	7.00	7.94	2.24	17.18
Lighting wiring	L4	7.00m	0.70	9.80	7.21	2.55	19.56
Power cable	L5	18.00m	2.70	37.80	19.80	8.64	66.24
Carried to summary			5.80	81.20	73.33	23.18	177.71
PART M **HEATING WORK**							
15mm copper pipe	M1	7.00m	1.54	19.25	13.16	4.86	37.27
Elbow	M2	4nr	1.12	14.00	4.40	2.76	21.16
Tee	M3	1nr	0.22	2.75	1.95	0.71	5.41
Radiator, double convector size 1400 x 520mm	M4	2nr	2.60	32.50	231.36	39.58	303.44
Break into existing pipe and insert tee	M5	1nr	0.75	9.38	3.95	2.00	15.32
Carried to summary			6.23	77.88	254.82	49.90	382.60

	Ref	Qty	Hours	Hours £	Mat'ls £	O & P £	Total £
PART N **ALTERATION WORK**							
Take out existing window size 1500 x 1000mm and lintel over, adapt opening to receive 1770 x 2000mm patio door and insert new lintel over (both measured separately) and make good	N1	1nr	20.00	190.00	20.00	31.50	241.50
Carried to summary			20.00	190.00	20.00	31.50	241.50

SUMMARY

	Hours	Hours £	Mat'ls £	O & P £	Total £
PART A PRELIMINARIES	0.00	0.00	0.00	0.00	1,610.25
PART B SUBSTRUCTURE TO DPC LEVEL	66.85	689.26	666.25	264.93	1,444.43
PART C EXTERNAL WALLS	90.66	861.27	1,576.00	365.59	2,802.86
PART D FLAT ROOF	0.00	0.00	0.00	0.00	0.00
PART E PITCHED ROOF	56.68	536.71	1,155.55	253.84	1,946.10
PART F WINDOWS AND EXTERNAL DOORS	19.14	181.83	916.96	164.82	1,263.61
PART G INTERNAL PARTITIONS AND DOORS	0.00	0.00	0.00	0.00	0.00
PART H WALL FINISHES	17.87	169.77	75.66	36.81	282.24
PART J FLOOR FINISHES	5.82	55.29	119.62	26.24	201.15
PART K CEILING FINISHES	13.31	87.87	43.49	19.70	151.06
PART L ELECTRICAL WORK	5.80	81.20	73.33	23.18	177.71
PART M HEATING WORK	6.23	77.88	254.82	49.90	382.60
PART N ALTERATION WORK	20.00	190.00	20.00	31.50	241.50
Final total	302.36	2,931.08	4,901.68	1,236.51	10,503.51

	Ref	Qty	Hours	Hours £	Mat'ls £	O & P £	Total £
PART A PRELIMINARIES							
Concrete mixer	A1	9 wks					405.00
Small tools	A2	10 wks					350.00
Scaffolding (m2/weeks)	A3	210.00					472.50
Skip	A4	6 wks					600.00
Clean up	A5	10 hrs					80.00
Carried to summary							1,907.50
PART B SUBSTRUCTURE TO DPC LEVEL							
Excavate topsoil 150mm thick by hand	B1	19.85m2	5.96	56.62	0.00	8.49	65.11
Excavate to reduce levels by hand	B2	5.95m3	14.88	141.36	0.00	21.20	162.56
Excavate for trench foundations by hand	B3	1.88m3	3.17	30.12	0.00	4.52	34.63
Earthwork support to sides of trenches	B4	16.64m2	4.90	46.55	19.97	9.98	76.50
Backfilling with excavated material	B5	0.70m3	0.28	2.66	0.00	0.40	3.06
Hardcore 225mm thick	B6	14.58m2	2.19	20.81	74.07	14.23	109.11
Hardcore filling to trench	B7	0.15m3	0.06	0.57	0.76	0.20	1.53
Concrete grade (1:3:6) in foundations	B8	1.54m3	1.53	14.54	104.63	17.87	137.04
Concrete grade (1:2:4) in bed 150mm thick	B9	14.58m2	4.37	41.52	158.63	30.02	230.17
Concrete (1:2:4) in cavity wall filling	B10	5.70m2	0.84	7.98	20.69	4.30	32.97
Carried forward			38.18	362.71	378.75	111.22	852.68

	Ref	Qty	Hours	Hours £	Mat'ls £	O & P £	Total £
Brought forward			38.18	362.71	378.75	111.22	852.68
Damp-proof membrane	B11	15.13m2	0.61	5.80	7.62	2.01	15.43
Reinforcement ref A193 in foundation	B12	5.70m2	0.68	6.46	7.41	2.08	15.95
Steel fabric reinforcement ref A193 in slab	B13	14.58m2	2.19	20.81	18.95	5.96	45.72
Solid blockwork 140mm thick in cavity wall	B14	7.40m2	9.63	91.49	94.70	27.93	214.11
Common bricks 112.5mm thick in cavity wall	B15	5.70m2	9.63	91.49	75.18	25.00	191.66
Facing bricks in 112.5mm thick in skin of cavity wall	B16	1.71m2	2.27	21.57	39.79	9.20	70.56
Form cavity 50mm wide in cavity wall	B17	7.41m2	0.22	2.09	4.45	0.98	7.52
DPC 112mm wide	B18	11.40m	0.57	5.42	9.23	2.20	16.84
DPC 140mm wide	B19	11.40m	0.68	6.46	11.06	2.63	20.15
Bond in block wall	B20	1.30m	0.58	5.51	2.83	1.25	9.59
Bond in half brick wall	B21	0.30m	0.46	4.37	0.61	0.75	5.73
50mm thick insulation board	B22	14.58m2	4.37	41.52	65.32	16.03	122.86
Carried to summary			70.07	665.67	715.90	207.23	1,588.80

PART C
EXTERNAL WALLS

	Ref	Qty	Hours	Hours £	Mat'ls £	O & P £	Total £
Solid blockwork 140mm thick in cavity wall	C1	25.90m2	33.67	319.87	329.45	97.40	746.71
Facing brickwork 112.5mm thick in cavity wall	C2	25.90m2	46.62	442.89	597.25	156.02	1,196.16
75mm thick insulation in cavity wall	C3	25.90m2	5.70	54.15	150.74	30.73	235.62
Carried forward			85.99	816.91	1,077.44	284.15	2,178.50

	Ref	Qty	Hours	Hours £	Mat'ls £	O & P £	Total £
Brought forward			85.99	816.91	1,077.44	284.15	2,178.50
Steel lintel 2400mm long	C4	2nr	0.50	4.75	196.92	30.25	231.92
Steel lintel 1500mm long	C5	3nr	0.60	5.70	322.41	49.22	377.33
Steel lintel 1150mm long	C6	1nr	0.15	1.43	47.28	0.21	1.64
Close cavity wall at jambs	C7	17.32m	0.87	8.27	35.33	6.54	50.13
Close cavity wall at cills	C8	5.71m	0.29	2.76	11.65	2.16	16.57
Close cavity wall at top	C9	11.40m	0.57	5.42	23.26	4.30	32.98
DPC 112mm wide at jambs	C10	17.32m	0.87	8.27	14.03	3.34	25.64
DPC 112mm wide at cills	C11	5.71m	0.29	2.76	4.63	1.11	8.49
Carried to summary			89.98	854.81	1,685.67	381.07	2,921.55
PART D FLAT ROOF			N/A	N/A	N/A	N/A	N/A
PART E PITCHED ROOF							
100 x 75mm sawn softwood wall plate	E1	6.00m	1.80	17.10	13.32	4.56	34.98
150 x 50mm sawn softwood pole plate plugged to brickwork	E2	6.30m	1.89	17.96	20.29	5.74	43.98
100 x 50mm sawn softwood rafters	E3	31.50m	6.30	56.70	103.32	24.00	184.02
125 x 25mm sawn softwood purlin	E4	6.30m	0.86	8.17	15.18	3.50	26.85
150 x 50mm sawn softwood joists	E5	24.10m	4.90	46.55	55.86	15.36	117.77
150 x 50mm sawn softwood sprockets	E6	18nr	2.16	20.52	20.52	6.16	47.20
100mm layer of insulation quilt between joists fixed with chicken wire and 150mm layer over joists	E7	18.00m2	8.64	82.08	196.38	41.77	320.23
6mm softwood soffit 150mm wide	E8	13.30m	5.32	50.54	23.28	11.07	84.89
Carried forward			31.87	299.62	448.15	112.16	859.93

	Ref	Qty	Hours	Hours £	Mat'ls £	O & P £	Total £
Brought forward			31.87	299.62	448.15	112.16	859.93
19mm wrought softwood fascia/ barge board 200mm high	E9	6.30m	6.65	63.18	12.41	11.34	86.92
Marley Plain roof tiles on felt and battens	E10	28.35m2	33.87	321.77	733.42	158.28	1,213.46
Double eaves course	E11	6.30m	2.21	21.00	20.29	6.19	47.48
Verge with plain tile undercloak	E12	9.00m	2.25	21.38	28.98	7.55	57.91
Lead flashing code 5, 200mm girth	E13	6.30m	3.78	35.91	42.80	11.81	90.52
Rake out joint for flashing	E14	6.30m	2.21	21.00	0.95	3.29	25.24
112mm diameter PVC-U gutter	E15	6.30m	1.64	15.58	28.16	6.56	50.30
Stop end	E16	1nr	0.14	1.33	1.11	0.37	2.81
Stop end outlet	E17	1nr	0.25	2.38	1.11	0.52	4.01
68mm diameter PVC-U down pipe	E18	2.50m	0.63	5.99	8.48	2.17	16.63
Shoe	E19	1nr	0.30	2.85	2.09	0.74	5.68
Paint fascia and soffit	E20	3.99m2	2.79	26.51	3.83	4.55	34.89
Carried to summary			88.59	838.46	1,331.78	325.54	2,495.77
PART F WINDOWS AND EXTERNAL DOORS							
PVC-U door size 840 x 1980mm complete (B)	F1	1nr	2.50	23.75	209.99	35.06	268.80
PVC-U sliding patio door size 1700 x 2075mm (C)	F2	2nr	14.00	133.00	587.04	108.01	828.05
Carried forward			16.50	156.75	797.03	143.07	1,096.85

	Ref	Qty	Hours	Hours £	Mat'ls £	O & P £	Total £
Brought forward			16.50	156.75	797.03	143.07	1,096.85
PVC-U window size 1200 x 1200mm complete (A)	F3	3nr	6.00	57.00	472.77	79.47	609.24
25 x 225mm wrought softwood window board	F4	3.60m	1.08	10.26	22.39	4.90	37.55
Paint window board	F5	3.60m	0.57	5.42	3.46	1.33	10.21
Carried to summary			24.15	229.43	1,295.65	228.76	1,753.84
PART G INTERNAL PARTITIONS AND DOORS		N/A	N/A	N/A	N/A	N/A	N/A
PART H WALL FINISHES							
19 x 100mm wrought softwood skirting	H1	8.70m	1.74	16.53	11.92	4.27	32.72
12mm plasterboard fixed to walls with dabs	H2	18.40m2	7.18	68.21	36.62	15.72	120.55
12mm plasterboard fixed to walls less than 300mm wide with dabs	H3	23.03m	4.15	39.43	21.19	9.09	69.71
Two coats emulsion paint to walls	H4	21.85m2	5.22	49.59	12.89	9.37	71.85
Paint skirting	H5	8.70m	1.31	12.45	4.18	2.49	19.12
Carried to summary			19.60	186.20	86.80	40.95	313.95
PART J FLOOR FINISHES							
Cement and sand floor screed 40mm thick	J1	14.58m2	3.65	34.68	44.84	11.93	91.44
Vinyl floor tiles, size 300 x 300mm	J2	14.58m2	3.50	33.25	101.77	20.25	155.27
Carried to summary			7.15	67.93	146.61	32.18	246.72

	Ref	Qty	Hours	Hours £	Mat'ls £	O & P £	Total £
PART K CEILING FINISHES							
Plasterboard with taped butt joints fixed to joists	K1	14.58m2	5.25	2.08	26.97	4.36	33.41
5mm skim coat to plasterboard ceilings	K2	14.58m2	7.29	69.26	17.79	13.06	100.10
Two coats emulsion paint to ceilings	K3	14.58m2	3.79	36.01	8.60	6.69	51.30
Carried to summary			16.33	107.34	53.36	24.11	184.81
PART L ELECTRICAL WORK							
13 amp double switched socket outlet with neon	L1	4nr	1.60	22.40	33.04	8.32	63.76
Lighting point	L2	2nr	0.70	9.80	13.60	3.51	26.91
Lighting switch	L3	3nr	0.75	10.50	11.91	3.36	25.77
Lighting wiring	L4	8.00m	0.80	11.20	8.24	2.92	22.36
Power cable	L5	20.00m	3.00	42.00	22.00	9.60	73.60
Carried to summary			6.85	95.90	88.79	27.70	212.39
PART M HEATING WORK							
15mm copper pipe	M1	8.00m	1.76	22.00	15.04	5.56	42.60
Elbow	M2	4nr	1.12	14.00	4.40	2.76	21.16
Tee	M3	1nr	0.22	2.75	1.95	0.71	5.41
Radiator, double convector size 1400 x 520mm	M4	2nr	2.60	32.50	231.36	39.58	303.44
Break into existing pipe and insert tee	M5	1nr	0.75	9.38	3.95	2.00	15.32
Carried to summary			6.45	80.63	256.70	50.60	387.92

	Ref	Qty	Hours	Hours £	Mat'ls £	O & P £	Total £
PART N ALTERATION WORK							
Take out existing window size 1500 x 1000mm and lintel over, adapt opening to receive 1770 x 2000mm patio door and insert new lintel over (both measured separately) and make good	N1	1nr	20.00	190.00	20.00	31.50	241.50
Carried to summary			20.00	190.00	20.00	31.50	241.50

SUMMARY

	Hours	Hours £	Mat'ls £	O & P £	Total £
PART A PRELIMINARIES	0.00	0.00	0.00	0.00	1,907.50
PART B SUBSTRUCTURE TO DPC LEVEL	70.07	665.67	715.90	207.23	1,588.80
PART C EXTERNAL WALLS	89.98	854.81	1,685.67	381.07	2,921.55
PART D FLAT ROOF	0.00	0.00	0.00	0.00	0.00
PART E PITCHED ROOF	88.59	838.46	1,331.78	325.54	2,495.77
PART F WINDOWS AND EXTERNAL DOORS	24.15	229.43	1,295.65	228.76	1,753.84
PART G INTERNAL PARTITIONS AND DOORS	0.00	0.00	0.00	0.00	0.00
PART H WALL FINISHES	19.60	186.20	86.80	40.95	313.95
PART J FLOOR FINISHES	7.15	67.93	146.61	32.18	246.72
PART K CEILING FINISHES	16.33	107.34	53.36	24.11	184.81
PART L ELECTRICAL WORK	6.85	95.90	88.79	27.70	212.39
PART M HEATING WORK	6.45	80.63	256.70	50.60	387.92
PART N ALTERATION WORK	20.00	190.00	20.00	31.50	241.50
Final total	349.17	3,316.37	5,681.26	1,349.64	12,254.75

	Ref	Qty	Hours	Hours £	Mat'ls £	O & P £	Total £
PART A PRELIMINARIES							
Concrete mixer	A1	8 wks					360.00
Small tools	A2	9 wks					315.00
Scaffolding (m2/weeks)	A3	270.00					607.50
Skip	A4	5 wks					500.00
Clean up	A5	10 hrs					80.00
Carried to summary							1,862.50
PART B SUBSTRUCTURE TO DPC LEVEL							
Excavate topsoil 150mm thick by hand	B1	17.85m2	5.36	50.92	0.00	7.64	58.56
Excavate to reduce levels by hand	B2	5.35m3	13.38	127.11	0.00	19.07	146.18
Excavate for trench foundations by hand	B3	1.88m3	4.89	46.46	0.00	6.97	53.42
Earthwork support to sides of trenches	B4	16.64m2	6.66	63.27	19.97	12.49	95.73
Backfilling with excavated material	B5	0.63m3	0.37	3.52	0.00	0.53	4.04
Hardcore 225mm thick	B6	13.69m2	2.74	26.03	69.54	14.34	109.91
Hardcore filling to trench	B7	0.14m3	0.07	0.67	0.70	0.20	1.57
Concrete grade (1:3:6) in foundations	B8	1.54m3	2.08	19.76	104.63	18.66	143.05
Concrete grade (1:2:4) in bed 150mm thick	B9	13.69m2	4.11	39.05	148.94	28.20	216.18
Concrete (1:2:4) in cavity wall filling	B10	5.70m2	1.14	10.83	20.69	4.73	36.25
Carried forward			40.80	387.60	364.47	112.81	864.88

	Ref	Qty	Hours	Hours £	Mat'ls £	O & P £	Total £
Brought forward			40.80	387.60	364.47	112.81	864.88
Damp-proof membrane	B11	13.13m2	0.53	5.04	6.62	1.75	13.40
Reinforcement ref A193 in foundation	B12	5.70m2	0.68	6.46	7.41	2.08	15.95
Steel fabric reinforcement ref A193 in slab	B13	13.69m2	2.05	19.48	17.80	5.59	42.87
Solid blockwork 140mm thick in cavity wall	B14	7.41m2	9.63	91.49	94.70	27.93	214.11
Common bricks 112.5mm thick in cavity wall	B15	5.70m2	9.63	91.49	75.18	25.00	191.66
Facing bricks in 112.5mm thick in skin of cavity wall	B16	1.71m2	2.27	21.57	39.79	9.20	70.56
Form cavity 50mm wide in cavity wall	B17	7.41m2	0.22	2.09	4.45	0.98	7.52
DPC 112mm wide	B18	11.40m	0.57	5.42	9.23	2.20	16.84
DPC 140mm wide	B19	11.40m	0.68	6.46	11.06	2.63	20.15
Bond in block wall	B20	1.30m	0.58	5.51	2.83	1.25	9.59
Bond in half brick wall	B21	0.30m	0.46	4.37	0.61	0.75	5.73
50mm thick insulation board	B22	13.69m2	4.11	39.05	61.33	15.06	115.43
Carried to summary			72.21	686.00	695.48	207.22	1,588.70
PART C EXTERNAL WALLS							
Solid blockwork 140mm thick in cavity wall	C1	31.90m2	41.47	393.97	405.77	119.96	919.70
Facing brickwork 112.5mm thick in cavity wall	C2	31.90m2	57.42	545.49	753.61	194.87	1,493.97
75mm thick insulation in cavity wall	C3	31.90m2	7.02	66.69	185.66	37.85	290.20
Carried forward			105.91	1,006.15	1,345.04	352.68	2,703.86

	Ref	Qty	Hours	Hours £	Mat'ls £	O & P £	Total £
Brought forward			105.91	1,006.15	1,345.04	352.68	2,703.86
Steel lintel 2400mm long	C4	2nr	0.50	4.75	196.92	30.25	231.92
Steel lintel 1500mm long	C5	3nr	0.60	5.70	322.41	49.22	377.33
Steel lintel 1150mm long	C6	1nr	0.15	1.43	47.28	7.31	56.01
Close cavity wall at jambs	C7	17.32m	0.87	8.27	35.33	6.54	50.13
Close cavity wall at cills	C8	5.71m	0.29	2.76	11.65	2.16	16.57
Close cavity wall at top	C9	11.40m	0.57	5.42	23.26	4.30	32.98
DPC 112mm wide at jambs	C10	17.32m	0.87	8.27	14.03	3.34	25.64
DPC 112mm wide at cills	C11	5.71m	0.29	2.76	4.63	1.11	8.49
Carried to summary			109.90	1,044.05	1,953.27	449.60	3,446.92
PART D FLAT ROOF			N/A	N/A	N/A	N/A	N/A
PART E PITCHED ROOF							
100 x 75mm sawn softwood wall plate	E1	4.00m	1.40	13.30	8.88	3.33	25.51
150 x 50mm sawn softwood pole plate plugged to brickwork	E2	4.30m	1.29	12.26	13.85	3.92	30.02
100 x 50mm sawn softwood rafters	E3	49.50m	9.90	89.10	162.36	37.72	289.18
125 x 25mm sawn softwood purlin	E4	4.30m	0.86	8.17	10.36	2.78	21.31
150 x 50mm sawn softwood joists	E5	40.50	8.10	76.95	92.34	25.39	194.68
150 x 50mm sawn softwood sprockets	E6	20nr	2.40	22.80	22.80	6.84	52.44
100mm layer of insulation quilt between joists fixed with chicken wire and 150mm layer over joists	E7	16.00m2	7.68	72.96	174.56	37.13	284.65
6mm softwood soffit 150mm wide	E8	13.30m	5.32	50.54	23.28	11.07	84.89
Carried forward			36.95	346.08	508.43	128.18	982.68

	Ref	Qty	Hours	Hours £	Mat'ls £	O & P £	Total £
Brought forward			36.95	346.08	508.43	128.18	982.68
19mm wrought softwood fascia/ barge board 200mm high	E9	4.30m	6.65	63.18	8.47	10.75	82.39
Marley Plain roof tiles on felt and battens	E10	23.65m2	44.94	426.93	611.83	155.81	1,194.57
Double eaves course	E11	4.30m	1.51	14.35	13.85	4.23	32.42
Verge with plain tile undercloak	E12	11.00m	2.75	26.13	35.42	9.23	70.78
Lead flashing code 5, 200mm girth	E13	4.30m	2.58	24.51	28.72	7.98	61.21
Rake out joint for flashing	E14	4.30m	1.51	14.35	0.65	2.25	17.24
112mm diameter PVC-U gutter	E15	4.30m	1.12	10.64	19.26	4.49	34.39
Stop end	E16	1nr	0.14	1.33	1.11	0.37	2.81
Stop end outlet	E17	1nr	0.25	2.38	1.11	0.52	4.01
68mm diameter PVC-U down pipe	E18	2.50m	0.63	5.99	8.48	2.17	16.63
Shoe	E19	1nr	0.30	2.85	2.09	0.74	5.68
Paint fascia and soffit	E20	3.99m2	2.79	26.51	3.83	4.55	34.89
Carried to summary			102.12	965.19	1,243.25	331.27	2,539.71
PART F WINDOWS AND EXTERNAL DOORS							
PVC-U door size 840 x 1980mm complete (B)	F1	1nr	2.50	23.75	209.99	35.06	268.80
PVC-U sliding patio door size 1700 x 2075mm (C)	F2	2nr	14.00	133.00	587.04	108.01	828.05
Carried forward			16.50	156.75	797.03	143.07	1,096.85

	Ref	Qty	Hours	Hours £	Mat'ls £	O & P £	Total £
Brought forward			16.50	156.75	797.03	143.07	1,096.85
PVC-U window size 1200 x 1200mm complete (A)	F3	3nr	6.00	57.00	472.77	79.47	609.24
25 x 225mm wrought softwood window board	F4	3.60m	1.08	10.26	22.39	4.90	37.55
Paint window board	F5	3.60m	0.57	5.42	3.46	1.33	10.21
Carried to summary			24.15	229.43	1,295.65	228.76	1,753.84
PART G INTERNAL PARTITIONS AND DOORS		N/A	N/A	N/A	N/A	N/A	N/A
PART H WALL FINISHES							
19 x 100mm wrought softwood skirting	H1	8.70m	1.74	16.53	11.92	4.27	32.72
12mm plasterboard fixed to walls with dabs	H2	18.40m2	7.18	68.21	36.62	15.72	120.55
12mm plasterboard fixed to walls less than 300mm wide with dabs	H3	23.03m	4.15	39.43	21.19	9.09	69.71
Two coats emulsion paint to walls	H4	21.85m2	5.22	49.59	12.89	9.37	71.85
Paint skirting	H5	8.70m	1.31	12.45	4.18	2.49	19.12
Carried to summary			19.60	186.20	86.80	40.95	313.95
PART J FLOOR FINISHES							
Cement and sand floor screed 40mm thick	J1	12.58m2	3.15	29.93	38.12	10.21	78.25
Vinyl floor tiles, size 300 x 300mm	J2	12.58m2	3.02	28.69	87.81	17.48	133.98
Carried to summary			6.17	58.62	125.93	27.68	212.23

	Ref	Qty	Hours	Hours £	Mat'ls £	O & P £	Total £
PART K CEILING FINISHES							
Plasterboard with taped butt joints fixed to joists	K1	12.58m2	4.53	2.08	23.27	3.80	29.15
5mm skim coat to plasterboard ceilings	K2	12.58m2	6.29	59.76	15.35	11.27	86.37
Two coats emulsion paint to ceilings	K3	12.58m2	3.27	31.07	7.42	5.77	44.26
Carried to summary			14.09	92.90	46.04	20.84	159.78
PART L ELECTRICAL WORK							
13 amp double switched socket outlet with neon	L1	3nr	1.20	16.80	24.78	6.24	47.82
Lighting point	L2	2nr	0.70	9.80	13.60	3.51	26.91
Lighting switch	L3	3nr	0.50	7.00	11.91	2.84	21.75
Lighting wiring	L4	8.00m	0.80	11.20	8.24	2.92	22.36
Power cable	L5	18.00m	2.70	37.80	19.80	8.64	66.24
Carried to summary			5.90	82.60	78.33	24.14	185.07
PART M HEATING WORK							
15mm copper pipe	M1	9.00m	1.98	24.75	16.92	6.25	47.92
Elbow	M2	4nr	1.12	14.00	4.40	2.76	21.16
Tee	M3	1nr	0.22	2.75	1.95	0.71	5.41
Radiator, double convector size 1400 x 520mm	M4	2nr	2.60	32.50	231.36	39.58	303.44
Break into existing pipe and insert tee	M5	1nr	0.75	9.38	3.95	2.00	15.32
Carried to summary			6.67	83.38	258.58	51.29	393.25

	Ref	Qty	Hours	Hours £	Mat'ls £	O & P £	Total £
PART N **ALTERATION WORK**							
Take out existing window size 1500 x 1000mm and lintel over, adapt opening to receive 1770 x 2000mm patio door and insert new lintel over (both measured separately) and make good	N1	1nr	20.00	190.00	20.00	31.50	241.50
Carried to summary			20.00	190.00	20.00	31.50	241.50

SUMMARY

	Hours	Hours £	Mat'ls £	O & P £	Total £
PART A PRELIMINARIES	0.00	0.00	0.00	0.00	1,862.50
PART B SUBSTRUCTURE TO DPC LEVEL	72.21	686.00	695.48	207.22	1,588.70
PART C EXTERNAL WALLS	109.90	1,044.05	1,953.27	449.60	3,446.92
PART D FLAT ROOF	0.00	0.00	0.00	0.00	0.00
PART E PITCHED ROOF	102.12	965.19	1,243.25	331.27	2,539.71
PART F WINDOWS AND EXTERNAL DOORS	24.15	229.43	1,295.65	228.76	1,753.84
PART G INTERNAL PARTITIONS AND DOORS	0.00	0.00	0.00	0.00	0.00
PART H WALL FINISHES	19.60	186.20	86.80	40.95	313.95
PART J FLOOR FINISHES	6.17	58.62	125.93	27.68	212.23
PART K CEILING FINISHES	14.09	92.90	46.04	20.84	159.78
PART L ELECTRICAL WORK	5.90	82.60	78.33	24.14	185.07
PART M HEATING WORK	6.67	83.38	258.58	51.29	393.25
PART N ALTERATION WORK	20.00	190.00	20.00	31.50	241.50
Final total	380.81	3,618.37	5,803.33	1,413.25	12,697.45

	Ref	Qty	Hours	Hours £	Mat'ls £	O & P £	Total £
PART A PRELIMINARIES							
Concrete mixer	A1	9 wks					405.00
Small tools	A2	10 wks					350.00
Scaffolding (m2/weeks)	A3	325.00					731.25
Skip	A4	6 wks					600.00
Clean up	A5	10 hrs					80.00
Carried to summary							2,166.25
PART B SUBSTRUCTURE TO DPC LEVEL							
Excavate topsoil 150mm thick by hand	B1	22.01m2	6.60	62.70	0.00	9.41	72.11
Excavate to reduce levels by hand	B2	6.60m3	16.60	157.70	0.00	23.66	181.36
Excavate for trench foundations by hand	B3	2.05m3	5.33	50.64	0.00	7.60	58.23
Earthwork support to sides of trenches	B4	18.10m2	7.24	68.78	21.72	13.58	104.08
Backfilling with excavated material	B5	0.63m3	0.42	3.99	0.00	0.60	4.59
Hardcore 225mm thick	B6	16.28m2	2.44	23.18	82.70	15.88	121.76
Hardcore filling to trench	B7	0.15m3	0.08	0.76	0.76	0.23	1.75
Concrete grade (1:3:6) in foundations	B8	1.67	2.25	21.38	113.46	20.23	155.06
Concrete grade (1:2:4) in bed 150mm thick	B9	16.28m2	4.88	46.36	177.13	33.52	257.01
Concrete (1:2:4) in cavity wall filling	B10	6.20m2	1.24	11.78	22.51	5.14	39.43
Carried forward			47.08	447.26	418.28	129.83	995.37

	Ref	Qty	Hours	Hours £	Mat'ls £	O & P £	Total £
Brought forward			47.08	447.26	418.28	129.83	995.37
Damp-proof membrane	B11	16.88m2	0.68	6.46	9.79	2.44	18.69
Reinforcement ref A193 in foundation	B12	6.20m2	0.74	7.03	8.06	2.26	17.35
Steel fabric reinforcement ref A193 in slab	B13	16.28m2	2.44	23.18	21.16	6.65	50.99
Solid blockwork 140mm thick in cavity wall	B14	8.06m2	10.48	99.56	103.00	30.38	232.94
Common bricks 112.5mm thick in cavity wall	B15	6.20m2	10.48	99.56	81.78	27.20	208.54
Facing bricks in 112.5mm thick in skin of cavity wall	B16	1.86m2	3.35	31.83	43.28	11.27	86.37
Form cavity 50mm wide in cavity wall	B17	8.06m2	0.24	2.28	4.84	1.07	8.19
DPC 112mm wide	B18	12.40m	0.62	5.89	10.06	2.39	18.34
DPC 140mm wide	B19	12.40m	0.74	7.03	12.03	2.86	21.92
Bond in block wall	B20	1.30m	0.58	5.51	2.83	1.25	9.59
Bond in half brick wall	B21	0.30m	0.46	4.37	0.61	0.75	5.73
50mm thick insulation board	B22	16.28m2	4.88	46.36	72.93	17.89	137.18
Carried to summary			82.77	786.32	788.65	236.24	1,811.21
PART C EXTERNAL WALLS							
Solid blockwork 140mm thick in cavity wall	C1	32.96m2	42.85	407.08	419.23	123.95	950.25
Facing brickwork 112.5mm thick in cavity wall	C2	32.96m2	59.33	563.64	760.06	198.55	1,522.25
75mm thick insulation in cavity wall	C3	32.96m2	7.25	68.88	191.83	39.11	299.81
Carried forward			109.43	1,039.59	1,371.12	361.61	2,772.31

	Ref	Qty	Hours	Hours £	Mat'ls £	O & P £	Total £
Brought forward			109.43	1,039.59	1,371.12	361.61	2,772.31
Steel lintel 2400mm long	C4	2nr	0.50	4.75	196.92	30.25	231.92
Steel lintel 1500mm long	C5	4nr	0.80	7.60	429.48	65.56	502.64
Steel lintel 1150mm long	C6	1nr	0.15	1.43	47.28	7.31	56.01
Close cavity wall at jambs	C7	19.72m	0.99	9.41	40.23	7.45	57.08
Close cavity wall at cills	C8	6.91m	0.35	3.33	14.10	2.61	20.04
Close cavity wall at top	C9	12.40m	0.62	5.89	25.30	4.68	35.87
DPC 112mm wide at jambs	C10	19.72m	0.99	9.41	15.97	3.81	29.18
DPC 112mm wide at cills	C11	6.91m	0.35	3.33	5.60	1.34	10.26
Carried to summary			114.03	1,083.29	2,098.72	477.30	3,659.31
PART D FLAT ROOF			N/A	N/A	N/A	N/A	N/A
PART E PITCHED ROOF							
100 x 75mm sawn softwood wall plate	E1	5.00m	1.75	16.63	11.10	4.16	31.88
150 x 50mm sawn softwood pole plate plugged to brickwork	E2	5.30m	1.59	15.11	17.07	4.83	37.00
100 x 50mm sawn softwood rafters	E3	49.50m	9.90	89.10	162.36	37.72	289.18
125 x 25mm sawn softwoo purlin	E4	5.30m	1.06	10.07	12.77	3.43	26.27
150 x 50mm sawn softwood joists	E5	40.50m	8.91	84.65	92.34	26.55	203.53
150 x 50mm sawn softwood sprockets	E6	20nr	2.40	22.80	22.80	6.84	52.44
100mm layer of insulation quilt between joists fixed with chicken wire and 150mm layer over joists	E7	20.00m2	9.60	91.20	218.20	46.41	355.81
6mm softwood soffit 150mm wide	E8	16.30m	5.72	54.34	25.03	11.91	91.28
Carried forward			40.93	383.89	561.67	141.83	1,087.39

	Ref	Qty	Hours	Hours £	Mat'ls £	O & P £	Total £
Brought forward			40.93	383.89	561.67	141.83	1,087.39
19mm wrought softwood fascia/ barge board 200mm high	E9	5.30m	7.15	67.93	10.46	11.76	90.14
Marley Plain roof tiles on felt and battens	E10	29.15m2	55.39	526.21	754.11	192.05	1,472.36
Double eaves course	E11	5.30m	1.86	17.67	17.07	5.21	39.95
Verge with plain tile undercloak	E12	11.00m	2.75	26.13	35.47	9.24	70.83
Lead flashing code 5, 200mm girth	E13	5.30m	3.18	30.21	35.40	9.84	75.45
Rake out joint for flashing	E14	5.30m	1.86	17.67	0.80	2.77	21.24
112mm diameter PVC-U gutter	E15	5.30m	1.38	13.11	23.69	5.52	42.32
Stop end	E16	1nr	0.14	1.33	1.11	0.37	2.81
Stop end outlet	E17	1nr	0.25	2.38	1.11	0.52	4.01
68mm diameter PVC-U down pipe	E18	2.50m	0.63	5.99	8.48	2.17	16.63
Shoe	E19	1nr	0.30	2.85	2.09	0.74	5.68
Paint fascia and soffit	E20	4.29m2	3.00	28.50	4.12	4.89	37.51
Carried to summary			118.82	1,123.84	1,455.58	386.91	2,966.33
PART F WINDOWS AND EXTERNAL DOORS							
PVC-U door size 840 x 1980mm complete (B)	F1	1nr	2.50	23.75	209.99	35.06	268.80
PVC-U sliding patio door size 1700 x 2075mm (C)	F2	2nr	14.00	133.00	587.04	108.01	828.05
Carried forward			16.50	156.75	797.03	143.07	1,096.85

	Ref	Qty	Hours	Hours £	Mat'ls £	O & P £	Total £
Brought forward			16.50	156.75	797.03	143.07	1,096.85
PVC-U window size 1200 x 1200mm complete (A)	F3	4nr	8.00	76.00	630.36	105.95	812.31
25 x 225mm wrought softwood window board	F4	4.80m	1.44	13.68	29.86	6.53	50.07
Paint window board	F5	4.80m	0.76	7.22	4.61	1.77	13.60
Carried to summary			26.70	253.65	1,461.86	257.33	1,972.84
PART G INTERNAL PARTITIONS AND DOORS		N/A	N/A	N/A	N/A	N/A	N/A
PART H WALL FINISHES							
19 x 100mm wrought softwood skirting	H1	9.70m	1.94	18.43	13.29	4.76	36.48
12mm plasterboard fixed to walls with dabs	H2	19.60m2	7.65	72.68	39.00	16.75	128.43
12mm plasterboard fixed to walls less than 300mm wide with dabs	H3	26.63m	4.79	45.51	24.50	10.50	80.51
Two coats emulsion paint to walls	H4	23.46m2	5.63	53.49	13.86	10.10	77.45
Paint skirting	H5	9.70m	1.46	13.87	4.66	2.78	21.31
Carried to summary			21.47	203.97	95.31	44.89	344.17
PART J FLOOR FINISHES							
Cement and sand floor screed 40mm thick	J1	16.28m2	4.07	38.67	49.33	13.20	101.19
Vinyl floor tiles, size 300 x 300mm	J2	16.28m2	3.90	37.05	113.63	22.60	173.28
Carried to summary			7.97	75.72	162.96	35.80	274.48

	Ref	Qty	Hours	Hours £	Mat'ls £	O & P £	Total £
PART K CEILING FINISHES							
Plasterboard with taped butt joints fixed to joists	K1	16.28m2	5.96	2.08	30.12	4.83	37.03
5mm skim coat to plasterboard ceilings	K2	16.28m2	8.14	77.33	19.86	14.58	111.77
Two coats emulsion paint to ceilings	K3	16.28m2	1.95	18.53	9.60	4.22	32.34
Carried to summary			16.05	97.94	59.58	23.63	181.14
PART L ELECTRICAL WORK							
13 amp double switched socket outlet with neon	L1	4nr	1.60	22.40	33.04	8.32	63.76
Lighting point	L2	3nr	1.05	14.70	20.40	5.27	40.37
Lighting switch	L3	3nr	0.75	10.50	11.91	3.36	25.77
Lighting wiring	L4	9.00m	0.70	9.80	27.00	5.52	42.32
Power cable	L5	20.00m	3.00	42.00	22.00	9.60	73.60
Carried to summary			7.10	99.40	114.35	32.06	245.81
PART M HEATING WORK							
15mm copper pipe	M1	10.00m	2.20	27.50	18.80	6.95	53.25
Elbow	M2	4nr	1.12	14.00	4.40	2.76	21.16
Tee	M3	1nr	0.22	2.75	1.95	0.71	5.41
Radiator, double convector size 1400 x 520mm	M4	3nr	3.90	48.75	347.04	59.37	455.16
Break into existing pipe and insert tee	M5	1nr	0.75	9.38	3.95	2.00	15.32
Carried to summary			8.19	102.38	376.14	71.78	550.29

	Ref	Qty	Hours	Hours £	Mat'ls £	O & P £	Total £
PART N							
ALTERATION WORK							
Take out existing window size 1500 x 1000mm and lintel over, adapt opening to receive 1770 x 2000mm patio door and insert new lintel over (both measured separately) and make good	N1	1nr	20.00	190.00	20.00	31.50	241.50
Carried to summary			20.00	190.00	20.00	31.50	241.50

SUMMARY

	Hours	Hours £	Mat'ls £	O & P £	Total £
PART A PRELIMINARIES	0.00	0.00	0.00	0.00	2,166.25
PART B SUBSTRUCTURE TO DPC LEVEL	82.77	786.32	788.65	236.24	1,811.21
PART C EXTERNAL WALLS	114.03	1,083.29	2,098.72	477.30	3,659.31
PART D FLAT ROOF	0.00	0.00	0.00	0.00	0.00
PART E PITCHED ROOF	118.82	1,123.84	1,455.58	386.91	2,966.33
PART F WINDOWS AND EXTERNAL DOORS	26.70	253.65	1,461.86	257.33	1,972.84
PART G INTERNAL PARTITIONS AND DOORS	0.00	0.00	0.00	0.00	0.00
PART H WALL FINISHES	21.47	203.97	95.31	44.89	344.17
PART J FLOOR FINISHES	7.97	75.72	162.96	35.80	274.48
PART K CEILING FINISHES	16.05	97.94	59.58	23.63	181.14
PART L ELECTRICAL WORK	7.10	99.40	114.35	32.06	245.81
PART M HEATING WORK	8.19	102.38	376.14	71.78	550.29
PART N ALTERATION WORK	20.00	190.00	20.00	31.50	241.50
Final total	423.10	4,016.51	6,633.15	1,597.44	14,413.33

	Ref	Qty	Hours	Hours £	Mat'ls £	O & P £	Total £
PART A PRELIMINARIES							
Concrete mixer	A1	10 wks					450.00
Small tools	A2	11 wks					385.00
Scaffolding (m2/weeks)	A3	385.00					866.25
Skip	A4	7 wks					700.00
Clean up	A5	10 hrs					80.00
Carried to summary							2,481.25
PART B SUBSTRUCTURE TO DPC LEVEL							
Excavate topsoil 150mm thick by hand	B1	26.15m2	7.85	74.58	0.00	11.19	85.76
Excavate to reduce levels by hand	B2	7.85m3	19.63	186.49	0.00	27.97	214.46
Excavate for trench foundations by hand	B3	2.21m3	5.75	54.63	0.00	8.19	62.82
Earthwork support to sides of trenches	B4	19.56m2	7.81	74.20	23.47	14.65	112.31
Backfilling with excavated material	B5	0.77m3	0.31	2.95	0.00	0.44	3.39
Hardcore 225mm thick	B6	19.98m2	3.00	28.50	101.50	19.50	149.50
Hardcore filling to trench	B7	0.17m3	0.09	0.86	0.86	0.26	1.97
Concrete grade (1:3:6) in foundations	B8	1.81m3	2.44	23.18	122.97	21.92	168.07
Concrete grade (1:2:4) in bed 150mm thick	B9	19.98m2	5.99	56.91	211.94	40.33	309.17
Concrete (1:2:4) in cavity wall filling	B10	6.70m2	1.34	12.73	24.32	5.56	42.61
Carried forward			54.21	515.00	485.06	150.01	1,150.06

	Ref	Qty	Hours	Hours £	Mat'ls £	O & P £	Total £
Brought forward			54.21	515.00	485.06	150.01	1,150.06
Damp-proof membrane	B11	20.63m2	0.83	7.89	11.97	2.98	22.83
Reinforcement ref A193 in foundation	B12	6.70m2	0.80	7.60	8.71	2.45	18.76
Steel fabric reinforcement ref A193 in slab	B13	19.98m2	3.00	28.50	25.97	8.17	62.64
Solid blockwork 140mm thick in cavity wall	B14	8.71m2	11.32	107.54	111.31	32.83	251.68
Common bricks 112.5mm thick in cavity wall	B15	6.70m2	11.32	107.54	88.37	29.39	225.30
Facing bricks in 112.5mm thick in skin of cavity wall	B16	2.01m2	3.62	34.39	46.77	12.17	93.33
Form cavity 50mm wide in cavity wall	B17	8.71m2	0.26	2.47	5.23	1.16	8.86
DPC 112mm wide	B18	13.40m	0.67	6.37	10.85	2.58	19.80
DPC 140mm wide	B19	13.40m	0.80	7.60	13.00	3.09	23.69
Bond in block wall	B20	1.30m	0.58	5.51	2.83	1.25	9.59
Bond in half brick wall	B21	0.30m	0.46	4.37	0.61	0.75	5.73
50mm thick insulation board	B22	19.98m2	5.99	56.91	89.51	21.96	168.38
Carried to summary			93.86	891.67	900.19	268.78	2,060.64
PART C EXTERNAL WALLS							
Solid blockwork 140mm thick in cavity wall	C1	34.46m2	44.80	425.60	438.33	129.59	993.52
Facing brickwork 112.5mm thick in cavity wall	C2	34.46m2	62.03	589.29	794.65	207.59	1,591.53
75mm thick insulation in cavity wall	C3	34.46m2	7.58	72.01	200.56	40.89	313.46
Carried forward			114.41	1,086.90	1,433.54	378.07	2,898.50

	Ref	Qty	Hours	Hours £	Mat'ls £	O & P £	Total £
Brought forward			114.41	1,086.90	1,433.54	378.07	2,898.50
Steel lintel 2400mm long	C4	2nr	0.50	4.75	196.92	30.25	231.92
Steel lintel 1500mm long	C5	4nr	0.80	7.60	429.48	65.56	502.64
Steel lintel 1150mm long	C6	1nr	0.15	1.43	47.28	7.31	56.01
Close cavity wall at jambs	C7	19.72m	0.99	9.41	40.23	7.45	57.08
Close cavity wall at cills	C8	6.91m	0.35	3.33	14.10	2.61	20.04
Close cavity wall at top	C9	13.40m	0.67	6.37	27.34	5.06	38.76
DPC 112mm wide at jambs	C10	19.72m	0.99	9.41	15.97	3.81	29.18
DPC 112mm wide at cills	C11	6.91m	0.35	3.33	5.60	1.34	10.26
Carried to summary			119.06	1,131.07	2,163.18	494.14	3,788.39
PART D FLAT ROOF			N/A	N/A	N/A	N/A	N/A
PART E PITCHED ROOF							
100 x 75mm sawn softwood wall plate	E1	6.00m	1.80	17.10	13.32	4.56	34.98
150 x 50mm sawn softwood pole plate plugged to brickwork	E2	6.30m	1.89	17.96	20.29	5.74	43.98
100 x 50mm sawn softwood rafters	E3	49.50m	9.90	89.10	162.36	37.72	289.18
125 x 25mm sawn softwood purlin	E4	6.30m	0.86	8.17	15.18	3.50	26.85
150 x 50mm sawn softwood joists	E5	40.50m	8.91	84.65	92.34	26.55	203.53
150 x 50mm sawn softwood sprockets	E6	20nr	2.40	22.80	22.80	6.84	52.44
100mm layer of insulation quilt between joists fixed with chicken wire and 150mm layer over joists	E7	24.00m2	11.52	109.44	261.84	55.69	426.97
6mm softwood soffit 150mm wide	E8	15.30m	6.12	58.14	26.78	12.74	97.66
Carried forward			43.40	407.35	614.91	153.34	1,175.60

	Ref	Qty	Hours	Hours £	Mat'ls £	O & P £	Total £
Brought forward			43.40	407.35	614.91	153.34	1,175.60
19mm wrought softwood fascia/ barge board 200mm high	E9	6.30m	7.65	72.68	12.41	12.76	97.85
Marley Plain roof tiles on felt and battens	E10	36.65m2	65.84	625.48	896.40	228.28	1,750.16
Double eaves course	E11	6.30m	2.21	21.00	20.29	6.19	47.48
Verge with plain tile undercloak	E12	11.00m	2.75	26.13	35.42	9.23	70.78
Lead flashing code 5, 200mm girth	E13	6.30m	3.78	35.91	42.08	11.70	89.69
Rake out joint for flashing	E14	6.30m	2.21	21.00	0.95	3.29	25.24
112mm diameter PVC-U gutter	E15	6.30m	1.64	15.58	28.16	6.56	50.30
Stop end	E16	1nr	0.14	1.33	1.11	0.37	2.81
Stop end outlet	E17	1nr	0.25	2.38	1.11	0.52	4.01
68mm diameter PVC-U down pipe	E18	2.50m	0.63	5.99	8.48	2.17	16.63
Shoe	E19	1nr	0.30	2.85	2.09	0.74	5.68
Paint fascia and soffit	E20	4.59m2	3.21	30.50	4.41	5.24	40.14
Carried to summary			134.01	1,268.15	1,667.82	440.39	3,376.36
PART F WINDOWS AND EXTERNAL DOORS							
PVC-U door size 840 x 1980mm complete (B)	F1	1nr	2.50	23.75	209.99	35.06	268.80
PVC-U sliding patio door size 1700 x 2075mm (C)	F2	2nr	14.00	133.00	587.04	108.01	828.05
Carried forward			16.50	156.75	797.03	143.07	1,096.85

	Ref	Qty	Hours	Hours £	Mat'ls £	O & P £	Total £
Brought forward			16.50	156.75	797.03	143.07	1,096.85
PVC-U window size 1200 x 1200mm complete (A)	F3	4nr	8.00	76.00	630.36	105.95	812.31
25 x 225mm wrought softwood window board	F4	4.80m	1.44	13.68	29.86	6.53	50.07
Paint window board	F5	4.80m	0.76	7.22	4.61	1.77	13.60
Carried to summary			26.70	253.65	1,461.86	257.33	1,972.84
PART G INTERNAL PARTITIONS AND DOORS		N/A	N/A	N/A	N/A	N/A	N/A
PART H WALL FINISHES							
19 x 100mm wrought softwood skirting	H1	10.07m	2.14	20.33	14.66	5.25	40.24
12mm plasterboard fixed to walls with dabs	H2	21.96m2	8.56	81.32	43.70	18.75	143.77
12mm plasterboard fixed to walls less than 300mm wide with dabs	H3	26.63m	4.79	45.51	24.50	10.50	80.51
Two coats emulsion paint to plastered walls	H4	25.96m2	6.23	59.19	15.32	11.18	85.68
Paint skirting	H5	10.70m	1.61	15.30	5.14	3.07	23.50
Carried to summary			23.33	221.64	103.32	48.74	373.70
PART J FLOOR FINISHES							
Cement and sand floor screed 40mm thick	J1	19.88m2	5.00	47.50	60.54	16.21	124.25
Vinyl floor tiles, size 300 x 300mm	J2	19.88m2	4.79	45.51	139.46	27.74	212.71
Carried to summary			9.79	93.01	200.00	43.95	336.96

	Ref	Qty	Hours	Hours £	Mat'ls £	O & P £	Total £
PART K **CEILING FINISHES**							
Plasterboard with taped butt joints fixed to joists	K1	19.88m2	7.19	2.08	36.96	5.86	44.90
5mm skim coat to plasterboard ceilings	K2	19.88m2	9.99	94.91	24.00	17.84	136.74
Two coats emulsion paint to ceilings	K3	19.88m2	5.19	49.31	11.79	9.16	70.26
Carried to summary			22.37	146.29	72.75	32.86	251.90
PART L **ELECTRICAL WORK**							
13 amp double switched socket outlet with neon	L1	5nr	2.00	28.00	41.30	10.40	79.70
Lighting point	L2	3nr	1.05	14.70	20.40	5.27	40.37
Lighting switch	L3	3nr	0.75	10.50	11.91	3.36	25.77
Lighting wiring	L4	11.00m	1.10	15.40	11.33	4.01	30.74
Power cable	L5	22.00m	3.30	46.20	24.20	10.56	80.96
Carried to summary			8.20	114.80	109.14	33.59	257.53
PART M **HEATING WORK**							
15mm copper pipe	M1	11.00m	2.42	30.25	26.68	8.54	65.47
Elbow	M2	4nr	1.12	14.00	4.40	2.76	21.16
Tee	M3	1nr	0.22	2.75	1.95	0.71	5.41
Radiator, double convector size 1400 x 520mm	M4	3nr	3.90	48.75	347.04	59.37	455.16
Break into existing pipe and insert tee	M5	1nr	0.75	9.38	3.95	2.00	15.32
Carried to summary			8.41	105.13	384.02	73.37	562.52

	Ref	Qty	Hours	Hours £	Mat'ls £	O & P £	Total £
PART N **ALTERATION WORK**							
Take out existing window size 1500 x 1000mm and lintel over, adapt opening to receive 1770 x 2000mm patio door and insert new lintel over (both measured separately) and make good	N1	1nr	20.00	190.00	20.00	31.50	241.50
Carried to summary			20.00	190.00	20.00	31.50	241.50

SUMMARY

	Hours	Hours £	Mat'ls £	O & P £	Total £
PART A PRELIMINARIES	0.00	0.00	0.00	0.00	2,481.25
PART B SUBSTRUCTURE TO DPC LEVEL	93.86	891.67	900.19	268.78	2,060.64
PART C EXTERNAL WALLS	119.06	1,131.07	2,163.18	494.14	3,788.39
PART D FLAT ROOF	0.00	0.00	0.00	0.00	0.00
PART E PITCHED ROOF	134.01	1,268.15	1,667.82	440.39	3,376.36
PART F WINDOWS AND EXTERNAL DOORS	26.70	253.65	1,461.86	257.33	1,972.84
PART G INTERNAL PARTITIONS AND DOORS	0.00	0.00	0.00	0.00	0.00
PART H WALL FINISHES	23.33	221.64	103.32	48.74	373.70
PART J FLOOR FINISHES	9.79	93.01	200.00	43.95	336.96
PART K CEILING FINISHES	22.37	146.29	72.75	32.86	251.90
PART L ELECTRICAL WORK	8.20	114.80	109.14	33.59	257.53
PART M HEATING WORK	8.41	105.13	384.02	73.37	562.52
PART N ALTERATION WORK	20.00	190.00	20.00	31.50	241.50
Final total	465.73	4,415.41	7,082.28	1,724.65	15,703.59

	Ref	Qty	Hours	Hours £	Mat'ls £	O & P £	Total £
PART A PRELIMINARIES							
Concrete mixer	A1	7 wks					315.00
Small tools	A2	8 wks					280.00
Scaffolding (m2/weeks)	A3	280.00					630.00
Skip	A4	4 wks					400.00
Clean up	A5	8 hrs					64.00
Carried to summary							1,689.00
PART B SUBSTRUCTURE TO DPC LEVEL							
Excavate topsoil 150mm thick by hand	B1	7.10m2	2.13	20.24	0.00	3.04	23.27
Excavate to reduce levels by hand	B2	2.13m3	5.33	50.64	0.00	7.60	58.23
Excavate for trench foundations by hand	B3	1.06m3	2.76	26.22	0.00	3.93	30.15
Earthwork support to sides of trenches	B4	9.34m2	3.74	35.53	11.21	7.01	53.75
Backfilling with excavated material	B5	0.40m3	0.16	1.52	0.00	0.23	1.75
Hardcore 225mm thick	B6	4.08m2	0.61	5.80	20.73	3.98	30.50
Hardcore filling to trench	B7	0.08m3	0.05	0.48	0.40	0.13	1.01
Concrete grade (1:3:6) in foundations	B8	0.86m3	1.16	11.02	58.43	10.42	79.87
Concrete grade (1:2:4) in bed 150mm thick	B9	4.08m2	1.22	11.59	44.39	8.40	64.38
Concrete (1:2:4) in cavity wall filling	B10	3.20m2	0.64	6.08	11.62	2.66	20.36
Carried forward			17.80	169.10	146.78	47.38	363.26

	Ref	Qty	Hours	Hours £	Mat'ls £	O & P £	Total £
Brought forward			17.80	169.10	146.78	47.38	363.26
Damp-proof membrane	B11	4.38m2	0.17	1.62	2.54	0.62	4.78
Reinforcement ref A193 in foundation	B12	3.22m2	0.38	3.61	4.16	1.17	8.94
Steel fabric reinforcement ref A193 in slab	B13	4.08m2	0.61	5.80	5.30	1.66	12.76
Solid blockwork 140mm thick in cavity wall	B14	4.16m2	5.41	51.40	53.16	15.68	120.24
Common bricks 112.5mm thick in cavity wall	B15	3.20m2	5.41	51.40	42.21	14.04	107.65
Facing bricks in 112.5mm thick in skin of cavity wall	B16	0.96m2	1.73	16.44	22.34	5.82	44.59
Form cavity 50mm wide in cavity wall	B17	4.16m2	0.13	1.24	2.50	0.56	4.30
DPC 112mm wide	B18	6.40m	0.32	3.04	5.18	1.23	9.45
DPC 140mm wide	B19	6.40m	0.38	3.61	6.21	1.47	11.29
Bond in block wall	B20	1.30m	0.58	5.51	2.83	1.25	9.59
Bond in half brick wall	B21	0.30m	0.46	4.37	0.61	0.75	5.73
50mm thick insulation board	B22	4.08m2	1.22	11.59	18.28	4.48	34.35
Carried to summary			34.60	328.70	312.10	96.12	736.92
PART C EXTERNAL WALLS							
Solid blockwork 140mm thick in cavity wall	C1	23.94m2	31.12	295.64	304.52	90.02	690.18
Facing brickwork 112.5mm thick in cavity wall	C2	23.94m2	43.09	409.36	552.06	144.21	1,105.63
75mm thick insulation in cavity wall	C3	23.94m2	5.27	50.07	139.33	28.41	217.80
Carried forward			79.48	755.06	995.91	262.65	2,013.62

	Ref	Qty	Hours	Hours £	Mat'ls £	O & P £	Total £
Brought forward			79.48	755.06	995.91	262.65	2,013.62
Steel lintel 2400mm long	C4	2nr	0.50	4.75	196.92	30.25	231.92
Steel lintel 1500mm long	C5	4nr	0.80	7.60	429.88	65.62	503.10
Close cavity wall at jambs	C7	13.76m	0.69	6.56	28.07	5.19	39.82
Close cavity wall at cills	C8	5.07m	0.25	2.38	10.34	1.91	14.62
Close cavity wall at top	C9	6.40m	0.32	3.04	13.06	2.42	18.52
DPC 112mm wide at jambs	C10	13.76m	0.69	6.56	11.15	2.66	20.36
DPC 112mm wide at cills	C11	5.07m	0.25	2.38	4.11	0.97	7.46
Carried to summary			82.98	788.31	1,689.44	371.66	2,849.41

PART D
FLAT ROOF

	Ref	Qty	Hours	Hours £	Mat'ls £	O & P £	Total £
200 x 50mm sawn softwood joists	D1	15.05m	3.76	35.72	46.66	12.36	94.74
200 x 50mm sawn softwood sprocket pieces	D2	8nr	1.12	10.64	10.72	3.20	24.56
18mm thick WPB grade decking	D3	7.10m2	6.39	57.51	55.81	17.00	130.32
50 x 50mm (avg) wide sawn softwood firrings	D4	15.05m	2.71	25.75	34.01	8.96	68.72
High density polyethylene vapour barrier 150mm thick	D5	6.00m2	1.12	10.64	56.82	10.12	77.58
100 x 75mm sawn softwood wall plate	D6	7.00m	2.10	19.95	15.54	5.32	40.81
100 x 75mm sawn softwood tilt fillet	D7	3.30m	0.83	7.89	8.71	2.49	19.08
Build in ends of 200 x 50mm joists	D8	7nr	2.45	23.28	1.05	3.65	27.97
Carried forward			20.48	191.37	229.32	63.10	483.79

	Ref	Qty	Hours	Hours £	Mat'ls £	O & P £	Total £
Brought forward			20.48	191.37	229.32	63.10	483.79
Rake out joint for flashing	D9	3.30m	1.16	11.02	0.50	1.73	13.25
6mm thick soffit 150mm wide	D10	7.30m	2.92	27.74	12.78	6.08	46.60
19mm wrought softwood fascia 200mm high	D11	7.30m	3.65	34.68	14.38	7.36	56.41
Three layer fibre-based roofing felt	D12	7.10m2	3.91	37.15	59.71	14.53	111.38
Felt turn-down 100mm girth	D13	7.30m	0.73	6.94	6.13	1.96	15.02
Felt flashing 150mm girth	D14	3.30m	0.33	3.14	5.54	1.30	9.98
112mm diameter PVC-U gutter	D15	3.30m	0.86	8.17	14.75	3.44	26.36
Stop end	D16	1nr	0.14	1.33	1.11	0.37	2.81
Stop end outlet	D17	1nr	0.25	2.38	1.11	0.52	4.01
68mm diameter PVC-U down pipe	D18	4.80m	1.20	11.40	16.27	4.15	31.82
Shoe	D19	1nr	0.30	2.85	2.09	0.74	5.68
Paint fascia and soffit	D20	2.19m2	1.53	14.54	2.10	2.50	19.13
Carried to summary			37.46	352.68	365.79	107.77	826.23
PART E PITCHED ROOF			N/A	N/A	N/A	N/A	N/A
PART F WINDOWS AND EXTERNAL DOORS							
PVC-U door size 840 x 1980mm complete (B)	F1	1nr	2.50	23.75	209.99	35.06	268.80
PVC-U sliding patio door size 1700 x 2075mm (C)	F2	2nr	14.00	133.00	587.04	108.01	828.05
Carried forward			16.50	156.75	797.03	143.07	1,096.85

	Ref	Qty	Hours	Hours £	Mat'ls £	O & P £	Total £
Brought forward			16.50	156.75	797.03	143.07	1,096.85
PVC-U window size 1200 x 1200mm complete (A)	F3	4nr	8.00	76.00	630.36	105.95	812.31
25 x 225mm wrought softwood window board	F4	4.80m	1.44	13.68	29.86	6.53	50.07
Paint window board	F5	4.80m	0.76	7.22	4.61	1.77	13.60
Carried to summary			26.70	253.65	1,461.86	257.33	1,972.84
PART G INTERNAL PARTITIONS AND DOORS		N/A	N/A	N/A	N/A	N/A	N/A
PART H WALL FINISHES							
19 x 100mm wrought softwood skirting	H1	10.34m	1.76	16.72	14.17	4.63	35.52
12mm plasterboard fixed to walls with dabs	H2	20.04m2	7.81	74.20	39.88	17.11	131.19
12mm plasterboard fixed to walls less than 300mm wide with dabs	H3	12.63m	2.27	21.57	11.62	4.98	38.16
Two coats emulsion paint to walls	H4	22.86m2	5.94	56.43	13.49	10.49	80.41
Paint skirting	H5	10.34m	2.07	19.67	4.96	3.69	28.32
Carried to summary			19.85	188.58	84.12	40.90	313.60
PART J FLOOR FINISHES							
Cement and sand floor screed 40mm thick	J1	4.08m2	1.02	9.69	12.36	3.31	25.36
Vinyl floor tiles, size 300 x 300mm	J2	4.08m2	0.70	6.65	28.47	5.27	40.39
Carried forward			1.72	16.34	40.83	8.58	65.75

	Ref	Qty	Hours	Hours £	Mat'ls £	O & P £	Total £
Brought forward			1.72	16.34	40.83	8.58	65.75
25mm thick tongued and grooved boarding	J3	4.08m2	3.02	28.69	38.35	10.06	77.10
150 x 50mm sawn softwood joists	J4	9.50m	2.09	19.86	21.66	6.23	47.74
Cut and pin ends of joists to existing brick wall	J5	5nr	0.90	8.55	0.00	1.28	9.83
Build in ends of joists to blockwork	J6	5nr	0.50	4.75	0.00	0.71	5.46
Carried to summary			8.23	78.19	100.84	26.85	205.88
PART K CEILING FINISHES							
Plasterboard with taped butt joints fixed to joists	K1	8.16m2	2.94	2.08	15.10	2.58	19.76
5mm skim coat to plasterboard ceilings	K2	8.16m2	4.08	38.76	9.96	7.31	56.03
Two coats emulsion paint to ceilings	K3	8.16m2	2.12	20.14	4.81	3.74	28.69
Carried to summary			9.14	60.98	29.87	13.63	104.48
PART L ELECTRICAL WORK							
13 amp double switched socket outlet with neon	L1	4nr	1.60	22.40	33.04	8.32	63.76
Lighting point	L2	2nr	0.70	9.80	13.60	3.51	26.91
Lighting switch	L3	4nr	1.00	14.00	15.88	4.48	34.36
Lighting wiring	L4	10.00m	1.00	14.00	10.30	3.65	27.95
Power cable	L5	24.00m	3.60	50.40	26.40	11.52	88.32
Carried to summary			7.90	110.60	99.22	31.47	241.29

	Ref	Qty	Hours	Hours £	Mat'ls £	O & P £	Total £
PART M **HEATING WORK**							
15mm copper pipe	M1	9.00m	1.98	24.75	16.92	6.25	47.92
Elbow	M2	8nr	2.24	28.00	8.80	5.52	42.32
Tee	M3	2nr	0.44	5.50	3.90	1.41	10.81
Radiator, double convector size 1400 x 520mm	M4	4nr	5.20	65.00	462.72	79.16	606.88
Break into existing pipe and insert tee	M5	1nr	0.75	9.38	3.95	2.00	15.32
Carried to summary			10.61	132.63	496.29	94.34	723.25
PART N **ALTERATION WORK**							
Take out existing window size 1500 x 1000mm and lintel over, adapt opening to receive 1770 x 2000mm patio door and insert new lintel over (both measured separately) and make good	N1	1nr	20.00	190.00	20.00	31.50	241.50
Take out existing window size 1500 x 1000mm, enlarge opening to receive new PVC-U door (measured separately)	N2	1nr	25.00	237.50	45.00	42.38	324.88
Carried to summary			45.00	427.50	65.00	73.88	566.38

SUMMARY

	Hours	Hours £	Mat'ls £	O & P £	Total £
PART A PRELIMINARIES	0.00	0.00	0.00	0.00	1,689.00
PART B SUBSTRUCTURE TO DPC LEVEL	34.60	328.70	312.10	96.12	736.92
PART C EXTERNAL WALLS	82.98	788.31	1,689.44	371.66	2,849.41
PART D FLAT ROOF	37.46	352.68	365.79	107.77	826.23
PART E PITCHED ROOF	0.00	0.00	0.00	0.00	0.00
PART F WINDOWS AND EXTERNAL DOORS	26.70	253.65	1,461.86	257.33	1,972.84
PART G INTERNAL PARTITIONS AND DOORS	0.00	0.00	0.00	0.00	0.00
PART H WALL FINISHES	19.85	188.58	84.12	40.90	313.60
PART J FLOOR FINISHES	8.23	78.19	100.84	26.85	205.88
PART K CEILING FINISHES	9.14	60.98	29.87	13.63	104.48
PART L ELECTRICAL WORK	7.90	110.60	99.22	31.47	241.29
PART M HEATING WORK	10.66	132.63	496.29	94.34	723.25
PART N ALTERATION WORK	45.00	427.50	65.00	73.88	566.38
Final total	282.52	2,721.82	4,704.53	1,113.95	10,229.28

	Ref	Qty	Hours	Hours £	Mat'ls £	O & P £	Total £
PART A PRELIMINARIES							
Concrete mixer	A1	8 wks					360.00
Small tools	A2	9 wks					315.00
Scaffolding (m2/weeks)	A3	360.00					810.00
Skip	A4	5 wks					500.00
Clean up	A5	10 hrs					80.00
Carried to summary							2,065.00
PART B SUBSTRUCTURE TO DPC LEVEL							
Excavate topsoil 150mm thick by hand	B1	9.25m2	2.78	26.41	0.00	3.96	30.37
Excavate to reduce levels by hand	B2	2.77m3	6.93	65.84	0.00	9.88	75.71
Excavate for trench foundations by hand	B3	1.22m3	3.17	30.12	0.00	4.52	34.63
Earthwork support to sides of trenches	B4	10.80m2	4.32	41.04	12.96	8.10	62.10
Backfilling with excavated material	B5	0.47m3	0.28	2.66	0.00	0.40	3.06
Hardcore 225mm thick	B6	5.78m2	1.16	11.02	29.36	6.06	46.44
Hardcore filling to trench	B7	0.09m3	0.05	0.48	0.45	0.14	1.06
Concrete grade (1:3:6) in foundations	B8	1.00m3	1.16	11.02	67.94	11.84	90.80
Concrete grade (1:2:4) in bed 150mm thick	B9	5.78m2	1.22	11.59	62.87	11.17	85.63
Concrete (1:2:4) in cavity wall filling	B10	3.70m2	0.64	6.08	13.43	2.93	22.44
Carried forward			21.71	206.25	187.01	58.99	452.24

	Ref	Qty	Hours	Hours £	Mat'ls £	O & P £	Total £
Brought forward			21.71	206.25	187.01	58.99	452.24
Damp-proof membrane	B11	6.13m2	0.25	2.38	3.56	0.89	6.83
Reinforcement ref A193 in foundation	B12	3.70m2	0.44	4.18	4.81	1.35	10.34
Steel fabric reinforcement ref A193 in slab	B13	5.78m2	0.87	8.27	7.51	2.37	18.14
Solid blockwork 140mm thick in cavity wall	B14	4.81m2	6.25	59.38	61.47	18.13	138.97
Common bricks 112.5mm thick in cavity wall	B15	3.70m2	6.25	59.38	48.80	16.23	124.40
Facing bricks in 112.5mm thick in skin of cavity wall	B16	1.11m2	2.00	19.00	25.83	6.72	51.55
Form cavity 50mm wide in cavity wall	B17	4.81m2	0.14	1.33	2.89	0.63	4.85
DPC 112mm wide	B18	7.40m	0.37	3.52	5.99	1.43	10.93
DPC 140mm wide	B19	7.40m	0.44	4.18	7.18	1.70	13.06
Bond in block wall	B20	1.30m	0.58	5.51	2.83	1.25	9.59
Bond in half brick wall	B21	0.30m	0.46	4.37	0.61	0.75	5.73
50mm thick insulation board	B22	5.78m2	1.73	16.44	25.89	6.35	48.67
Carried to summary			41.49	394.16	384.38	116.78	895.32
PART C **EXTERNAL WALLS**							
Solid blockwork 140mm thick in cavity wall	C1	28.99m2	37.69	358.06	368.75	109.02	835.83
Facing brickwork 112.5mm thick in cavity wall	C2	28.99m2	52.18	495.71	668.51	174.63	1,338.85
75mm thick insulation in cavity wall	C3	28.99m2	6.38	60.61	168.72	34.40	263.73
Carried forward			96.25	914.38	1,205.98	318.05	2,438.41

	Ref	Qty	Hours	Hours £	Mat'ls £	O & P £	Total £
Brought forward			96.25	914.38	1,205.98	318.05	2,438.41
Steel lintel 2400mm long	C4	2nr	0.50	4.75	196.92	30.25	231.92
Steel lintel 1500mm long	C5	4nr	0.80	7.60	429.88	65.62	503.10
Close cavity wall at jambs	C7	13.76m	0.69	6.56	28.07	5.19	39.82
Close cavity wall at cills	C8	5.07m	0.25	2.38	10.34	1.91	14.62
Close cavity wall at top	C9	7.60m	0.37	3.52	15.10	2.79	21.41
DPC 112mm wide at jambs	C10	13.76m	0.69	6.56	11.15	2.66	20.36
DPC 112mm wide at cills	C11	5.07m	0.25	2.38	4.11	0.97	7.46
Carried to summary			99.80	948.10	1,901.55	427.45	3,277.10

PART D
FLAT ROOF

	Ref	Qty	Hours	Hours £	Mat'ls £	O & P £	Total £
200 x 150mm sawn softwood joists	D1	19.35m	4.84	45.98	59.98	15.89	121.85
200 x 50mm sawn softwood sprocket pieces	D2	8nr	1.12	10.64	10.72	3.20	24.56
18mm thick WPB grade decking	D3	9.25m2	8.33	74.97	72.71	22.15	169.83
50 x 50mm (avg) wide sawn softwood firrings	D4	19.25m	3.47	32.97	43.51	11.47	87.95
High density polyethylene vapour barrier 150mm thick	D5	8.00m2	1.16	11.02	75.76	13.02	99.80
100 x 75mm sawn softwood wall plate	D6	8.00m	2.40	22.80	17.76	6.08	46.64
100 x 75mm sawn softwood tilt fillet	D7	4.30m	1.08	10.26	11.35	3.24	24.85
Build in ends of 200 x 50mm joists	D8	9nr	3.15	29.93	1.35	4.69	35.97
Carried forward			25.55	238.56	293.14	79.76	611.46

	Ref	Qty	Hours	Hours £	Mat'ls £	O & P £	Total £
Brought forward			25.55	238.56	293.14	79.76	611.46
Rake out joint for flashing	D9	4.30m	1.51	14.35	0.65	2.25	17.24
6mm thick soffit 150mm wide	D10	8.30m	3.32	31.54	14.53	6.91	52.98
19mm wrought softwood fascia 200mm high	D11	8.30m	4.15	39.43	16.35	8.37	64.14
Three layer fibre-based roofing felt	D12	9.25m2	4.63	43.99	77.79	18.27	140.04
Felt turn-down 100mm girth	D13	8.30m	0.83	7.89	6.97	2.23	17.08
Felt flashing 150mm girth	D14	4.30m	0.52	4.94	7.22	1.82	13.98
112mm diameter PVC-U gutter	D15	4.30m	1.12	10.64	19.22	4.48	34.34
Stop end	D16	1nr	0.14	1.33	1.11	0.37	2.81
Stop end outlet	D17	1nr	0.25	2.38	1.11	0.52	4.01
68mm diameter PVC-U down pipe	D18	4.80m	1.20	11.40	16.27	4.15	31.82
Shoe	D19	1nr	0.30	2.85	2.09	0.74	5.68
Paint fascia and soffit	D20	2.49m2	1.74	16.53	2.39	2.84	21.76
Carried to summary			45.26	425.81	458.84	132.70	1,017.34
PART E PITCHED ROOF			N/A	N/A	N/A	N/A	N/A
PART F WINDOWS AND EXTERNAL DOORS							
PVC-U door size 840 x 1980mm complete (B)	F1	1nr	2.50	23.75	209.99	35.06	268.80
PVC-U sliding patio door size 1700 x 2075mm (C)	F2	2nr	14.00	133.00	587.04	108.01	828.05
Carried forward			16.50	156.75	797.03	143.07	1,096.85

	Ref	Qty	Hours	Hours £	Mat'ls £	O & P £	Total £
Brought forward			16.50	156.75	797.03	143.07	1,096.85
PVC-U window size 1200 x 1200mm complete (A)	F3	4nr	8.00	76.00	630.36	105.95	812.31
25 x 225mm wrought softwood window board	F4	4.80m	1.44	13.68	29.86	6.53	50.07
Paint window board	F5	4.80m	0.76	7.22	4.61	1.77	13.60
Carried to summary			26.70	253.65	1,461.86	257.33	1,972.84
PART G INTERNAL PARTITIONS AND DOORS		N/A	N/A	N/A	N/A	N/A	N/A
PART H WALL FINISHES							
19 x 100mm wrought softwood skirting	H1	12.34m	2.07	19.67	16.91	5.49	42.06
12mm plasterboard fixed to walls with dabs	H2	24.94m2	9.73	92.44	49.63	21.31	163.37
12mm plasterboard fixed to walls less than 300mm wide with dabs	H3	18.33m	3.39	32.21	32.37	9.69	74.26
Two coats emulsion paint to walls	H4	26.76m2	5.49	52.16	15.79	10.19	78.14
Paint skirting	H5	12.34m	2.47	23.47	9.92	5.01	38.39
Carried to summary			23.15	219.93	124.62	51.68	396.23
PART J FLOOR FINISHES							
Cement and sand floor screed 40mm thick	J1	5.78m2	1.45	13.78	17.51	4.69	35.98
Vinyl floor tiles, size 300 x 300mm	J2	5.78m2	0.98	9.31	40.03	7.40	56.74
Carried forward			2.43	23.09	57.54	12.09	92.72

	Ref	Qty	Hours	Hours £	Mat'ls £	O & P £	Total £
Brought forward			2.43	23.09	57.54	12.09	92.72
25mm thick tongued and grooved boarding	J3	5.78m2	4.28	40.66	54.33	14.25	109.24
150 x 50mm sawn softwood joists	J4	13.30m	2.93	27.84	30.32	8.72	66.88
Cut and pin ends of joists to existing brick wall	J5	7nr	1.25	11.88	1.75	2.04	15.67
Build in ends of joists to blockwork	J6	7nr	0.70	6.65	1.05	1.16	8.86
Carried to summary			11.59	110.11	144.99	38.26	293.36

PART K
CEILING FINISHES

	Ref	Qty	Hours	Hours £	Mat'ls £	O & P £	Total £
Plasterboard with taped butt joints fixed to joists	K1	11.56m2	4.16	39.52	21.39	9.14	70.05
5mm skim coat to plasterboard ceilings	K2	11.56m2	5.78	54.91	14.10	10.35	79.36
Two coats emulsion paint to ceilings	K3	11.56m2	3.00	28.50	4.81	5.00	38.31
Carried to summary			12.94	122.93	40.30	24.48	187.71

PART L
ELECTRICAL WORK

	Ref	Qty	Hours	Hours £	Mat'ls £	O & P £	Total £
13 amp double switched socket outlet with neon	L1	4nr	1.60	22.40	33.04	8.32	63.76
Lighting point	L2	4nr	0.70	9.80	27.20	5.55	42.55
Lighting switch	L3	4nr	1.00	14.00	15.88	4.48	34.36
Lighting wiring	L4	12.00m	1.00	14.00	12.36	3.95	30.31
Power cable	L5	28.00m	3.60	50.40	30.80	12.18	93.38
Carried to summary			7.90	110.60	119.28	34.48	264.36

	Ref	Qty	Hours	Hours £	Mat'ls £	O & P £	Total £
PART M **HEATING WORK**							
15mm copper pipe	M1	11.00m	1.98	24.75	20.68	6.81	52.24
Elbow	M2	8nr	2.24	28.00	8.80	5.52	42.32
Tee	M3	2nr	0.44	5.50	3.90	1.41	10.81
Radiator, double convector size 1400 x 520mm	M4	4nr	5.20	65.00	462.72	79.16	606.88
Break into existing pipe and insert tee	M5	1nr	0.75	9.38	3.95	2.00	15.32
Carried to summary			10.61	132.63	500.05	94.90	727.58
PART N **ALTERATION WORK**							
Take out existing window size 1500 x 1000mm and lintel over, adapt opening to receive 1770 x 2000mm patio door and insert new lintel over (both measured separately) and make good	N1	1nr	20.00	190.00	20.00	31.50	241.50
Take out existing window size 1500 x 1000mm, enlarge opening to receive new PVC-U door (measured separately)	N2	1nr	25.00	237.50	45.00	42.38	324.88
Carried to summary			45.00	427.50	65.00	73.88	566.38

SUMMARY

	Hours	Hours £	Mat'ls £	O & P £	Total £
PART A PRELIMINARIES	0.00	0.00	0.00	0.00	2,065.00
PART B SUBSTRUCTURE TO DPC LEVEL	41.49	394.16	384.38	116.78	895.32
PART C EXTERNAL WALLS	99.80	948.10	1,901.55	427.45	3,277.10
PART D FLAT ROOF	45.26	425.81	458.84	132.70	1,017.34
PART E PITCHED ROOF	0.00	0.00	0.00	0.00	0.00
PART F WINDOWS AND EXTERNAL DOORS	26.70	253.65	1,461.86	257.33	1,972.84
PART G INTERNAL PARTITIONS AND DOORS	0.00	0.00	0.00	0.00	0.00
PART H WALL FINISHES	23.15	219.93	124.62	51.68	396.23
PART J FLOOR FINISHES	11.59	110.11	144.99	38.26	293.36
PART K CEILING FINISHES	12.94	122.93	40.30	24.48	187.71
PART L ELECTRICAL WORK	7.90	110.60	119.28	34.48	264.36
PART M HEATING WORK	10.61	132.63	500.05	94.90	727.58
PART N ALTERATION WORK	45.00	427.50	65.00	73.88	566.38
Final total	324.44	3,145.42	5,200.87	1,251.94	11,663.22

	Ref	Qty	Hours	Hours £	Mat'ls £	O & P £	Total £
PART A PRELIMINARIES							
Concrete mixer	A1	9 wks					405.00
Small tools	A2	10 wks					350.00
Scaffolding (m2/weeks)	A3	450.00					1,012.50
Skip	A4	6 wks					600.00
Clean up	A5	12 hrs					96.00
Carried to summary							2,463.50
PART B SUBSTRUCTURE TO DPC LEVEL							
Excavate topsoil 150mm thick by hand	B1	11.4m2	3.42	32.49	0.00	4.87	37.36
Excavate to reduce levels by hand	B2	3.42m3	8.55	81.23	0.00	12.18	93.41
Excavate for trench foundations by hand	B3	1.39m3	3.61	34.30	0.00	5.14	39.44
Earthwork support to sides of trenches	B4	12.26m2	4.90	46.55	14.71	9.19	70.45
Backfilling with excavated material	B5	0.55m3	0.33	3.14	0.00	0.47	3.61
Hardcore 225mm thick	B6	7.48m2	1.50	14.25	38.00	7.84	60.09
Hardcore filling to trench	B7	0.12m3	0.06	0.57	0.60	0.18	1.35
Concrete grade (1:3:6) in foundations	B8	1.13m3	1.53	14.54	76.77	13.70	105.00
Concrete grade (1:2:4) in bed 150mm thick	B9	7.48m2	2.24	21.28	81.38	15.40	118.06
Concrete (1:2:4) in cavity wall filling	B10	4.20m2	0.84	7.98	15.25	3.48	26.71
Carried forward			26.98	256.31	226.71	72.45	555.47

	Ref	Qty	Hours	Hours £	Mat'ls £	O & P £	Total £
Brought forward			26.98	256.31	226.71	72.45	555.47
Damp-proof membrane	B11	7.88m2	0.32	3.04	4.57	1.14	8.75
Reinforcement ref A193 in foundation	B12	4.20m2	0.50	4.75	5.46	1.53	11.74
Steel fabric reinforcement ref A193 in slab	B13	7.48m2	1.12	10.64	9.72	3.05	23.41
Solid blockwork 140mm thick in cavity wall	B14	5.46m2	7.10	67.45	69.78	20.58	157.81
Common bricks 112.5mm thick in cavity wall	B15	4.20m2	7.10	67.45	55.40	18.43	141.28
Facing bricks in 112.5mm thick in skin of cavity wall	B16	1.26m2	2.27	21.57	29.32	7.63	58.52
Form cavity 50mm wide in cavity wall	B17	5.46m2	0.16	1.52	3.28	0.72	5.52
DPC 112mm wide	B18	8.40m	0.42	3.99	6.80	1.62	12.41
DPC 140mm wide	B19	8.40m	0.50	4.75	8.15	1.94	14.84
Bond in block wall	B20	1.30m	0.58	5.51	2.83	1.25	9.59
Bond in half brick wall	B21	0.30m	0.46	4.37	0.61	0.75	5.73
50mm thick insulation board	B22	7.48m2	2.24	21.28	33.51	8.22	63.01
Carried to summary			49.75	472.63	456.14	139.31	1,068.08
PART C EXTERNAL WALLS							
Solid blockwork 140mm thick in cavity wall	C1	34.04m2	44.25	420.38	432.99	128.00	981.37
Facing brickwork 112.5mm thick in cavity wall	C2	34.04m2	61.27	582.07	784.96	205.05	1,572.08
75mm thick insulation in cavity wall	C3	34.04m2	7.49	71.16	198.11	40.39	309.65
Carried forward			113.01	1,073.60	1,416.06	373.45	2,863.10

	Ref	Qty	Hours	Hours £	Mat'ls £	O & P £	Total £
Brought forward			113.01	1,073.60	1,416.06	373.45	2,863.10
Steel lintel 2400mm long	C4	2nr	0.50	4.75	196.92	30.25	231.92
Steel lintel 1500mm long	C5	4nr	0.80	7.60	429.88	65.62	503.10
Close cavity wall at jambs	C7	13.76m	0.69	6.56	28.07	5.19	39.82
Close cavity wall at cills	C8	5.07m	0.25	2.38	10.34	1.91	14.62
Close cavity wall at top	C9	8.40m	0.42	3.99	17.16	3.17	24.32
DPC 112mm wide at jambs	C10	13.76m	0.69	6.56	11.15	2.66	20.36
DPC 112mm wide at cills	C11	5.07m	0.25	2.38	4.11	0.97	7.46
Carried to summary			116.61	1,107.80	2,113.69	483.22	3,704.71

PART D
FLAT ROOF

	Ref	Qty	Hours	Hours £	Mat'ls £	O & P £	Total £
200 x 50mm sawn softwood joists	D1	23.65m	5.91	56.15	73.32	19.42	148.88
200 x 50mm sawn softwood sprocket pieces	D2	8nr	1.12	10.64	10.72	3.20	24.56
18mm thick WPB grade decking	D3	11.40m2	10.26	92.34	89.60	27.29	209.23
50 x 50mm (avg) wide sawn softwood firrings	D4	23.65m	4.26	40.47	53.45	14.09	108.01
High density polyethylene vapour barrier 150mm thick	D5	10.00m2	2.00	19.00	94.70	17.06	130.76
100 x 75mm sawn softwood wall plate	D6	9.00m	2.70	25.65	19.98	6.84	52.47
100 x 75mm sawn softwood tilt fillet	D7	5.30m	1.33	12.64	13.39	3.90	29.93
Build in ends of 200 x 50mm joists	D8	11nr	3.85	36.58	1.65	5.73	43.96
Carried forward			31.43	293.46	356.81	97.54	747.80

	Ref	Qty	Hours	Hours £	Mat'ls £	O & P £	Total £
Brought forward			31.43	293.46	356.81	97.54	747.80
Rake out joint for flashing	D9	5.30m	1.86	17.67	0.80	2.77	21.24
6mm thick soffit 150mm wide	D10	9.30m	3.72	35.34	16.28	7.74	59.36
19mm wrought softwood fascia 200mm high	D11	9.30m	4.65	44.18	18.32	9.37	71.87
Three layer fibre-based roofing felt	D12	11.40m2	6.27	59.57	95.87	23.32	178.75
Felt turn-down 100mm girth	D13	9.30m	0.93	8.84	7.81	2.50	19.14
Felt flashing 150mm girth	D14	5.30m	0.63	5.99	8.90	2.23	17.12
112mm diameter PVC-U gutter	D15	5.30m	1.12	10.64	23.69	5.15	39.48
Stop end	D16	1nr	0.14	1.33	1.11	0.37	2.81
Stop end outlet	D17	1nr	0.25	2.38	1.11	0.52	4.01
68mm diameter PVC-U down pipe	D18	4.80m	1.20	11.40	16.27	4.15	31.82
Shoe	D19	1nr	0.30	2.85	2.09	0.74	5.68
Paint fascia and soffit	D20	2.79m2	1.95	18.53	2.68	3.18	24.39
Carried to summary			54.45	512.15	551.74	159.58	1,223.47
PART E PITCHED ROOF			N/A	N/A	N/A	N/A	N/A
PART F WINDOWS AND EXTERNAL DOORS							
PVC-U door size 840 x 1980mm complete (B)	F1	1nr	2.50	23.75	209.99	35.06	268.80
PVC-U sliding patio door size 1700 x 2075mm (C)	F2	2nr	14.00	133.00	587.04	108.01	828.05
Carried forward			16.50	156.75	797.03	143.07	1,096.85

	Ref	Qty	Hours	Hours £	Mat'ls £	O & P £	Total £
Brought forward			16.50	156.75	797.03	143.07	1,096.85
PVC-U window size 1200 x 1200mm complete (A)	F3	4nr	8.00	76.00	630.36	105.95	812.31
25 x 225mm wrought softwood window board	F4	4.80m	1.44	13.68	29.86	6.53	50.07
Paint window board	F5	4.80m	0.76	7.22	4.61	1.77	13.60
Carried to summary			26.70	253.65	1,461.86	257.33	1,972.84
PART G INTERNAL PARTITIONS AND DOORS		N/A	N/A	N/A	N/A	N/A	N/A
PART H WALL FINISHES							
19 x 100mm wrought softwood skirting	H1	14.34m	2.41	22.90	19.65	6.38	48.93
12mm plasterboard fixed to walls with dabs	H2	29.84m2	11.64	110.58	59.38	25.49	195.45
12mm plasterboard fixed to walls less than 300mm wide with dabs	H3	18.33m	3.39	32.21	32.37	9.69	74.26
Two coats emulsion paint to plastered walls	H4	32.66m2	6.68	63.46	19.27	12.41	95.14
Paint skirting	H5	14.34m	1.80	17.10	6.88	3.60	27.58
Carried to summary			25.92	246.24	137.55	57.57	441.36
PART J FLOOR FINISHES							
Cement and sand floor screed 40mm thick	J1	7.48m2	1.87	17.77	22.66	6.06	46.49
Vinyl floor tiles, size 300 x 300mm	J2	7.48m2	1.80	17.10	52.22	10.40	79.72
Carried forward			3.67	34.87	74.88	16.46	126.21

	Ref	Qty	Hours	Hours £	Mat'ls £	O & P £	Total £
Brought forward			3.67	34.87	74.88	16.46	126.21
25mm thick tongued and grooved boarding	J3	7.48m2	5.54	52.63	70.30	18.44	141.37
150 x 50mm sawn softwood joists	J4	15.20m	3.34	31.73	34.66	9.96	76.35
Cut and pin ends of joists to existing brick wall	J5	8nr	1.49	14.16	2.00	2.42	18.58
Build in ends of joists to blockwork	J6	8nr	0.80	7.60	1.20	1.32	10.12
Carried to summary			14.84	140.98	183.04	48.60	372.62
PART K CEILING FINISHES							
Plasterboard with taped butt joints fixed to joists	K1	14.96m2	5.38	51.11	27.68	11.82	90.61
5mm skim coat to plasterboard ceilings	K2	14.96m2	7.48	71.06	18.25	13.40	102.71
Two coats emulsion paint to ceilings	K3	14.96m2	3.89	36.96	8.83	6.87	52.65
Carried to summary			16.75	159.13	54.76	32.08	245.97
PART L ELECTRICAL WORK							
13 amp double switched socket outlet with neon	L1	6nr	2.40	33.60	49.56	12.47	95.63
Lighting point	L2	4nr	1.40	19.60	27.20	7.02	53.82
Lighting switch	L3	4nr	1.00	14.00	15.88	4.48	34.36
Lighting wiring	L4	14.00m	1.40	19.60	14.42	5.10	39.12
Power cable	L5	32.00m	4.80	67.20	35.20	15.36	117.76
Carried to summary			11.00	154.00	142.26	44.44	340.70

	Ref	Qty	Hours	Hours £	Mat'ls £	O & P £	Total £
PART M **HEATING WORK**							
15mm copper pipe	M1	13.00m	2.86	35.75	24.44	9.03	69.22
Elbow	M2	8nr	2.24	28.00	8.80	5.52	42.32
Tee	M3	2nr	0.44	5.50	3.90	1.41	10.81
Radiator, double convector size 1400 x 520mm	M4	4nr	5.20	65.00	462.72	79.16	606.88
Break into existing pipe and insert tee	M5	1nr	0.75	9.38	3.95	2.00	15.32
Carried to summary			11.49	143.63	503.81	97.12	744.55
PART N **ALTERATION WORK**							
Take out existing window size 1500 x 1000mm and lintel over, adapt opening to receive 1770 x 2000mm patio door and insert new lintel over (both measured separately) and make good	N1	1nr	20.00	190.00	20.00	31.50	241.50
Take out existing window size 1500 x 1000mm, enlarge opening to receive new PVC-U door (measured separately)	N2	1nr	25.00	237.50	45.00	42.38	324.88
Carried to summary			45.00	427.50	65.00	73.88	566.38

SUMMARY

	Hours	Hours £	Mat'ls £	O & P £	Total £
PART A PRELIMINARIES	0.00	0.00	0.00	0.00	2,463.50
PART B SUBSTRUCTURE TO DPC LEVEL	49.75	472.63	456.14	139.31	1,068.08
PART C EXTERNAL WALLS	116.61	1,107.80	2,113.69	483.22	3,704.71
PART D FLAT ROOF	54.45	512.15	551.74	159.58	1,223.47
PART E PITCHED ROOF	0.00	0.00	0.00	0.00	0.00
PART F WINDOWS AND EXTERNAL DOORS	26.70	253.65	1,461.86	257.33	1,972.84
PART G INTERNAL PARTITIONS AND DOORS	0.00	0.00	0.00	0.00	0.00
PART H WALL FINISHES	25.92	246.24	137.55	57.57	441.36
PART J FLOOR FINISHES	14.84	140.98	183.04	48.60	372.62
PART K CEILING FINISHES	16.75	159.13	54.76	32.08	245.97
PART L ELECTRICAL WORK	11.00	154.00	142.26	44.44	340.70
PART M HEATING WORK	11.49	143.63	503.81	97.12	744.55
PART N ALTERATION WORK	45.00	427.50	65.00	73.88	566.38
Final total	372.51	3,617.71	5,669.85	1,393.13	13,144.18

	Ref	Qty	Hours	Hours £	Mat'ls £	O & P £	Total £
PART A PRELIMINARIES							
Concrete mixer	A1	9 wks					405.00
Small tools	A2	9 wks					315.00
Scaffolding (m2/weeks)	A3	360.00					810.00
Skip	A4	5 wks					500.00
Clean up	A5	10 hrs					80.00
Carried to summary							2,110.00
PART B SUBSTRUCTURE TO DPC LEVEL							
Excavate topsoil 150mm thick by hand	B1	10.40m2	3.12	29.64	0.00	4.45	34.09
Excavate to reduce levels by hand	B2	3.12m3	7.80	74.10	0.00	11.12	85.22
Excavate for trench foundations by hand	B3	1.39m3	3.61	34.30	0.00	5.14	39.44
Earthwork support to sides of trenches	B4	12.26m2	4.90	46.55	14.71	9.19	70.45
Backfilling with excavated material	B5	0.47m3	0.28	2.66	0.00	0.40	3.06
Hardcore 225mm thick	B6	6.48m2	1.30	12.35	32.92	6.79	52.06
Hardcore filling to trench	B7	0.09m3	0.05	0.48	0.45	0.14	1.06
Concrete grade (1:3:6) in foundations	B8	1.13m3	1.53	14.54	76.77	13.70	105.00
Concrete grade (1:2:4) in bed 150mm thick	B9	6.48m2	1.94	18.43	70.50	13.34	102.27
Concrete (1:2:4) in cavity wall filling	B10	4.20m2	0.84	7.98	15.25	3.48	26.71
Carried forward			25.37	241.02	210.60	67.74	519.36

	Ref	Qty	Hours	Hours £	Mat'ls £	O & P £	Total £
Brought forward			25.37	241.02	210.60	67.74	519.36
Damp-proof membrane	B11	6.88m2	0.28	2.66	3.99	1.00	7.65
Reinforcement ref A193 in foundation	B12	4.20m2	0.50	4.75	5.46	1.53	11.74
Steel fabric reinforcement ref A193 in slab	B13	6.48m2	0.97	9.22	8.42	2.65	20.28
Solid blockwork 140mm thick in cavity wall	B14	5.46m2	7.10	67.45	69.78	20.58	157.81
Common bricks 112.5mm thick in cavity wall	B15	4.20m2	7.10	67.45	55.40	18.43	141.28
Facing bricks in 112.5mm thick in skin of cavity wall	B16	1.26m2	2.27	21.57	29.32	7.63	58.52
Form cavity 50mm wide in cavity wall	B17	5.46m2	0.16	1.52	3.28	0.72	5.52
DPC 112mm wide	B18	8.40m	0.42	3.99	6.80	1.62	12.41
DPC 140mm wide	B19	8.40m	0.50	4.75	8.15	1.94	14.84
Bond in block wall	B20	1.30m	0.58	5.51	2.83	1.25	9.59
Bond in half brick wall	B21	0.30m	0.46	4.37	0.61	0.75	5.73
50mm thick insulation board	B22	6.48m2	1.94	18.43	29.03	7.12	54.58
Carried to summary			47.65	452.68	433.67	132.95	1,019.30
PART C EXTERNAL WALLS							
Solid blockwork 140mm thick in cavity wall	C1	34.04m2	44.25	420.38	432.99	128.00	981.37
Facing brickwork 112.5mm thick in cavity wall	C2	34.04m2	61.27	582.07	784.96	205.05	1,572.08
75mm thick insulation in cavity wall	C3	34.04m2	7.49	71.16	198.11	40.39	309.65
Carried forward			113.01	1,073.60	1,416.06	373.45	2,863.10

	Ref	Qty	Hours	Hours £	Mat'ls £	O & P £	Total £
Brought forward			113.01	1,073.60	1,416.06	373.45	2,863.10
Steel lintel 2400mm long	C4	2nr	0.50	4.75	196.92	30.25	231.92
Steel lintel 1500mm long	C5	4nr	0.80	7.60	429.88	65.62	503.10
Close cavity wall at jambs	C7	13.76m	0.69	6.56	28.07	5.19	39.82
Close cavity wall at cills	C8	5.07m	0.25	2.38	10.34	1.91	14.62
Close cavity wall at top	C9	8.40m	0.42	3.99	17.16	3.17	24.32
DPC 112mm wide at jambs	C10	13.76m	0.69	6.56	11.15	2.66	20.36
DPC 112mm wide at cills	C11	5.07m	0.25	2.38	4.11	0.97	7.46
Carried to summary			116.61	1,107.80	2,113.69	483.22	3,704.71

PART D
FLAT ROOF

	Ref	Qty	Hours	Hours £	Mat'ls £	O & P £	Total £
200 x 50mm sawn softwood joists	D1	22.04m	5.51	52.35	68.36	18.11	138.81
200 x 50mm sawn softwood sprocket pieces	D2	12nr	1.68	15.96	16.08	4.81	36.85
18mm thick WPB grade decking	D3	10.40m2	9.36	84.24	81.74	24.90	190.88
50 x 50mm (avg) wide sawn softwood firrings	D4	22.05m	3.97	37.72	49.83	13.13	100.68
High density polyethylene vapour barrier 150mm thick	D5	9.00m2	1.80	17.10	85.23	15.35	117.68
100 x 75mm sawn softwood wall plate	D6	9.00m	2.70	25.65	19.98	6.84	52.47
100 x 75mm sawn softwood tilt fillet	D7	3.30m	0.83	7.89	8.71	2.49	19.08
Build in ends of 200 x 50mm joists	D8	7nr	2.46	23.37	1.05	3.66	28.08
Carried forward			28.31	264.27	330.98	89.29	684.53

	Ref	Qty	Hours	Hours £	Mat'ls £	O & P £	Total £
Brought forward			28.31	264.27	330.98	89.29	684.53
Rake out joint for flashing	D9	3.30m	1.16	11.02	0.50	1.73	13.25
6mm thick soffit 150mm wide	D10	9.30m	3.72	35.34	16.28	7.74	59.36
19mm wrought softwood fascia 200mm high	D11	9.30m	4.65	44.18	18.32	9.37	71.87
Three layer fibre-based roofing felt	D12	10.40m2	5.72	54.34	89.99	21.65	165.98
Felt turn-down 100mm girth	D13	9.30m	0.93	8.84	7.81	2.50	19.14
Felt flashing 150mm girth	D14	3.30m	0.40	3.80	5.54	1.40	10.74
112mm diameter PVC-U gutter	D15	3.30m	0.86	8.17	14.75	3.44	26.36
Stop end	D16	1nr	0.14	1.33	1.11	0.37	2.81
Stop end outlet	D17	1nr	0.25	2.38	1.11	0.52	4.01
68mm diameter PVC-U down pipe	D18	4.80m	1.20	11.40	16.27	4.15	31.82
Shoe	D19	1nr	0.30	2.85	2.09	0.74	5.68
Paint fascia and soffit	D20	2.79m2	1.95	18.53	2.68	3.18	24.39
Carried to summary			49.59	466.43	507.43	146.08	1,119.93
PART E PITCHED ROOF			N/A	N/A	N/A	N/A	N/A
PART F WINDOWS AND EXTERNAL DOORS							
PVC-U door size 840 x 1980mm complete (B)	F1	1nr	2.50	23.75	209.99	35.06	268.80
PVC-U sliding patio door size 1700 x 2075mm (C)	F2	2nr	14.00	133.00	587.04	108.01	828.05
Carried forward			16.50	156.75	797.03	143.07	1,096.85

	Ref	Qty	Hours	Hours £	Mat'ls £	O & P £	Total £
Brought forward			16.50	156.75	797.03	143.07	1,096.85
PVC-U window size 1200 x 1200mm complete (A)	F3	4nr	8.00	76.00	630.36	105.95	812.31
25 x 225mm wrought softwood window board	F4	4.80m	1.44	13.68	29.86	6.53	50.07
Paint window board	F5	4.80m	0.76	7.22	4.61	1.77	13.60
Carried to summary			26.70	253.65	1,461.86	257.33	1,972.84
PART G INTERNAL PARTITIONS AND DOORS		N/A	N/A	N/A	N/A	N/A	N/A
PART H WALL FINISHES							
19 x 100mm wrought softwood skirting	H1	14.34m	2.44	23.18	19.65	6.42	49.25
12mm plasterboard fixed to walls with dabs	H2	29.84m2	11.64	110.58	59.38	25.49	195.45
12mm plasterboard fixed to walls less than 300mm wide with dabs	H3	18.33m	3.39	32.21	32.37	9.69	74.26
Two coats emulsion paint to walls	H4	32.66m2	6.68	63.46	19.27	12.41	95.14
Paint skirting	H5	14.34m	1.80	17.10	6.88	3.60	27.58
Carried to summary			25.95	246.53	137.55	57.61	441.69
PART J FLOOR FINISHES							
Cement and sand floor screed 40mm thick	J1	6.48m2	1.62	15.39	19.63	5.25	40.27
Vinyl floor tiles, size 300 x 300mm	J2	6.48m2	1.56	14.82	45.23	9.01	69.06
Carried forward			3.18	30.21	64.86	14.26	109.33

	Ref	Qty	Hours	Hours £	Mat'ls £	O & P £	Total £
Brought forward			3.18	30.21	64.86	14.26	109.33
25mm thick tongued and grooved boarding	J3	6.48m2	4.79	45.51	60.91	15.96	122.38
150 x 50mm sawn softwood joists	J4	14.50m	2.03	19.29	33.06	7.85	60.20
Cut and pin ends of joists to existing brick wall	J5	5nr	0.90	8.55	1.25	1.47	11.27
Build in ends of joists to blockwork	J6	5nr	0.50	4.75	0.75	0.83	6.33
Carried to summary			11.40	108.30	160.83	40.37	309.50

PART K
CEILING FINISHES

	Ref	Qty	Hours	Hours £	Mat'ls £	O & P £	Total £
Plasterboard with taped butt joints fixed to joists	K1	12.96m2	4.67	44.37	23.98	10.25	78.60
5mm skim coat to plasterboard ceilings	K2	12.96m2	6.48	61.56	15.81	11.61	88.98
Two coats emulsion paint to ceilings	K3	12.96m2	3.37	32.02	7.65	5.95	45.61
Carried to summary			14.52	137.94	47.44	27.81	213.19

PART L
ELECTRICAL WORK

	Ref	Qty	Hours	Hours £	Mat'ls £	O & P £	Total £
13 amp double switched socket outlet with neon	L1	4nr	1.60	22.40	33.04	8.32	63.76
Lighting point	L2	4nr	1.40	19.60	27.20	7.02	53.82
Lighting switch	L3	4nr	1.00	14.00	15.88	4.48	34.36
Lighting wiring	L4	12.00m	1.20	16.80	12.36	4.37	33.53
Power cable	L5	28.00m	4.20	58.80	30.80	13.44	103.04
Carried to summary			9.40	131.60	119.28	37.63	288.51

	Ref	Qty	Hours	Hours £	Mat'ls £	O & P £	Total £
PART M **HEATING WORK**							
15mm copper pipe	M1	11.00m	2.42	30.25	20.68	7.64	58.57
Elbow	M2	8nr	2.24	28.00	8.80	5.52	42.32
Tee	M3	2nr	0.68	8.50	3.90	1.86	14.26
Radiator, double convector size 1400 x 520mm	M4	4nr	5.20	65.00	462.72	79.16	606.88
Break into existing pipe and insert tee	M5	1nr	0.75	9.38	3.95	2.00	15.32
Carried to summary			11.29	141.13	500.05	96.18	737.35
PART N **ALTERATION WORK**							
Take out existing window size 1500 x 1000mm and lintel over, adapt opening to receive 1770 x 2000mm patio door and insert new lintel over (both measured separately) and make good	N1	1nr	20.00	190.00	20.00	31.50	241.50
Take out existing window size 1500 x 1000mm, enlarge opening to receive new PVC-U door (measured separately)	N2	1nr	25.00	237.50	45.00	42.38	324.88
Carried to summary			45.00	427.50	65.00	73.88	566.38

SUMMARY

	Hours	Hours £	Mat'ls £	O & P £	Total £
PART A PRELIMINARIES	0.00	0.00	0.00	0.00	2,110.00
PART B SUBSTRUCTURE TO DPC LEVEL	47.65	452.68	433.67	132.95	1,019.30
PART C EXTERNAL WALLS	116.61	1,107.80	2,113.69	483.22	3,704.71
PART D FLAT ROOF	49.59	466.43	507.43	146.08	1,119.93
PART E PITCHED ROOF	0.00	0.00	0.00	0.00	0.00
PART F WINDOWS AND EXTERNAL DOORS	26.70	253.65	1,461.86	257.33	1,972.84
PART G INTERNAL PARTITIONS AND DOORS	0.00	0.00	0.00	0.00	0.00
PART H WALL FINISHES	25.95	246.53	137.55	57.61	441.69
PART J FLOOR FINISHES	11.40	108.30	160.83	40.37	309.50
PART K CEILING FINISHES	14.52	137.94	47.44	27.81	213.19
PART L ELECTRICAL WORK	9.40	131.60	119.28	37.63	288.51
PART M HEATING WORK	11.29	141.13	500.05	96.18	737.35
PART N ALTERATION WORK	45.00	427.50	65.00	73.88	566.38
Final total	358.11	3,473.56	5,546.80	1,353.06	12,483.40

	Ref	Qty	Hours	Hours £	Mat'ls £	O & P £	Total £
PART A PRELIMINARIES							
Concrete mixer	A1	10 wks					450.00
Small tools	A2	10 wks					350.00
Scaffolding (m2/weeks)	A3	450.00					1,012.50
Skip	A4	6 wks					600.00
Clean up	A5	12 hrs					96.00
Carried to summary							2,508.50
PART B SUBSTRUCTURE TO DPC LEVEL							
Excavate topsoil 150mm thick by hand	B1	14.19m2	4.26	40.47	0.00	6.07	46.54
Excavate to reduce levels by hand	B2	4.06m3	10.15	96.43	0.00	14.46	110.89
Excavate for trench foundations by hand	B3	1.39m3	3.61	34.30	0.00	5.14	39.44
Earthwork support to sides of trenches	B4	13.72m2	5.49	52.16	16.46	10.29	78.91
Backfilling with excavated material	B5	0.55m3	0.22	2.09	0.00	0.31	2.40
Hardcore 225mm thick	B6	9.18m2	1.38	13.11	46.63	8.96	68.70
Hardcore filling to trench	B7	0.12m3	0.06	0.57	0.60	0.18	1.35
Concrete grade (1:3:6) in foundations	B8	1.27m3	1.71	16.25	86.28	15.38	117.90
Concrete grade (1:2:4) in bed 150mm thick	B9	9.18m2	2.75	26.13	99.88	18.90	144.91
Concrete (1:2:4) in cavity wall filling	B10	4.70m2	0.94	8.93	17.06	3.90	29.89
Carried forward			30.57	290.42	266.91	83.60	640.92

	Ref	Qty	Hours	Hours £	Mat'ls £	O & P £	Total £
Brought forward			30.57	290.42	266.91	83.60	640.92
Damp-proof membrane	B11	9.63m2	0.37	3.52	5.59	1.37	10.47
Reinforcement ref A193 in foundation	B12	4.70m2	0.44	4.18	6.11	1.54	11.83
Steel fabric reinforcement ref A193 in slab	B13	9.18m2	1.38	13.11	11.93	3.76	28.80
Solid blockwork 140mm thick in cavity wall	B14	6.11m2	7.94	75.43	78.08	23.03	176.54
Common bricks 112.5mm thick in cavity wall	B15	4.70m2	7.94	75.43	61.99	20.61	158.03
Facing bricks in 112.5mm thick in skin of cavity wall	B16	1.41m2	2.54	24.13	32.81	8.54	65.48
Form cavity 50mm wide in cavity wall	B17	1.41m2	0.18	1.71	3.67	0.81	6.19
DPC 112mm wide	B18	9.40m	0.47	4.47	7.61	1.81	13.89
DPC 140mm wide	B19	9.40m	0.56	5.32	9.12	2.17	16.61
Bond in block wall	B20	1.30m	0.58	5.51	2.83	1.25	9.59
Bond in half brick wall	B21	0.30m	0.46	4.37	0.61	0.75	5.73
50mm thick insulation board	B22	9.18m2	2.75	26.13	41.13	10.09	77.34
Carried to summary			56.18	533.71	528.39	159.32	1,221.42
PART C EXTERNAL WALLS							
Solid blockwork 140mm thick in cavity wall	C1	39.09m2	50.81	482.70	497.22	146.99	1,126.90
Facing brickwork 112.5mm thick in cavity wall	C2	39.09m2	70.36	668.42	901.42	235.48	1,805.32
75mm thick insulation in cavity wall	C3	39.09m2	8.60	81.70	227.50	46.38	355.58
Carried forward			129.77	1,232.82	1,626.14	428.84	3,287.80

	Ref	Qty	Hours	Hours £	Mat'ls £	O & P £	Total £
Brought forward			129.77	1,232.82	1,626.14	428.84	3,287.80
Steel lintel 2400mm long	C4	2nr	0.50	4.75	196.92	30.25	231.92
Steel lintel 1500mm long	C5	4nr	0.80	7.60	429.88	65.62	503.10
Close cavity wall at jambs	C7	13.76m	0.69	6.56	28.07	5.19	39.82
Close cavity wall at cills	C8	5.07m	0.25	2.38	10.34	1.91	14.62
Close cavity wall at top	C9	9.40m	0.47	4.47	19.18	3.55	27.19
DPC 112mm wide at jambs	C10	13.76m	0.69	6.56	11.15	2.66	20.36
DPC 112mm wide at cills	C11	5.07m	0.25	2.38	4.11	0.97	7.46
Carried to summary			133.42	1,267.49	2,325.79	538.99	4,132.27
PART D FLAT ROOF							
200 x 50mm sawn softwood joists	D1	28.35m	7.09	67.36	87.89	23.29	178.53
200 x 50mm sawn softwood sprocket pieces	D2	12nr	1.68	15.96	16.08	4.81	36.85
18mm thick WPB grade decking	D3	13.55m2	12.20	109.80	106.50	32.45	248.75
50 x 50mm (avg) wide sawn softwood firrings	D4	28.35m	5.10	48.45	64.07	16.88	129.40
High density polyethylene vapour barrier 150mm thick	D5	12.00m2	2.40	22.80	113.64	20.47	156.91
100 x 75mm sawn softwood wall plate	D6	10.00m	3.00	28.50	22.20	7.61	58.31
100 x 75mm sawn softwood tilt fillet	D7	4.30m	1.08	10.26	11.35	3.24	24.85
Build in ends of 200 x 50mm joists	D8	9nr	3.15	29.93	1.35	4.69	35.97
Carried forward			35.70	333.05	423.08	113.42	869.55

	Ref	Qty	Hours	Hours £	Mat'ls £	O & P £	Total £
Brought forward			35.70	333.05	423.08	113.42	869.55
Rake out joint for flashing	D9	4.30m	1.51	14.35	0.65	2.25	17.24
6mm thick soffit 150mm wide	D10	10.30m	4.12	39.14	18.03	8.58	65.75
19mm wrought softwood fascia 200mm high	D11	10.30m	5.15	48.93	20.29	10.38	79.60
Three layer fibre-based roofing felt	D12	13.55m2	7.45	70.78	113.96	27.71	212.45
Felt turn-down 100mm girth	D13	10.30m	1.03	9.79	8.65	2.77	21.20
Felt flashing 150mm girth	D14	4.30m	0.52	4.94	7.22	1.82	13.98
112mm diameter PVC-U gutter	D15	4.30m	1.12	10.64	19.22	4.48	34.34
Stop end	D16	1nr	0.14	1.33	1.11	0.37	2.81
Stop end outlet	D17	1nr	0.25	2.38	1.11	0.52	4.01
68mm diameter PVC-U down pipe	D18	4.80m	1.20	11.40	16.27	4.15	31.82
Shoe	D19	1nr	0.30	2.85	2.09	0.74	5.68
Paint fascia and soffit	D20	3.09m2	2.16	20.52	2.97	3.52	27.01
Carried to summary			60.65	570.08	634.65	180.71	1,385.43
PART E PITCHED ROOF			N/A	N/A	N/A	N/A	N/A
PART F WINDOWS AND EXTERNAL DOORS							
PVC-U door size 840 x 1980mm complete (B)	F1	1nr	2.50	23.75	209.99	35.06	268.80
PVC-U sliding patio door size 1700 x 2075mm (C)	F2	2nr	14.00	133.00	587.04	108.01	828.05
Carried forward			16.50	156.75	797.03	143.07	1,096.85

	Ref	Qty	Hours	Hours £	Mat'ls £	O & P £	Total £
Brought forward			16.50	156.75	797.03	143.07	1,096.85
PVC-U window size 1200 x 1200mm complete (A)	F3	4nr	8.00	76.00	630.36	105.95	812.31
25 x 225mm wrought softwood window board	F4	4.80m	1.44	13.68	29.86	6.53	50.07
Paint window board	F5	4.80m	0.76	7.22	4.61	1.77	13.60
Carried to summary			26.70	253.65	1,461.86	257.33	1,972.84
PART G INTERNAL PARTITIONS AND DOORS		N/A	N/A	N/A	N/A	N/A	N/A
PART H WALL FINISHES							
19 x 100mm wrought softwood skirting	H1	16.34m	2.75	26.13	22.39	7.28	55.79
12mm plasterboard fixed to walls with dabs	H2	34.74m2	13.55	128.73	69.13	29.68	227.53
12mm plasterboard fixed to walls less than 300mm wide with dabs	H3	18.33m	3.34	31.73	32.37	9.62	73.72
Two coats emulsion paint to walls	H4	38.56m2	7.89	74.96	22.75	14.66	112.36
Paint skirting	H5	16.34m	2.05	19.48	7.84	4.10	31.41
Carried to summary			29.58	281.01	154.48	65.32	500.81
PART J FLOOR FINISHES							
Cement and sand floor screed 40mm thick	J1	9.18m2	2.30	21.85	27.82	7.45	57.12
Vinyl floor tiles, size 300 x 300mm	J2	9.18m2	2.20	20.90	64.07	12.75	97.72
Carried forward			4.50	42.75	91.89	20.20	154.84

	Ref	Qty	Hours	Hours £	Mat'ls £	O & P £	Total £
Brought forward			4.50	42.75	91.89	20.20	154.84
25mm thick tongued and grooved boarding	J3	9.18m2	6.79	64.51	86.29	22.62	173.41
150 x 50mm sawn softwood joists	J4	20.30m	2.87	27.27	46.28	11.03	84.58
Cut and pin ends of joists to existing brick wall	J5	7nr	1.26	11.97	1.75	2.06	15.78
Build in ends of joists to blockwork	J6	7nr	0.70	6.65	1.05	1.16	8.86
Carried to summary			16.12	153.14	227.26	57.06	437.46
PART K CEILING FINISHES							
Plasterboard with taped butt joints fixed to joists	K1	18.36m2	6.60	62.70	33.97	14.50	111.17
5mm skim coat to plasterboard ceilings	K2	18.36m2	9.18	87.21	22.40	16.44	126.05
Two coats emulsion paint to ceilings	K3	18.36m2	4.77	45.32	10.83	8.42	64.57
Carried to summary			20.55	195.23	67.20	39.36	301.79
PART L ELECTRICAL WORK							
13 amp double switched socket outlet with neon	L1	6nr	2.40	33.60	49.66	12.49	95.75
Lighting point	L2	4nr	1.40	19.60	27.20	7.02	53.82
Lighting switch	L3	4nr	1.00	14.00	15.88	4.48	34.36
Lighting wiring	L4	12.00m	1.20	16.80	12.36	4.37	33.53
Power cable	L5	32.00m	4.80	67.20	35.20	15.36	117.76
Carried to summary			10.80	151.20	140.30	43.73	335.23

	Ref	Qty	Hours	Hours £	Mat'ls £	O & P £	Total £
PART M HEATING WORK							
15mm copper pipe	M1	13.00m	2.86	35.75	24.46	9.03	69.24
Elbow	M2	8nr	2.24	28.00	8.80	5.52	42.32
Tee	M3	2nr	0.68	8.50	3.90	1.86	14.26
Radiator, double convector size 1400 x 520mm	M4	4nr	5.20	65.00	462.72	79.16	606.88
Break into existing pipe and insert tee	M5	1nr	0.75	9.38	3.95	2.00	15.32
Carried to summary			11.73	146.63	503.83	97.57	748.02
PART N ALTERATION WORK							
Take out existing window size 1500 x 1000mm and lintel over, adapt opening to receive 1770 x 2000mm patio door and insert new lintel over (both measured separately) and make good	N1	1nr	20.00	190.00	20.00	31.50	241.50
Take out existing window size 1500 x 1000mm, enlarge opening to receive new PVC-U door (measured separately)	N2	1nr	25.00	237.50	45.00	42.38	324.88
Carried to summary			45.00	427.50	65.00	73.88	566.38

SUMMARY

	Hours	Hours £	Mat'ls £	O & P £	Total £
PART A PRELIMINARIES	0.00	0.00	0.00	0.00	2,508.50
PART B SUBSTRUCTURE TO DPC LEVEL	56.18	533.71	528.39	159.32	1,221.42
PART C EXTERNAL WALLS	133.42	1,267.49	2,325.79	538.99	4,132.27
PART D FLAT ROOF	60.65	570.08	634.65	180.71	1,385.43
PART E PITCHED ROOF	0.00	0.00	0.00	0.00	0.00
PART F WINDOWS AND EXTERNAL DOORS	26.70	253.65	1,461.86	257.33	1,972.84
PART G INTERNAL PARTITIONS AND DOORS	0.00	0.00	0.00	0.00	0.00
PART H WALL FINISHES	29.58	281.01	154.48	65.32	500.81
PART J FLOOR FINISHES	16.12	153.14	227.26	57.06	437.46
PART K CEILING FINISHES	20.55	195.23	67.20	39.36	301.79
PART L ELECTRICAL WORK	10.80	151.20	140.30	43.73	335.23
PART M HEATING WORK	11.73	146.63	503.83	97.57	748.02
PART N ALTERATION WORK	45.00	427.50	65.00	73.88	566.38
Final total	410.73	3,979.64	6,108.76	1,513.27	14,110.15

	Ref	Qty	Hours	Hours £	Mat'ls £	O & P £	Total £
PART A PRELIMINARIES							
Concrete mixer	A1	11 wks					495.00
Small tools	A2	11 wks					385.00
Scaffolding (m2/weeks)	A3	550.00					1,237.50
Skip	A4	7 wks					700.00
Clean up	A5	12 hrs					96.00
Carried to summary							2,913.50
PART B SUBSTRUCTURE TO DPC LEVEL							
Excavate topsoil 150mm thick by hand	B1	17.50m2	5.25	49.88	0.00	7.48	57.36
Excavate to reduce levels by hand	B2	5.00m3	12.50	118.75	0.00	17.81	136.56
Excavate for trench foundations by hand	B3	1.72m3	4.47	42.47	0.00	6.37	48.83
Earthwork support to sides of trenches	B4	15.18m2	6.07	57.67	18.22	11.38	87.27
Backfilling with excavated material	B5	0.62m3	0.37	3.52	0.00	0.53	4.04
Hardcore 225mm thick	B6	11.88m2	2.38	22.61	60.35	12.44	95.40
Hardcore filling to trench	B7	0.14m3	0.07	0.67	0.70	0.20	1.57
Concrete grade (1:3:6) in foundations	B8	1.40m3	1.89	17.96	95.11	16.96	130.02
Concrete grade (1:2:4) in bed 150mm thick	B9	11.88m2	3.56	33.82	129.25	24.46	187.53
Concrete (1:2:4) in cavity wall filling	B10	5.20m2	1.04	9.88	18.88	4.31	33.07
Carried forward			37.60	357.20	322.51	101.96	781.67

	Ref	Qty	Hours	Hours £	Mat'ls £	O & P £	Total £
Brought forward			37.60	357.20	322.51	101.96	781.67
Damp-proof membrane	B11	12.38m2	0.52	4.94	7.18	1.82	13.94
Reinforcement ref A193 in foundation	B12	5.20m2	0.62	5.89	6.76	1.90	14.55
Steel fabric reinforcement ref A193 in slab	B13	11.38m2	1.78	16.91	15.46	4.86	37.23
Solid blockwork 140mm thick in cavity wall	B14	6.76m2	8.79	83.51	86.39	25.48	195.38
Common bricks 112.5mm thick in cavity wall	B15	5.20m2	8.79	83.51	68.59	22.81	174.91
Facing bricks in 112.5mm thick in skin of cavity wall	B16	1.56m2	2.81	26.70	36.30	9.45	72.44
Form cavity 50mm wide in cavity wall	B17	6.76m2	0.20	1.90	4.06	0.89	6.85
DPC 112mm wide	B18	10.40m	0.52	4.94	8.42	2.00	15.36
DPC 140mm wide	B19	10.40m	0.62	5.89	10.09	2.40	18.38
Bond in block wall	B20	1.30m	0.59	5.61	2.83	1.27	9.70
Bond in half brick wall	B21	0.30m	0.46	4.37	0.61	0.75	5.73
50mm thick insulation board	B22	11.88m2	3.56	33.82	53.22	13.06	100.10
Carried to summary			63.30	623.35	591.20	197.58	1,368.13
PART C EXTERNAL WALLS							
Solid blockwork 140mm thick in cavity wall	C1	44.14m2	57.38	545.11	561.46	165.99	1,272.56
Facing brickwork 112.5mm thick in cavity wall	C2	44.14m2	79.45	754.78	1,017.87	265.90	2,038.54
75mm thick insulation in cavity wall	C3	44.14m2	9.71	92.25	256.89	52.37	401.51
Carried forward			146.54	1,392.13	1,836.22	484.25	3,712.60

	Ref	Qty	Hours	Hours £	Mat'ls £	O & P £	Total £
Brought forward			146.54	1,392.13	1,836.22	484.25	3,712.60
Steel lintel 2400mm long	C4	2nr	0.50	4.75	196.92	30.25	231.92
Steel lintel 1500mm long	C5	4nr	0.80	7.60	429.88	65.62	503.10
Close cavity wall at jambs	C7	13.76m	0.69	6.56	28.07	5.19	39.82
Close cavity wall at cills	C8	5.07m	0.25	2.38	10.34	1.91	14.62
Close cavity wall at top	C9	10.40m	0.51	4.85	21.22	3.91	29.97
DPC 112mm wide at jambs	C10	13.76m	0.69	6.56	11.15	2.66	20.36
DPC 112mm wide at cills	C11	5.07m	0.25	2.38	4.11	0.97	7.46
Carried to summary			150.23	1,427.19	2,537.91	594.76	4,559.86
PART D **FLAT ROOF**							
200 x 50mm sawn softwood joists	D1	34.65m	8.66	82.27	107.42	28.45	218.14
200 x 50mm sawn softwood sprocket pieces	D2	12nr	1.68	15.96	16.08	4.81	36.85
18mm thick WPB grade decking	D3	16.70m2	15.03	135.27	131.26	39.98	306.51
50 x 50mm (avg) wide sawn softwood firrings	D4	34.65m	6.24	59.28	78.31	20.64	158.23
High density polyethylene vapour barrier 150mm thick	D5	15.00m2	3.00	28.50	142.05	25.58	196.13
100 x 75mm sawn softwood wall plate	D6	11.00m	3.30	31.35	24.42	8.37	64.14
100 x 75mm sawn softwood tilt fillet	D7	5.30m	1.08	10.26	13.99	3.64	27.89
Build in ends of 200 x 50mm joists	D8	11nr	3.85	36.58	1.65	5.73	43.96
Carried forward			42.84	399.47	515.18	137.20	1,051.84

	Ref	Qty	Hours	Hours £	Mat'ls £	O & P £	Total £
Brought forward			42.84	399.47	515.18	137.20	1,051.84
Rake out joint for flashing	D9	5.30m	1.86	17.67	0.80	2.77	21.24
6mm thick soffit 150mm wide	D10	11.30m	4.52	42.94	19.78	9.41	72.13
19mm wrought softwood fascia 200mm high	D11	11.30m	5.65	53.68	22.26	11.39	87.33
Three layer fibre-based roofing felt	D12	16.70m2	9.19	87.31	140.45	34.16	261.92
Felt turn-down 100mm girth	D13	11.30m	1.13	10.74	9.49	3.03	23.26
Felt flashing 150mm girth	D14	5.30m	0.64	6.08	8.90	2.25	17.23
112mm diameter PVC-U gutter	D15	5.30m	1.38	13.11	23.69	5.52	42.32
Stop end	D16	1nr	0.14	1.33	1.11	0.37	2.81
Stop end outlet	D17	1nr	0.25	2.38	1.11	0.52	4.01
68mm diameter PVC-U down pipe	D18	4.80m	1.20	11.40	16.27	4.15	31.82
Shoe	D19	1nr	0.30	2.85	2.09	0.74	5.68
Paint fascia and soffit	D20	3.39m2	2.37	22.52	3.25	3.86	29.63
Carried to summary			71.47	671.45	764.38	215.37	1,651.20
PART E PITCHED ROOF			N/A	N/A	N/A	N/A	N/A
PART F WINDOWS AND EXTERNAL DOORS							
PVC-U door size 840 x 1980mm complete (B)	F1	1nr	2.50	23.75	209.99	35.06	268.80
PVC-U sliding patio door size 1700 x 2075mm (C)	F2	2nr	14.00	133.00	587.04	108.01	828.05
Carried forward			16.50	156.75	797.03	143.07	1,096.85

	Ref	Qty	Hours	Hours £	Mat'ls £	O & P £	Total £
Brought forward			16.50	156.75	797.03	143.07	1,096.85
PVC-U window size 1200 x 1200mm complete (A)	F3	4nr	8.00	76.00	630.36	105.95	812.31
25 x 225mm wrought softwood window board	F4	4.80m	1.44	13.68	29.86	6.53	50.07
Paint window board	F5	4.80m	0.76	7.22	4.61	1.77	13.60
Carried to summary			26.70	253.65	1,461.86	257.33	1,972.84
PART G INTERNAL PARTITIONS AND DOORS		N/A	N/A	N/A	N/A	N/A	N/A
PART H WALL FINISHES							
19 x 100mm wrought softwood skirting	H1	18.34m	3.12	29.64	25.13	8.22	62.99
12mm plasterboard fixed to walls with dabs	H2	39.64m2	15.46	146.87	78.88	33.86	259.61
12mm plasterboard fixed to walls less than 300mm wide with dabs	H3	18.33m	3.39	32.21	32.37	9.69	74.26
Two coats emulsion paint to walls	H4	42.46m2	8.49	80.66	25.05	15.86	121.56
Paint skirting	H5	18.34m	2.39	22.71	8.80	4.73	36.23
Carried to summary			32.85	312.08	170.23	72.35	554.65
PART J FLOOR FINISHES							
Cement and sand floor screed 40mm thick	J1	11.88m2	2.97	28.22	36.00	9.63	73.85
Vinyl floor tiles, size 300 x 300mm	J2	11.88m2	2.85	27.08	83.62	16.60	127.30
Carried forward			5.82	55.29	119.62	26.24	201.15

	Ref	Qty	Hours	Hours £	Mat'ls £	O & P £	Total £
Brought forward			5.82	55.29	119.62	26.24	201.15
25mm thick tongued and grooved boarding	J3	11.88m2	8.79	83.51	111.67	29.28	224.45
150 x 50mm sawn softwood joists	J4	23.20m	3.25	30.88	52.90	12.57	96.34
Cut and pin ends of joists to existing brick wall	J5	8nr	1.44	13.68	2.00	2.35	18.03
Build in ends of joists to blockwork	J6	8nr	0.80	7.60	1.20	1.32	10.12
Carried to summary			20.10	190.95	287.39	71.75	550.09

PART K
CEILING FINISHES

	Ref	Qty	Hours	Hours £	Mat'ls £	O & P £	Total £
Plasterboard with taped butt joints fixed to joists	K1	23.76m2	6.60	62.70	43.96	16.00	122.66
5mm skim coat to plasterboard ceilings	K2	23.76m2	9.18	87.21	28.99	17.43	133.63
Two coats emulsion paint to ceilings	K3	23.76m2	4.77	45.32	14.02	8.90	68.24
Carried to summary			20.55	195.23	86.97	42.33	324.52

PART L
ELECTRICAL WORK

	Ref	Qty	Hours	Hours £	Mat'ls £	O & P £	Total £
13 amp double switched socket outlet with neon	L1	6nr	2.40	33.60	49.66	12.49	95.75
Lighting point	L2	4nr	1.40	19.60	27.20	7.02	53.82
Lighting switch	L3	4nr	1.00	14.00	15.88	4.48	34.36
Lighting wiring	L4	14.00m	1.40	19.60	14.42	5.10	39.12
Power cable	L5	36.00m	5.40	75.60	39.60	17.28	132.48
Carried to summary			11.60	162.40	146.76	46.37	355.53

	Ref	Qty	Hours	Hours £	Mat'ls £	O & P £	Total £
PART M **HEATING WORK**							
15mm copper pipe	M1	15.00m	3.30	41.25	28.20	10.42	79.87
Elbow	M2	8nr	2.24	28.00	8.80	5.52	42.32
Tee	M3	2nr	0.68	8.50	3.90	1.86	14.26
Radiator, double convector size 1400 x 520mm	M4	4nr	5.20	65.00	462.72	79.16	606.88
Break into existing pipe and insert tee	M5	1nr	0.75	9.38	3.95	2.00	15.32
Carried to summary			12.17	152.13	507.57	98.95	758.65
PART N **ALTERATION WORK**							
Take out existing window size 1500 x 1000mm and lintel over, adapt opening to receive 1770 x 2000mm patio door and insert new lintel over (both measured separately) and make good	N1	1nr	20.00	190.00	20.00	31.50	241.50
Take out existing window size 1500 x 1000mm, enlarge opening to receive new PVC-U door (measured separately)	N2	1nr	25.00	237.50	45.00	42.38	324.88
Carried to summary			45.00	427.50	65.00	73.88	566.38

SUMMARY

	Hours	Hours £	Mat'ls £	O & P £	Total £
PART A PRELIMINARIES	0.00	0.00	0.00	0.00	2,913.50
PART B SUBSTRUCTURE TO DPC LEVEL	63.30	623.35	591.20	197.58	1,368.13
PART C EXTERNAL WALLS	150.23	1,427.19	2,537.91	594.76	4,559.86
PART D FLAT ROOF	71.47	671.45	764.38	215.37	1,651.20
PART E PITCHED ROOF	0.00	0.00	0.00	0.00	0.00
PART F WINDOWS AND EXTERNAL DOORS	26.70	253.65	1,461.86	257.33	1,972.84
PART G INTERNAL PARTITIONS AND DOORS	0.00	0.00	0.00	0.00	0.00
PART H WALL FINISHES	32.85	312.08	170.23	72.35	554.65
PART J FLOOR FINISHES	20.10	190.95	287.39	71.75	550.09
PART K CEILING FINISHES	20.55	195.23	86.97	42.33	324.52
PART L ELECTRICAL WORK	11.60	162.40	146.76	46.37	355.53
PART M HEATING WORK	12.17	152.13	507.57	98.95	758.65
PART N ALTERATION WORK	45.00	427.50	65.00	73.88	566.38
Final total	453.97	4,415.93	6,619.27	1,670.67	15,575.35

	Ref	Qty	Hours	Hours £	Mat'ls £	O & P £	Total £
PART A PRELIMINARIES							
Concrete mixer	A1	12 wks					540.00
Small tools	A2	12 wks					420.00
Scaffolding (m2/weeks)	A3	720.00					1,620.00
Skip	A4	8 wks					800.00
Clean up	A5	12 hrs					96.00
Carried to summary							3,476.00
PART B SUBSTRUCTURE TO DPC LEVEL							
Excavate topsoil 150mm thick by hand	B1	19.85m2	5.96	56.62	0.00	8.49	65.11
Excavate to reduce levels by hand	B2	5.95m3	14.88	141.36	0.00	21.20	162.56
Excavate for trench foundations by hand	B3	1.88m3	4.89	46.46	0.00	6.97	53.42
Earthwork support to sides of trenches	B4	16.64m2	6.66	63.27	19.97	12.49	95.73
Backfilling with excavated material	B5	0.70m3	0.42	3.99	0.00	0.60	4.59
Hardcore 225mm thick	B6	14.58m2	2.92	27.74	74.07	15.27	117.08
Hardcore filling to trench	B7	0.15m3	0.08	0.76	0.76	0.23	1.75
Concrete grade (1:3:6) in foundations	B8	1.54m3	2.08	19.76	104.63	18.66	143.05
Concrete grade (1:2:4) in bed 150mm thick	B9	14.58m2	4.37	41.52	158.94	30.07	230.52
Concrete (1:2:4) in cavity wall filling	B10	5.70m2	1.14	10.83	20.69	4.73	36.25
Carried forward			43.40	412.30	379.06	118.70	910.06

	Ref	Qty	Hours	Hours £	Mat'ls £	O & P £	Total £
Brought forward			43.40	412.30	379.06	118.70	910.06
Damp-proof membrane	B11	15.13m2	0.61	5.80	8.88	2.20	16.88
Reinforcement ref A193 in foundation	B12	5.70m2	0.68	6.46	7.41	2.08	15.95
Steel fabric reinforcement ref A193 in slab	B13	14.58m2	2.19	20.81	18.95	5.96	45.72
Solid blockwork 140mm thick in cavity wall	B14	7.41m2	9.63	91.49	94.70	27.93	214.11
Common bricks 112.5mm thick in cavity wall	B15	5.70m2	9.63	91.49	75.18	25.00	191.66
Facing bricks in 112.5mm thick in skin of cavity wall	B16	1.71m2	3.08	29.26	39.79	10.36	79.41
Form cavity 50mm wide in cavity wall	B17	7.41m2	0.22	2.09	4.45	0.98	7.52
DPC 112mm wide	B18	11.40m	0.57	5.42	9.23	2.20	16.84
DPC 140mm wide	B19	11.40m	0.68	6.46	11.06	2.63	20.15
Bond in block wall	B20	1.30m	0.58	5.51	2.83	1.25	9.59
Bond in half brick wall	B21	0.30m	0.46	4.37	0.61	0.75	5.73
50mm thick insulation board	B22	14.58m2	4.37	41.52	65.32	16.03	122.86
Carried to summary			76.10	722.95	717.47	216.06	1,656.48
PART C EXTERNAL WALLS							
Solid blockwork 140mm thick in cavity wall	C1	46.65m2	60.65	576.18	593.39	175.43	1,345.00
Facing brickwork 112.5mm thick in cavity wall	C2	46.65m2	83.97	797.72	1,075.75	281.02	2,154.48
75mm thick insulation in cavity wall	C3	44.65m2	10.26	97.47	271.50	55.35	424.32
Carried forward			154.88	1,471.36	1,940.64	511.80	3,923.80

	Ref	Qty	Hours	Hours £	Mat'ls £	O & P £	Total £
Brought forward			154.88	1,471.36	1,940.64	511.80	3,923.80
Steel lintel 2400mm long	C4	2nr	0.50	4.75	196.92	30.25	231.92
Steel lintel 1500mm long	C5	6nr	1.20	11.40	644.82	98.43	754.65
Steel lintel 1150mm long	C6	1nr	0.15	1.43	47.18	7.29	55.90
Close cavity wall at jambs	C7	24.52m	1.23	11.69	50.02	9.26	70.96
Close cavity wall at cills	C8	9.31m	0.47	4.47	18.99	3.52	26.97
Close cavity wall at top	C9	11.40m	0.57	5.42	23.26	4.30	32.98
DPC 112mm wide at jambs	C10	24.52m	1.23	11.69	19.86	4.73	36.28
DPC 112mm wide at cills	C11	9.31m	0.47	4.47	7.54	1.80	13.81
Carried to summary			160.55	1,525.23	2,902.05	664.09	5,091.37
PART D FLAT ROOF							
200 x 50mm sawn softwood joists	D1	40.95m	10.24	97.28	126.95	33.63	257.86
200 x 50mm sawn softwood sprocket pieces	D2	12nr	1.68	15.96	16.08	4.81	36.85
18mm thick WPB grade decking	D3	19.85m2	17.86	160.74	156.02	47.51	364.27
50 x 50mm (avg) wide sawn softwood firrings	D4	40.95m	7.37	70.02	92.95	24.44	187.41
High density polyethylene vapour barrier 150mm thick	D5	18.00m2	3.60	34.20	170.46	30.70	235.36
100 x 75mm sawn softwood wall plate	D6	12.00m	3.69	35.06	26.64	9.25	70.95
100 x 75mm sawn softwood tilt fillet	D7	6.30m	2.21	21.00	16.63	5.64	43.27
Build in ends of 200 x 50mm joists	D8	13nr	4.53	43.04	0.95	6.60	50.58
Carried forward			51.18	477.28	606.68	162.59	1,246.55

	Ref	Qty	Hours	Hours £	Mat'ls £	O & P £	Total £
Brought forward			51.18	477.28	606.68	162.59	1,246.55
Rake out joint for flashing	D9	6.30m	2.21	21.00	0.95	3.29	25.24
6mm thick soffit 150mm wide	D10	12.30m	6.15	58.43	21.53	11.99	91.95
19mm wrought softwood fascia 200mm high	D11	12.30m	6.15	58.43	24.23	12.40	95.05
Three layer fibre-based roofing felt	D12	19.85m2	10.92	103.74	166.93	40.60	311.27
Felt turn-down 100mm girth	D13	12.30m	1.23	11.69	10.33	3.30	25.32
Felt flashing 150mm girth	D14	6.30m	0.76	7.22	10.58	2.67	20.47
112mm diameter PVC-U gutter	D15	6.30m	1.64	15.58	28.16	6.56	50.30
Stop end	D16	1nr	0.14	1.33	1.11	0.37	2.81
Stop end outlet	D17	1nr	0.25	2.38	1.11	0.52	4.01
68mm diameter PVC-U down pipe	D18	4.80m	1.20	11.40	16.27	4.15	31.82
Shoe	D19	1nr	0.30	2.85	2.09	0.74	5.68
Paint fascia and soffit	D20	3.69m2	2.58	24.51	3.54	4.21	32.26
Carried to summary			84.71	795.82	893.51	253.40	1,942.72
PART E PITCHED ROOF			N/A	N/A	N/A	N/A	N/A
PART F WINDOWS AND EXTERNAL DOORS							
PVC-U door size 840 x 1980mm complete (B)	F1	2nr	5.00	47.50	419.98	70.12	537.60
PVC-U sliding patio door size 1700 x 2075mm (C)	F2	2nr	14.00	133.00	587.04	108.01	828.05
Carried forward			19.00	180.50	1,007.02	178.13	1,365.65

	Ref	Qty	Hours	Hours £	Mat'ls £	O & P £	Total £
Brought forward			19.00	180.50	1,007.02	178.13	1,365.65
PVC-U window size 1200 x 1200mm complete (A)	F3	6nr	12.00	114.00	945.54	158.93	1,218.47
25 x 225mm wrought softwood window board	F4	7.20m	2.16	20.52	44.79	9.80	75.11
Paint window board	F5	7.20m	1.22	11.59	6.91	2.78	21.28
Carried to summary			34.38	326.61	2,004.26	349.63	2,680.50
PART G INTERNAL PARTITIONS AND DOORS		N/A	N/A	N/A	N/A	N/A	N/A
PART H WALL FINISHES							
19 x 100mm wrought softwood skirting	H1	19.56m	3.33	31.64	26.80	8.77	67.20
12mm plasterboard fixed to walls with dabs	H2	40.00m2	15.60	148.20	79.60	34.17	261.97
12mm plasterboard fixed to walls less than 300mm wide with dabs	H3	13.83m	6.09	57.86	57.85	17.36	133.06
Two coats emulsion paint to walls	H4	25.07m2	9.01	85.60	26.59	16.83	129.01
Paint skirting	H5	19.56m	2.54	24.13	9.39	5.03	38.55
Carried to summary			36.57	347.42	200.23	82.15	629.79
PART J FLOOR FINISHES							
Cement and sand floor screed 40mm thick	J1	14.58m2	3.65	34.68	44.84	11.93	91.44
Vinyl floor tiles, size 300 x 300mm	J2	14.58m2	3.50	33.25	101.77	20.25	155.27
Carried forward			7.15	67.93	146.61	32.18	246.72

	Ref	Qty	Hours	Hours £	Mat'ls £	O & P £	Total £
Brought forward			7.15	67.93	146.61	32.18	246.72
25mm thick tongued and grooved boarding	J3	14.58m2	10.79	102.51	137.05	35.93	275.49
150 x 50mm sawn softwood joists	J4	29.00m	4.06	38.57	66.12	15.70	120.39
Cut and pin ends of joists to existing brick wall	J5	10nr	1.80	17.10	2.50	2.94	22.54
Build in ends of joists to blockwork	J6	10nr	1.00	9.50	1.50	1.65	12.65
Carried to summary			24.80	235.60	353.78	88.41	677.79
PART K CEILING FINISHES							
Plasterboard with taped butt joints fixed to joists	K1	29.18m2	10.51	99.85	53.98	23.07	176.90
5mm skim coat to plasterboard ceilings	K2	29.18m2	14.59	138.61	35.60	26.13	200.34
Two coats emulsion paint to ceilings	K3	29.18m2	7.59	72.11	17.22	13.40	102.72
Carried to summary			32.69	310.56	106.80	62.60	479.96
PART L ELECTRICAL WORK							
13 amp double switched socket outlet with neon	L1	8nr	3.20	44.80	66.08	16.63	127.51
Lighting point	L2	4nr	1.40	19.60	27.20	7.02	53.82
Lighting switch	L3	5nr	1.25	17.50	19.85	5.60	42.95
Lighting wiring	L4	16.00m	1.60	22.40	16.48	5.83	44.71
Power cable	L5	40.00m	6.00	84.00	44.00	19.20	147.20
Carried to summary			13.45	188.30	173.61	54.29	416.20

	Ref	Qty	Hours	Hours £	Mat'ls £	O & P £	Total £
PART M HEATING WORK							
15mm copper pipe	M1	18.00m	3.96	49.50	33.84	12.50	95.84
Elbow	M2	8nr	2.24	28.00	8.80	5.52	42.32
Tee	M3	2nr	0.68	8.50	3.90	1.86	14.26
Radiator, double convector size 1400 x 520mm	M4	4nr	5.20	65.00	462.72	79.16	606.88
Break into existing pipe and insert tee	M5	1nr	0.75	9.38	3.95	2.00	15.32
Carried to summary			12.83	160.38	513.21	101.04	774.62
PART N ALTERATION WORK							
Take out existing window size 1500 x 1000mm and lintel over, adapt opening to receive 1770 x 2000mm patio door and insert new lintel over (both measured separately) and make good	N1	1nr	20.00	190.00	20.00	31.50	241.50
Take out existing window size 1500 x 1000mm, enlarge opening to receive new PVC-U door (measured separately)	N2	1nr	25.00	237.50	45.00	42.38	324.88
Carried to summary			45.00	427.50	65.00	73.88	566.38

SUMMARY

	Hours	Hours £	Mat'ls £	O & P £	Total £
PART A PRELIMINARIES	0.00	0.00	0.00	0.00	3,476.00
PART B SUBSTRUCTURE TO DPC LEVEL	76.10	722.95	717.47	216.06	1,656.48
PART C EXTERNAL WALLS	160.55	1,525.23	2,902.05	664.09	5,091.37
PART D FLAT ROOF	84.71	795.82	893.51	253.40	1,942.72
PART E PITCHED ROOF	0.00	0.00	0.00	0.00	0.00
PART F WINDOWS AND EXTERNAL DOORS	34.38	326.61	2,044.26	349.63	2,680.49
PART G INTERNAL PARTITIONS AND DOORS	0.00	0.00	0.00	0.00	0.00
PART H WALL FINISHES	36.57	347.42	200.23	82.15	629.79
PART J FLOOR FINISHES	24.80	235.60	353.78	88.41	677.79
PART K CEILING FINISHES	32.69	310.56	106.80	62.60	479.96
PART L ELECTRICAL WORK	13.45	188.30	173.61	54.29	416.20
PART M HEATING WORK	12.83	160.38	513.21	101.04	774.62
PART N ALTERATION WORK	45.00	427.50	65.00	73.88	566.38
Final total	521.08	5,040.37	7,969.92	1,945.55	18,391.80

	Ref	Qty	Hours	Hours £	Mat'ls £	O & P £	Total £
PART A PRELIMINARIES							
Concrete mixer	A1	11 wks					495.00
Small tools	A2	11 wks					385.00
Scaffolding (m2/weeks)	A3	720.00					1,620.00
Skip	A4	7 wks					700.00
Clean up	A5	12 hrs					96.00
Carried to summary							3,296.00
PART B SUBSTRUCTURE TO DPC LEVEL							
Excavate topsoil 150mm thick by hand	B1	17.85m2	5.36	50.92	0.00	7.64	58.56
Excavate to reduce levels by hand	B2	5.35m3	13.38	127.11	0.00	19.07	146.18
Excavate for trench foundations by hand	B3	1.88m3	4.89	46.46	0.00	6.97	53.42
Earthwork support to sides of trenches	B4	16.64m2	6.66	63.27	19.97	12.49	95.73
Backfilling with excavated material	B5	0.62m3	0.37	3.52	0.00	0.53	4.04
Hardcore 225mm thick	B6	13.69m2	2.74	26.03	69.54	14.34	109.91
Hardcore filling to trench	B7	0.14m3	0.07	0.67	0.70	0.20	1.57
Concrete grade (1:3:6) in foundations	B8	1.54m3	2.08	19.76	104.63	18.66	143.05
Concrete grade (1:2:4) in bed 150mm thick	B9	14.58m2	4.11	39.05	148.94	28.20	216.18
Concrete (1:2:4) in cavity wall filling	B10	5.70m2	1.14	10.83	20.69	4.73	36.25
Carried forward			40.80	387.60	364.47	112.81	864.88

	Ref	Qty	Hours	Hours £	Mat'ls £	O & P £	Total £
Brought forward			40.80	387.60	364.47	112.81	864.88
Damp-proof membrane	B11	13.13m2	0.53	5.04	7.62	1.90	14.55
Reinforcement ref A193 in foundation	B12	5.70m2	0.68	6.46	7.41	2.08	15.95
Steel fabric reinforcement ref A193 in slab	B13	13.69m2	2.05	19.48	17.80	5.59	42.87
Solid blockwork 140mm thick in cavity wall	B14	7.41m2	9.63	91.49	94.70	27.93	214.11
Common bricks 112.5mm thick in cavity wall	B15	5.70m2	9.63	91.49	75.18	25.00	191.66
Facing bricks in 112.5mm thick in skin of cavity wall	B16	1.71m2	2.27	21.57	39.79	9.20	70.56
Form cavity 50mm wide in cavity wall	B17	7.41m2	0.22	2.09	4.45	0.98	7.52
DPC 112mm wide	B18	11.40m	0.57	5.42	9.23	2.20	16.84
DPC 140mm wide	B19	11.40m	0.68	6.46	11.06	2.63	20.15
Bond in block wall	B20	1.30m	0.58	5.51	2.83	1.25	9.59
Bond in half brick wall	B21	0.30m	0.46	4.37	0.61	0.75	5.73
50mm thick insulation board	B22	13.69m2	4.11	39.05	61.33	15.06	115.43
Carried to summary			72.21	686.00	696.48	207.37	1,589.85
PART C EXTERNAL WALLS							
Solid blockwork 140mm thick in cavity wall	C1	44.65m2	58.05	551.48	567.95	167.91	1,287.34
Facing brickwork 112.5mm thick in cavity wall	C2	44.65m2	80.37	763.52	1,029.63	268.97	2,062.12
75mm thick insulation in cavity wall	C3	44.65m2	10.26	97.47	271.50	55.35	424.32
Carried forward			148.68	1,412.46	1,869.08	492.23	3,773.77

	Ref	Qty	Hours	Hours £	Mat'ls £	O & P £	Total £
Brought forward			148.68	1,412.46	1,869.08	492.23	3,773.77
Steel lintel 2400mm long	C4	2nr	0.50	4.75	196.92	30.25	231.92
Steel lintel 1500mm long	C5	6nr	1.20	11.40	644.82	98.43	754.65
Steel lintel 1150mm long	C6	1nr	0.15	1.43	47.18	7.29	55.90
Close cavity wall at jambs	C7	24.52m	1.23	11.69	50.02	9.26	70.96
Close cavity wall at cills	C8	9.31m	0.47	4.47	18.99	3.52	26.97
Close cavity wall at top	C9	11.40m	0.57	5.42	23.26	4.30	32.98
DPC 112mm wide at jambs	C10	24.52m	1.23	11.69	19.86	4.73	36.28
DPC 112mm wide at cills	C11	9.31m	0.47	4.47	7.54	1.80	13.81
Carried to summary			154.35	1,466.33	2,830.49	644.52	4,941.34
PART D FLAT ROOF							
200 x 50mm sawn softwood joists	D1	37.35m	9.34	88.73	115.79	30.68	235.20
200 x 50mm sawn softwood sprocket pieces	D2	16nr	2.24	21.28	21.44	6.41	49.13
18mm thick WPB grade decking	D3	17.85m2	16.07	144.63	140.30	42.74	327.67
50 x 50mm (avg) wide sawn softwood firrings	D4	37.35m	6.72	63.84	84.41	22.24	170.49
High density polyethylene vapour barrier 150mm thick	D5	16.00m2	3.20	30.40	151.52	27.29	209.21
100 x 75mm sawn softwood wall plate	D6	12.00m	3.60	34.20	26.64	9.13	69.97
100 x 75mm sawn softwood tilt fillet	D7	4.30m	.083.15	0.00	11.33	1.70	13.03
Build in ends of 200 x 50mm joists	D8	9nr	4.53	43.04	1.35	6.66	51.04
Carried forward			45.70	426.12	552.78	146.83	1,125.73

	Ref	Qty	Hours	Hours £	Mat'ls £	O & P £	Total £
Brought forward			45.70	426.12	552.78	146.83	1,125.73
Rake out joint for flashing	D9	4.30m	1.51	14.35	0.65	2.25	17.24
6mm thick soffit 150mm wide	D10	12.30m	4.92	46.74	21.53	10.24	78.51
19mm wrought softwood fascia 200mm high	D11	12.30m	6.15	58.43	24.23	12.40	95.05
Three layer fibre-based roofing felt	D12	17.85m2	9.82	93.29	150.12	36.51	279.92
Felt turn-down 100mm girth	D13	12.30m	1.23	11.69	10.33	3.30	25.32
Felt flashing 150mm girth	D14	4.30m	0.52	4.94	7.22	1.82	13.98
112mm diameter PVC-U gutter	D15	4.30m	1.12	10.64	19.22	4.48	34.34
Stop end	D16	1nr	0.14	1.33	1.11	0.37	2.81
Stop end outlet	D17	1nr	0.25	2.38	1.11	0.52	4.01
68mm diameter PVC-U down pipe	D18	4.80m	1.20	11.40	16.27	4.15	31.82
Shoe	D19	1nr	0.30	2.85	2.09	0.74	5.68
Paint fascia and soffit	D20	3.69m2	2.58	24.51	3.54	4.21	32.26
Carried to summary			75.44	708.65	810.20	227.83	1,746.67
PART E PITCHED ROOF			N/A	N/A	N/A	N/A	N/A
PART F WINDOWS AND EXTERNAL DOORS							
PVC-U door size 840 x 1980mm complete (B)	F1	2nr	5.00	47.50	419.98	70.12	537.60
PVC-U sliding patio door size 1700 x 2075mm (C)	F2	2nr	14.00	133.00	587.04	108.01	828.05
Carried forward			19.00	180.50	1,007.02	178.13	1,365.65

	Ref	Qty	Hours	Hours £	Mat'ls £	O & P £	Total £
Brought forward			19.00	180.50	1,007.02	178.13	1,365.65
PVC-U window size 1200 x 1200mm complete (A)	F3	6nr	12.00	114.00	945.54	158.93	1,218.47
25 x 225mm wrought softwood window board	F4	7.20m	2.16	20.52	44.78	9.80	75.10
Paint window board	F5	7.20m	1.22	11.59	6.91	2.78	21.28
Carried to summary			34.38	326.61	2,004.25	349.63	2,680.49
PART G INTERNAL PARTITIONS AND DOORS		N/A	N/A	N/A	N/A	N/A	N/A
PART H WALL FINISHES							
19 x 100mm wrought softwood skirting	H1	19.56m	3.33	31.64	26.80	8.77	67.20
12mm plasterboard fixed to walls with dabs	H2	40.00m2	15.60	148.20	79.60	34.17	261.97
12mm plasterboard fixed to walls less than 300mm wide with dabs	H3	33.83m	6.09	57.86	57.85	17.36	133.06
Two coats emulsion paint to walls	H4	45.07m2	9.01	85.60	26.59	16.83	129.01
Paint skirting	H5	19.56m	2.54	24.13	9.39	5.03	38.55
Carried forward			36.57	347.42	200.23	82.15	629.79
PART J FLOOR FINISHES							
Cement and sand floor screed 40mm thick	J1	12.58m2	3.15	29.93	38.12	10.21	78.25
Vinyl floor tiles, size 300 x 300mm	J2	12.58m2	3.02	28.69	87.81	17.48	133.98
Carried forward			6.17	58.62	125.93	27.68	212.23

	Ref	Qty	Hours	Hours £	Mat'ls £	O & P £	Total £
Brought forward			6.17	58.62	125.93	27.68	212.23
25mm thick tongued and grooved boarding	J3	12.58m2	9.31	88.45	118.25	31.00	237.70
150 x 50mm sawn softwood joists	J4	27.30m	3.82	36.29	62.24	14.78	113.31
Cut and pin ends of joists to existing brick wall	J5	7nr	1.26	11.97	1.75	2.06	15.78
Build in ends of joists to blockwork	J6	7nr	0.70	6.65	1.05	1.16	8.86
Carried to summary			21.26	201.97	309.22	76.68	587.87
PART K CEILING FINISHES							
Plasterboard with taped butt joints fixed to joists	K1	25.16m2	9.06	86.07	46.55	19.89	152.51
5mm skim coat to plasterboard ceilings	K2	25.16m2	12.58	119.51	30.70	22.53	172.74
Two coats emulsion paint to ceilings	K3	25.16m2	6.54	62.13	14.84	11.55	88.52
Carried to summary			28.18	267.71	92.09	53.97	413.77
PART L ELECTRICAL WORK							
13 amp double switched socket outlet with neon	L1	6nr	2.40	33.60	49.36	12.44	95.40
Lighting point	L2	4nr	1.40	19.60	27.20	7.02	53.82
Lighting switch	L3	5nr	1.25	17.50	19.85	5.60	42.95
Lighting wiring	L4	16.00m	1.60	22.40	16.48	5.83	44.71
Power cable	L5	36.00m	5.40	75.60	39.60	17.28	132.48
Carried to summary			12.05	168.70	152.49	48.18	369.37

	Ref	Qty	Hours	Hours £	Mat'ls £	O & P £	Total £
PART M HEATING WORK							
15mm copper pipe	M1	19.00m	4.18	52.25	35.72	13.20	101.17
Elbow	M2	8nr	2.24	28.00	8.80	5.52	42.32
Tee	M3	2nr	0.68	8.50	3.90	1.86	14.26
Radiator, double convector size 1400 x 520mm	M4	4nr	5.20	65.00	462.72	79.16	606.88
Break into existing pipe and insert tee	M5	1nr	0.75	9.38	3.95	2.00	15.32
Carried to summary			13.05	163.13	515.09	101.73	779.95
PART N ALTERATION WORK							
Take out existing window size 1500 x 1000mm and lintel over, adapt opening to receive 1770 x 2000mm patio door and insert new lintel over (both measured separately) and make good	N1	1nr	20.00	190.00	20.00	31.50	241.50
Take out existing window size 1500 x 1000mm, enlarge opening to receive new PVC-U door (measured separately)	N2	1nr	25.00	237.50	45.00	42.38	324.88
Carried to summary			45.00	427.50	65.00	73.88	566.38

SUMMARY

	Hours	Hours £	Mat'ls £	O & P £	Total £
PART A PRELIMINARIES	0.00	0.00	0.00	0.00	3,296.00
PART B SUBSTRUCTURE TO DPC LEVEL	72.21	686.00	696.48	207.37	1,589.85
PART C EXTERNAL WALLS	154.35	1,466.33	2,830.49	644.52	4,941.34
PART D FLAT ROOF	75.44	708.65	810.20	227.83	1,746.67
PART E PITCHED ROOF	0.00	0.00	0.00	0.00	0.00
PART F WINDOWS AND EXTERNAL DOORS	34.88	326.61	2,044.25	349.63	2,680.49
PART G INTERNAL PARTITIONS AND DOORS	0.00	0.00	0.00	0.00	0.00
PART H WALL FINISHES	36.57	347.42	200.23	82.15	629.79
PART J FLOOR FINISHES	21.26	201.97	309.22	76.68	587.87
PART K CEILING FINISHES	28.18	267.71	92.09	53.97	413.77
PART L ELECTRICAL WORK	12.05	168.70	152.49	48.18	369.37
PART M HEATING WORK	13.05	163.13	515.09	101.73	779.95
PART N ALTERATION WORK	45.00	427.50	65.00	73.88	566.38
Final total	492.99	4,764.02	7,715.54	1,865.94	17,601.48

	Ref	Qty	Hours	Hours £	Mat'ls £	O & P £	Total £
PART A PRELIMINARIES							
Concrete mixer	A1	12 wks					540.00
Small tools	A2	12 wks					420.00
Scaffolding (m2/weeks)	A3	780.00					1,755.00
Skip	A4	8 wks					800.00
Clean up	A5	12 hrs					96.00
Carried to summary							3,611.00
PART B SUBSTRUCTURE TO DPC LEVEL							
Excavate topsoil 150mm thick by hand	B1	22.00m2	6.60	62.70	0.00	9.41	72.11
Excavate to reduce levels by hand	B2	6.60m3	16.60	157.70	0.00	23.66	181.36
Excavate for trench foundations by hand	B3	2.05m3	5.33	50.64	0.00	7.60	58.23
Earthwork support to sides of trenches	B4	18.10m2	7.24	68.78	21.72	13.58	104.08
Backfilling with excavated material	B5	0.70m3	0.42	3.99	0.00	0.60	4.59
Hardcore 225mm thick	B6	16.28m2	2.44	23.18	82.70	15.88	121.76
Hardcore filling to trench	B7	0.15m3	0.08	0.76	0.76	0.23	1.75
Concrete grade (1:3:6) in foundations	B8	1.67m3	2.25	21.38	113.46	20.23	155.06
Concrete grade (1:2:4) in bed 150mm thick	B9	16.28m2	4.88	46.36	177.13	33.52	257.01
Concrete (1:2:4) in cavity wall filling	B10	6.20m2	1.24	11.78	22.51	5.14	39.43
Carried forward			47.08	447.26	418.28	129.83	995.37

	Ref	Qty	Hours	Hours £	Mat'ls £	O & P £	Total £
Brought forward			47.08	447.26	418.28	129.83	995.37
Damp-proof membrane	B11	16.88m2	0.68	6.46	9.79	2.44	18.69
Reinforcement ref A193 in foundation	B12	6.20m2	0.74	7.03	8.06	2.26	17.35
Steel fabric reinforcement ref A193 in slab	B13	16.28m2	2.44	23.18	21.16	6.65	50.99
Solid blockwork 140mm thick in cavity wall	B14	8.06m2	10.48	99.56	103.00	30.38	232.94
Common bricks 112.5mm thick in cavity wall	B15	6.20m2	10.48	99.56	81.78	27.20	208.54
Facing bricks in 112.5mm thick in skin of cavity wall	B16	1.86m2	3.35	31.83	43.28	11.27	86.37
Form cavity 50mm wide in cavity wall	B17	8.06m2	0.24	2.28	4.84	1.07	8.19
DPC 112mm wide	B18	12.40m	0.62	5.89	10.04	2.39	18.32
DPC 140mm wide	B19	12.40m	0.74	7.03	12.03	2.86	21.92
Bond in block wall	B20	1.30m	0.58	5.51	2.83	1.25	9.59
Bond in half brick wall	B21	0.30m	0.46	4.37	0.61	0.75	5.73
50mm thick insulation board	B22	16.28m2	4.88	46.36	72.93	17.89	137.18
Carried to summary			82.77	786.32	788.63	236.24	1,811.19
PART C EXTERNAL WALLS							
Solid blockwork 140mm thick in cavity wall	C1	46.82m2	60.87	578.27	595.55	176.07	1,349.89
Facing brickwork 112.5mm thick in cavity wall	C2	46.82m2	84.28	800.66	1,079.67	282.05	2,162.38
75mm thick insulation in cavity wall	C3	46.82m2	10.30	97.85	272.49	55.55	425.89
Carried forward			155.45	1,476.78	1,947.71	513.67	3,938.16

	Ref	Qty	Hours	Hours £	Mat'ls £	O & P £	Total £
Brought forward			155.45	1,476.78	1,947.71	513.67	3,938.16
Steel lintel 2400mm long	C4	2nr	0.50	4.75	196.92	30.25	231.92
Steel lintel 1500mm long	C5	8nr	1.60	15.20	859.76	131.24	1,006.20
Steel lintel 1150mm long	C6	1nr	0.15	1.43	47.18	7.29	55.90
Close cavity wall at jambs	C7	29.32m	1.47	13.97	59.81	11.07	84.84
Close cavity wall at cills	C8	11.71m	0.59	5.61	23.89	4.42	33.92
Close cavity wall at top	C9	12.40m	0.62	5.89	25.30	4.68	35.87
DPC 112mm wide at jambs	C10	29.32m	1.47	13.97	23.75	5.66	43.37
DPC 112mm wide at cills	C11	11.71m	0.59	5.61	9.48	2.26	17.35
Carried to summary			162.29	1,541.76	3,146.62	703.26	5,391.63
PART D FLAT ROOF							
200 x 50mm sawn softwood joists	D1	45.65m	11.41	108.40	141.52	37.49	287.40
200 x 50mm sawn softwood sprocket pieces	D2	16nr	2.24	21.28	21.44	6.41	49.13
18mm thick WPB grade decking	D3	22.00m2	19.18	172.62	172.92	51.83	397.37
50 x 50mm (avg) wide sawn softwood firrings	D4	49.65m	8.22	78.09	103.17	27.19	208.45
High density polyethylene vapour barrier 150mm thick	D5	20.00m2	4.00	38.00	189.40	34.11	261.51
100 x 75mm sawn softwood wall plate	D6	13.00m	3.90	37.05	28.86	9.89	75.80
100 x 75mm sawn softwood tilt fillet	D7	5.30m	1.33	12.64	13.99	3.99	30.62
Build in ends of 200 x 50mm joists	D8	11nr	3.85	36.58	1.65	5.73	43.96
Carried forward			54.13	504.65	672.95	176.64	1,354.23

	Ref	Qty	Hours	Hours £	Mat'ls £	O & P £	Total £
Brought forward			54.13	504.65	672.95	176.64	1,354.23
Rake out joint for flashing	D9	5.30m	1.86	17.67	0.80	2.77	21.24
6mm thick soffit 150mm wide	D10	13.30m	5.32	50.54	23.28	11.07	84.89
19mm wrought softwood fascia 200mm high	D11	13.30m	6.65	63.18	26.20	13.41	102.78
Three layer fibre-based roofing felt	D12	22.00m2	12.10	114.95	185.02	45.00	344.97
Felt turn-down 100mm girth	D13	13.30m	1.33	12.64	11.17	3.57	27.38
Felt flashing 150mm girth	D14	5.30m	0.64	6.08	8.90	2.25	17.23
112mm diameter PVC-U gutter	D15	5.30m	1.38	13.11	23.69	5.52	42.32
Stop end	D16	1nr	0.14	1.33	1.11	0.37	2.81
Stop end outlet	D17	1nr	0.25	2.38	1.11	0.52	4.01
68mm diameter PVC-U down pipe	D18	4.80m	1.20	11.40	16.27	4.15	31.82
Shoe	D19	1nr	0.30	2.85	2.09	0.74	5.68
Paint fascia and soffit	D20	3.99m2	2.79	26.51	3.83	4.55	34.89
Carried to summary			88.09	827.27	976.42	270.55	2,074.24
PART E PITCHED ROOF			N/A	N/A	N/A	N/A	N/A
PART F WINDOWS AND EXTERNAL DOORS							
PVC-U door size 840 x 1980mm complete (B)	F1	2nr	5.00	47.50	419.98	70.12	537.60
PVC-U sliding patio door size 1700 x 2075mm (C)	F2	2nr	14.00	133.00	587.04	108.01	828.05
Carried forward			19.00	180.50	1,007.02	178.13	1,365.65

	Ref	Qty	Hours	Hours £	Mat'ls £	O & P £	Total £
Brought forward			19.00	180.50	1,007.02	178.13	1,365.65
PVC-U window size 1200 x 1200mm complete (A)	F3	8nr	16.00	152.00	1,260.72	211.91	1,624.63
25 x 225mm wrought softwood window board	F4	9.60m	2.88	27.36	59.71	13.06	100.13
Paint window board	F5	9.60m	1.52	14.44	9.22	3.55	27.21
Carried to summary			39.40	374.30	2,336.67	406.65	3,117.62
PART G INTERNAL PARTITIONS AND DOORS		N/A	N/A	N/A	N/A	N/A	N/A
PART H WALL FINISHES							
19 x 100mm wrought softwood skirting	H1	21.50m	3.66	34.77	29.46	9.63	73.86
12mm plasterboard fixed to walls with dabs	H2	42.02m2	16.39	155.71	83.62	35.90	275.22
12mm plasterboard fixed to walls less than 300mm wide with dabs	H3	41.03m	7.38	70.11	70.16	21.04	161.31
Two coats emulsion paint to walls	H4	48.17m2	9.63	91.49	28.42	17.99	137.89
Paint skirting	H5	21.50m	2.80	26.60	10.32	5.54	42.46
Carried forward			39.86	378.67	221.98	90.10	690.75
PART J FLOOR FINISHES							
Cement and sand floor screed 40mm thick	J1	16.28m2	4.07	38.67	49.33	13.20	101.19
Vinyl floor tiles, size 300 x 300mm	J2	16.28m2	3.90	37.05	113.63	22.60	173.28
Carried forward			7.97	75.72	162.96	35.80	274.48

	Ref	Qty	Hours	Hours £	Mat'ls £	O & P £	Total £
Brought forward			7.97	75.72	162.96	35.80	274.48
25mm thick tongued and grooved boarding	J3	16.28m2	12.04	114.38	153.03	40.11	307.52
150 x 50mm sawn softwood joists	J4	31.20m	4.37	41.52	71.14	16.90	129.55
Cut and pin ends of joists to existing brick wall	J5	8nr	1.44	13.68	2.04	2.36	18.08
Build in ends of joists to blockwork	J6	8nr	8.00	76.00	1.20	11.58	88.78
Carried to summary			33.82	321.29	390.37	106.75	818.41

PART K
CEILING FINISHES

	Ref	Qty	Hours	Hours £	Mat'ls £	O & P £	Total £
Plasterboard with taped butt joints fixed to joists	K1	33.56m2	12.08	114.76	62.09	26.53	203.38
5mm skim coat to plasterboard ceilings	K2	33.56m2	16.78	159.41	40.96	30.06	230.43
Two coats emulsion paint to ceilings	K3	33.56m2	8.73	82.94	19.80	15.41	118.15
Carried to summary			37.59	357.11	122.85	71.99	551.95

PART L
ELECTRICAL WORK

	Ref	Qty	Hours	Hours £	Mat'ls £	O & P £	Total £
13 amp double switched socket outlet with neon	L1	8nr	3.20	44.80	66.08	16.63	127.51
Lighting point	L2	6nr	2.10	29.40	40.80	10.53	80.73
Lighting switch	L3	5nr	1.25	17.50	19.85	5.60	42.95
Lighting wiring	L4	18.00m	1.80	25.20	18.54	6.56	50.30
Power cable	L5	40.00m	6.00	84.00	44.00	19.20	147.20
Carried to summary			14.35	200.90	189.27	58.53	448.70

	Ref	Qty	Hours	Hours £	Mat'ls £	O & P £	Total £
PART M HEATING WORK							
15mm copper pipe	M1	21.00m	4.62	57.75	39.48	14.58	111.81
Elbow	M2	8nr	2.24	28.00	8.80	5.52	42.32
Tee	M3	2nr	0.68	8.50	3.90	1.86	14.26
Radiator, double convector size 1400 x 520mm	M4	6nr	7.80	97.50	674.08	115.74	887.32
Break into existing pipe and insert tee	M5	1nr	0.75	9.38	3.95	2.00	15.32
Carried to summary			16.09	201.13	730.21	139.70	1,071.04
PART N ALTERATION WORK							
Take out existing window size 1500 x 1000mm and lintel over, adapt opening to receive 1770 x 2000mm patio door and insert new lintel over (both measured separately) and make good	N1	1nr	20.00	190.00	20.00	31.50	241.50
Take out existing window size 1500 x 1000mm, enlarge opening to receive new PVC-U door (measured separately)	N2	1nr	25.00	237.50	45.00	42.38	324.88
Carried to summary			45.00	427.50	65.00	73.88	566.38

SUMMARY

	Hours	Hours £	Mat'ls £	O & P £	Total £
PART A PRELIMINARIES	0.00	0.00	0.00	0.00	3,611.00
PART B SUBSTRUCTURE TO DPC LEVEL	82.77	786.32	788.63	236.24	1,811.19
PART C EXTERNAL WALLS	162.29	1,541.76	3,146.62	703.26	5,391.63
PART D FLAT ROOF	88.09	827.27	976.42	270.55	2,074.24
PART E PITCHED ROOF	0.00	0.00	0.00	0.00	0.00
PART F WINDOWS AND EXTERNAL DOORS	39.40	374.30	2,336.67	406.65	3,117.62
PART G INTERNAL PARTITIONS AND DOORS	0.00	0.00	0.00	0.00	0.00
PART H WALL FINISHES	39.86	378.67	221.98	90.10	690.75
PART J FLOOR FINISHES	33.82	321.29	390.37	106.75	818.41
PART K CEILING FINISHES	37.59	357.11	122.85	71.99	551.95
PART L ELECTRICAL WORK	14.35	200.90	189.27	58.53	448.70
PART M HEATING WORK	16.09	201.13	730.21	139.70	1,071.04
PART N ALTERATION WORK	45.00	427.50	65.00	73.88	566.38
Final total	559.26	5,416.25	8,968.02	2,157.65	20,152.91

	Ref	Qty	Hours	Hours £	Mat'ls £	O & P £	Total £
PART A PRELIMINARIES							
Concrete mixer	A1	12 wks					540.00
Small tools	A2	12 wks					420.00
Scaffolding (m2/weeks)	A3	780.00					1,755.00
Skip	A4	9 wks					900.00
Clean up	A5	12 hrs					96.00
Carried to summary							3,711.00
PART B SUBSTRUCTURE TO DPC LEVEL							
Excavate topsoil 150mm thick by hand	B1	26.15m2	7.85	74.58	0.00	11.19	85.76
Excavate to reduce levels by hand	B2	7.85m3	19.63	186.49	0.00	27.97	214.46
Excavate for trench foundations by hand	B3	2.21m3	5.75	54.63	0.00	8.19	62.82
Earthwork support to sides of trenches	B4	19.56m2	7.81	74.20	23.47	14.65	112.31
Backfilling with excavated material	B5	0.77m3	0.31	2.95	0.00	0.44	3.39
Hardcore 225mm thick	B6	19.98m2	3.00	28.50	101.50	19.50	149.50
Hardcore filling to trench	B7	0.17m3	0.09	0.86	0.86	0.26	1.97
Concrete grade (1:3:6) in foundations	B8	1.81m3	2.44	23.18	122.97	21.92	168.07
Concrete grade (1:2:4) in bed 150mm thick	B9	19.98m2	5.99	56.91	211.94	40.33	309.17
Concrete (1:2:4) in cavity wall filling	B10	6.70m2	1.34	12.73	24.32	5.56	42.61
Carried forward			54.21	515.00	485.06	150.01	1,150.06

	Ref	Qty	Hours	Hours £	Mat'ls £	O & P £	Total £
Brought forward			54.21	515.00	485.06	150.01	1,150.06
Damp-proof membrane	B11	20.63m2	0.83	7.89	11.97	2.98	22.83
Reinforcement ref A193 in foundation	B12	6.70m2	0.80	7.60	8.71	2.45	18.76
Steel fabric reinforcement ref A193 in slab	B13	19.98m2	3.00	28.50	25.97	8.17	62.64
Solid blockwork 140mm thick in cavity wall	B14	8.71m2	11.32	107.54	111.31	32.83	251.68
Common bricks 112.5mm thick in cavity wall	B15	6.70m2	11.32	107.54	88.37	29.39	225.30
Facing bricks in 112.5mm thick in skin of cavity wall	B16	2.01m2	3.62	34.39	46.77	12.17	93.33
Form cavity 50mm wide in cavity wall	B17	8.71m2	0.26	2.47	5.23	1.16	8.86
DPC 112mm wide	B18	13.40m	0.67	6.37	10.85	2.58	19.80
DPC 140mm wide	B19	13.40m	0.80	7.60	13.00	3.09	23.69
Bond in block wall	B20	1.30m	0.58	5.51	2.83	1.25	9.59
Bond in half brick wall	B21	0.30m	0.46	4.37	0.61	0.75	5.73
50mm thick insulation board	B22	19.98m2	5.99	56.91	89.51	21.96	168.38
Carried to summary			93.86	891.67	900.19	268.78	2,060.64
PART C EXTERNAL WALLS							
Solid blockwork 140mm thick in cavity wall	C1	51.87m2	60.87	578.27	659.79	185.71	1,423.76
Facing brickwork 112.5mm thick in cavity wall	C2	51.87m2	84.28	800.66	1,196.12	299.52	2,296.30
75mm thick insulation in cavity wall	C3	51.87m2	11.41	108.40	301.88	61.54	471.82
Carried forward			156.56	1,487.32	2,157.79	546.77	4,191.88

	Ref	Qty	Hours	Hours £	Mat'ls £	O & P £	Total £
Brought forward			156.56	1,487.32	2,157.79	546.77	4,191.88
Steel lintel 2400mm long	C4	2nr	0.50	4.75	196.92	30.25	231.92
Steel lintel 1500mm long	C5	8nr	1.60	15.20	859.76	131.24	1,006.20
Steel lintel 1150mm long	C6	1nr	0.15	1.43	47.18	7.29	55.90
Close cavity wall at jambs	C7	29.32m	1.47	13.97	59.81	11.07	84.84
Close cavity wall at cills	C8	11.71m	0.59	5.61	23.89	4.42	33.92
Close cavity of hollow wall at top	C9	13.40m		0.00	2.00	0.30	2.30
DPC 112mm wide at jambs	C10	29.32m	1.47	13.97	23.75	5.66	43.37
DPC 112mm wide at cills	C11	11.71m	0.59	5.61	9.48	2.26	17.35
Carried to summary			162.78	1,546.41	3,333.40	731.97	5,611.78
PART D FLAT ROOF							
200 x 50mm sawn softwood joists	D1	53.95m	13.49	128.16	167.25	44.31	339.72
200 x 50mm sawn softwood sprocket pieces	D2	16nr	2.24	21.28	21.44	6.41	49.13
18mm thick WPB grade decking	D3	22.00m2	19.18	172.62	172.92	51.83	397.37
50 x 50mm (avg) wide sawn softwood firrings	D4	49.65m	8.22	78.09	103.17	27.19	208.45
High density polyethylene vapour barrier 150mm thick	D5	24.00m2	4.80	45.60	227.28	40.93	313.81
100 x 75mm sawn softwood wall plate	D6	13.00m	3.90	37.05	28.86	9.89	75.80
100 x 75mm sawn softwood tilt fillet	D7	5.30m	1.33	12.64	13.99	3.99	30.62
Build in ends of 200 x 50mm joists	D8	11nr	3.85	36.58	1.65	5.73	43.96
Carried forward			57.01	532.01	736.56	190.28	1,458.85

	Ref	Qty	Hours	Hours £	Mat'ls £	O & P £	Total £
Brought forward			57.01	532.01	736.56	190.28	1,458.85
Rake out joint for flashing	D9	5.30m	1.86	17.67	0.80	2.77	21.24
6mm thick soffit 150mm wide	D10	13.30m	5.32	50.54	23.28	11.07	84.89
19mm wrought softwood fascia 200mm high	D11	13.30m	6.65	63.18	26.20	13.41	102.78
Three layer fibre-based roofing felt	D12	22.00m2	12.10	114.95	185.02	45.00	344.97
Felt turn-down 100mm girth	D13	13.30m	1.33	12.64	11.17	3.57	27.38
Felt flashing 150mm girth	D14	5.30m	0.64	6.08	8.90	2.25	17.23
112mm diameter PVC-U gutter	D15	5.30m	1.38	13.11	23.69	5.52	42.32
Stop end	D16	1nr	0.14	1.33	1.11	0.37	2.81
Stop end outlet	D17	1nr	0.25	2.38	1.11	0.52	4.01
68mm diameter PVC-U down pipe	D18	4.80m	1.20	11.40	16.27	4.15	31.82
Shoe	D19	1nr	0.30	2.85	2.09	0.74	5.68
Paint fascia and soffit	D20	3.99m2	2.79	26.51	3.83	4.55	34.89
Carried to summary			90.97	854.63	1,040.03	284.20	2,178.85
PART E PITCHED ROOF			N/A	N/A	N/A	N/A	N/A
PART F WINDOWS AND EXTERNAL DOORS							
PVC-U door size 840 x 1980mm complete (B)	F1	2nr	5.00	47.50	419.98	70.12	537.60
PVC-U sliding patio door size 1700 x 2075mm (C)	F2	2nr	14.00	133.00	587.04	108.01	828.05
Carried forward			19.00	180.50	1,007.02	178.13	1,365.65

	Ref	Qty	Hours	Hours £	Mat'ls £	O & P £	Total £
Brought forward			19.00	180.50	1,007.02	178.13	1,365.65
PVC-U window size 1200 x 1200mm complete (A)	F3	8nr	16.00	152.00	1,260.72	211.91	1,624.63
25 x 225mm wrought softwood window board	F4	9.60m	2.88	27.36	59.71	13.06	100.13
Paint window board	F5	9.60m	1.52	14.44	9.22	3.55	27.21
Carried to summary			39.40	374.30	2,336.67	406.65	3,117.62
PART G INTERNAL PARTITIONS AND DOORS							
50 x 75mm sawn softwood sole plate	G1	3.70m	0.81	7.70	4.14	1.24	9.52
50 x 75mm sawn softwood head	G2	3.70m	0.81	7.70	4.14	1.24	9.52
50 x 75mm sawn softwood studs	G3	17.15m	4.80	45.60	19.21	5.76	44.18
50 x 75mm sawn softwood noggings	G4	3.70m	1.04	9.88	4.14	1.24	9.52
Plasterboard	G5	14.80m2	5.33	50.64	27.38	8.21	62.97
35mm thick veneered internal door size 762 x 1981mm	G6	1nr	1.25	11.88	55.32	16.60	127.24
38 x 150mm wrought softwood lining	G7	4.87m	1.07	10.17	42.81	12.84	98.46
13 x 38mm wrought softwood stop	G8	4.87m	0.73	6.94	2.39	0.72	5.50
19 x 50mm wrought softwood architrave	G9	9.74m	1.46	13.87	2.24	0.67	5.15
19 x 100mm wrought softwood skirting	G10	5.88m	1.00	9.50	8.06	2.42	18.54
100mm rising steel butts	G11	1pr	0.30	2.85	3.95	1.19	9.09
Carried forward			18.60	176.70	173.78	52.13	399.69

	Ref	Qty	Hours	Hours £	Mat'ls £	O & P £	Total £
Brought forward			18.60	176.70	173.78	52.13	399.69
SAA mortice latch with lever furniture	G12	1nr	0.80	7.60	14.10	4.23	32.43
Two coats emulsion on plasterboard walls	G13	14.80m2	3.85	36.58	8.76	2.63	20.15
Paint general surfaces	G14	4.70m2	3.29	31.26	1.40	0.42	3.22
Carried to summary			26.54	252.13	198.04	59.41	455.49
PART H WALL FINISHES							
19 x 100mm wrought softwood skirting	H1	28.50m	4.91	46.65	35.39	12.31	94.34
12mm plasterboard fixed to walls with dabs	H2	46.22m2	18.03	171.29	92.77	39.61	303.66
12mm plasterboard fixed to walls less than 300mm wide with dabs	H3	41.03m	7.38	70.11	70.16	21.04	161.31
Two coats emulsion paint to walls	H4	48.17m2	9.63	91.49	28.42	17.99	137.89
Paint skirting	H5	21.50m	2.80	26.60	10.32	5.54	42.46
Carried to summary			42.75	406.13	237.06	96.48	739.66
PART J FLOOR FINISHES							
Cement and sand floor screed 40mm thick	J1	16.28m2	4.07	38.67	49.33	13.20	101.19
Vinyl floor tiles, size 300 x 300mm	J2	16.28m2	3.90	37.05	113.63	22.60	173.28
Carried forward			7.97	75.72	162.96	35.80	274.48

	Ref	Qty	Hours	Hours £	Mat'ls £	O & P £	Total £
Brought forward			7.97	75.72	162.96	35.80	274.48
25mm thick tongued and grooved boarding	J3	16.28m2	12.04	114.38	153.03	40.11	307.52
150 x 50mm sawn softwood joists	J4	31.20m	4.37	41.52	71.14	16.90	129.55
Cut and pin ends of joists to existing brick wall	J5	8nr	1.44	13.68	2.04	2.36	18.08
Build in ends of joists to blockwork	J6	8nr	8.00	76.00	1.20	11.58	88.78
Carried to summary			33.82	321.29	390.37	106.75	818.41
PART K CEILING FINISHES							
Plasterboard with taped butt joints fixed to joists	K1	33.56m2	12.08	114.76	62.09	26.53	203.38
5mm skim coat to plasterboard ceilings	K2	33.56m2	16.78	159.41	40.96	30.06	230.43
Two coats emulsion paint to ceilings	K3	33.56m2	8.73	82.94	19.80	15.41	118.15
Carried to summary			37.59	357.11	122.85	71.99	551.95
PART L ELECTRICAL WORK							
13 amp double switched socket outlet with neon	L1	8nr	3.20	44.80	66.08	16.63	127.51
Lighting point	L2	6nr	2.10	29.40	40.80	10.53	80.73
Lighting switch	L3	5nr	1.25	17.50	19.85	5.60	42.95
Lighting wiring	L4	18.00m	1.80	25.20	18.54	6.56	50.30
Power cable	L5	40.00m	6.00	84.00	44.00	19.20	147.20
Carried to summary			14.35	200.90	189.27	58.53	448.70

	Ref	Qty	Hours	Hours £	Mat'ls £	O & P £	Total £
PART M HEATING WORK							
15mm copper pipe	M1	21.00m	4.62	57.75	39.48	14.58	111.81
Elbow	M2	8nr	2.24	28.00	8.80	5.52	42.32
Tee	M3	2nr	0.68	8.50	3.90	1.86	14.26
Radiator, double convector size 1400 x 520mm	M4	6nr	7.80	97.50	674.08	115.74	887.32
Break into existing pipe and insert tee	M5	1nr	0.75	9.38	3.95	2.00	15.32
Carried to summary			16.09	201.13	730.21	139.70	1,071.04
PART N ALTERATION WORK							
Take out existing window size 1500 x 1000mm and lintel over, adapt opening to receive 1770 x 2000mm patio door and insert new lintel over (both measured separately) and make good	N1	1nr	20.00	190.00	20.00	31.50	241.50
Take out existing window size 1500 x 1000mm, enlarge opening to receive new PVC-U door (measured separately)	N2	1nr	25.00	237.50	45.00	42.38	324.88
Carried to summary			45.00	427.50	65.00	73.88	566.38

SUMMARY

	Hours	Hours £	Mat'ls £	O & P £	Total £
PART A PRELIMINARIES	0.00	0.00	0.00	0.00	3,711.00
PART B SUBSTRUCTURE TO DPC LEVEL	93.86	891.67	900.19	268.78	2,060.64
PART C EXTERNAL WALLS	162.78	1,546.41	3,333.40	731.97	5,611.78
PART D FLAT ROOF	90.97	854.63	1,040.03	284.20	2,178.85
PART E PITCHED ROOF	0.00	0.00	0.00	0.00	0.00
PART F WINDOWS AND EXTERNAL DOORS	39.40	374.30	2,336.67	406.65	3,117.62
PART G INTERNAL PARTITIONS AND DOORS	26.54	252.13	198.04	59.41	455.49
PART H WALL FINISHES	42.75	406.13	237.06	96.48	739.66
PART J FLOOR FINISHES	33.82	321.29	390.37	106.75	818.41
PART K CEILING FINISHES	37.59	357.11	122.85	71.99	551.95
PART L ELECTRICAL WORK	14.35	200.90	189.27	58.53	448.70
PART M HEATING WORK	16.09	201.13	730.21	139.70	1,071.04
PART N ALTERATION WORK	45.00	427.50	65.00	73.88	566.38
Final total	603.15	5,833.20	9,543.09	2,298.34	21,331.52

	Ref	Qty	Hours	Hours £	Mat'ls £	O & P £	Total £
PART A PRELIMINARIES							
Concrete mixer	A1	7 wks					315.00
Small tools	A2	8 wks					280.00
Scaffolding (m2/weeks)	A3	280.00					630.00
Skip	A4	4 wks					400.00
Clean up	A5	8 hrs					64.00
Carried to summary							1,689.00
PART B SUBSTRUCTURE TO DPC LEVEL							
Excavate topsoil 150mm thick by hand	B1	7.10m2	2.13	20.24	0.00	3.04	23.27
Excavate to reduce levels by hand	B2	2.13m3	5.33	50.64	0.00	7.60	58.23
Excavate for trench foundations by hand	B3	1.06m3	2.76	26.22	0.00	3.93	30.15
Earthwork support to sides of trenches	B4	9.34m2	3.74	35.53	11.21	7.01	53.75
Backfilling with excavated material	B5	0.40m3	0.16	1.52	0.00	0.23	1.75
Hardcore 225mm thick	B6	4.08m2	0.61	5.80	20.73	3.98	30.50
Hardcore filling to trench	B7	0.08m3	0.05	0.48	0.40	0.13	1.01
Concrete grade (1:3:6) in foundations	B8	0.86m3	1.16	11.02	58.43	10.42	79.87
Concrete grade (1:2:4) in bed 150mm thick	B9	4.08m2	1.22	11.59	44.39	8.40	64.38
Concrete (1:2:4) in cavity wall filling	B10	3.20m2	0.64	6.08	11.62	2.66	20.36
Carried forward			17.80	169.10	146.78	47.38	363.26

	Ref	Qty	Hours	Hours £	Mat'ls £	O & P £	Total £
Brought forward			17.80	169.10	146.78	47.38	363.26
Damp-proof membrane	B11	4.38m2	0.17	1.62	2.54	0.62	4.78
Reinforcement ref A193 in foundation	B12	3.22m2	0.38	3.61	4.16	1.17	8.94
Steel fabric reinforcement ref A193 in slab	B13	4.08m2	0.61	5.80	5.30	1.66	12.76
Solid blockwork 140mm thick in cavity wall	B14	4.16m2	5.41	51.40	53.16	15.68	120.24
Common bricks 112.5mm thick in cavity wall	B15	3.20m2	5.41	51.40	42.21	14.04	107.65
Facing bricks in 112.5mm thick in skin of cavity wall	B16	0.96m2	1.73	16.44	22.34	5.82	44.59
Form cavity 50mm wide in cavity wall	B17	4.16m2	0.13	1.24	2.50	0.56	4.30
DPC 112mm wide	B18	6.40m	0.32	3.04	5.18	1.23	9.45
DPC 140mm wide	B19	6.40m	0.38	3.61	6.21	1.47	11.29
Bond in block wall	B20	1.30m	0.58	5.51	2.83	1.25	9.59
Bond in half brick wall	B21	0.30m	0.46	4.37	0.61	0.75	5.73
50mm thick insulation board	B22	4.08m2	1.22	11.59	18.28	4.48	34.35
Carried to summary			34.60	328.70	312.10	96.12	736.92
PART C EXTERNAL WALLS							
Solid blockwork 140mm thick in cavity wall	C1	23.94m2	31.12	295.64	304.52	90.02	690.18
Facing brickwork 112.5mm thick in cavity wall	C2	23.94m2	43.09	409.36	552.06	144.21	1,105.63
75mm thick insulation in cavity wall	C3	23.94m2	5.27	50.07	139.33	28.41	217.80
Carried forward			79.48	755.06	995.91	262.65	2,013.62

	Ref	Qty	Hours	Hours £	Mat'ls £	O & P £	Total £
Brought forward			79.48	755.06	995.91	262.65	2,013.62
Steel lintel 2400mm long	C4	2nr	0.50	4.75	196.92	30.25	231.92
Steel lintel 1500mm long	C5	4nr	0.80	7.60	429.88	65.62	503.10
Close cavity wall at jambs	C7	13.76m	0.69	6.56	28.07	5.19	39.82
Close cavity wall at cills	C8	5.07m	0.25	2.38	10.34	1.91	14.62
Close cavity wall at top	C9	6.40m	0.32	3.04	13.06	2.42	18.52
DPC 112mm wide at jambs	C10	13.76m	0.69	6.56	11.15	2.66	20.36
DPC 112mm wide at cills	C11	5.07m	0.25	2.38	4.11	0.97	7.46
Carried to summary			82.98	788.31	1,689.44	371.66	2,849.41
PART D FLAT ROOF			N/A	N/A	N/A	N/A	N/A
PART E PITCHED ROOF							
100 x 75mm sawn softwood wall plate	E1	3.00m	1.05	9.98	6.66	2.50	19.13
150 x 50mm sawn softwood pole plate plugged to brickwork	E2	3.30m	0.99	9.41	10.63	3.01	23.04
100 x 50mm sawn softwood rafters	E3	17.50m	3.50	31.50	57.40	13.34	102.24
125 x 25mm sawn softwood purlin	E4	3.30m	0.66	6.27	7.91	2.13	16.31
150 x 50mm sawn softwood joists	E5	12.50m	2.75	26.13	28.50	8.19	62.82
150 x 50mm sawn softwood sprockets	E6	14nr	1.68	15.96	15.96	4.79	36.71
100mm layer of insulation quilt between joists fixed with chicken wire and 150mm layer over joists	E7	6.00m2	2.88	27.36	65.46	13.92	106.74
6mm softwood soffit 150mm wide	E8	8.30m	3.32	31.54	14.53	6.91	52.98
Carried forward			16.83	158.14	207.05	54.78	419.96

	Ref	Qty	Hours	Hours £	Mat'ls £	O & P £	Total £
Brought forward			16.83	158.14	207.05	54.78	419.96
19mm wrought softwood fascia/ barge board 200mm high	E9	3.30m	4.15	39.43	6.50	6.89	52.81
Marley Plain roof tiles on felt and battens	E10	11.55m2	21.95	208.53	298.79	76.10	583.41
Double eaves course	E11	3.30m	0.83	7.89	10.63	2.78	21.29
Verge with plain tile undercloak	E12	7.00m	1.75	16.63	22.54	5.87	45.04
Lead flashing code 5, 200mm girth	E13	3.30m	1.98	18.81	22.04	6.13	46.98
Rake out joint for flashing	E14	3.30m	1.16	11.02	0.50	1.73	13.25
112mm diameter PVC-U gutter	E15	3.30m	0.86	8.17	14.75	3.44	26.36
Stop end	E16	1nr	0.14	1.33	1.11	0.37	2.81
Stop end outlet	E17	1nr	0.25	2.38	1.11	0.52	4.01
68mm diameter PVC-U down pipe	E18	2.50m	0.63	5.99	8.48	2.17	16.63
Shoe	E19	1nr	0.30	2.85	2.09	0.74	5.68
Paint fascia and soffit	E20	2.59m2	1.53	14.54	2.49	2.55	19.58
Carried to summary			52.36	495.67	598.08	164.06	1,257.81

PART F
WINDOWS AND
EXTERNAL DOORS

	Ref	Qty	Hours	Hours £	Mat'ls £	O & P £	Total £
PVC-U door size 840 x 1980mm complete (B)	F1	1nr	2.50	23.75	209.99	35.06	268.80
PVC-U sliding patio door size 1700 x 2075mm (C)	F2	2nr	14.00	133.00	587.04	108.01	828.05
Carried forward			16.50	156.75	797.03	143.07	1,096.85

	Ref	Qty	Hours	Hours £	Mat'ls £	O & P £	Total £
Brought forward			16.50	156.75	797.03	143.07	1,096.85
PVC-U window size 1200 x 1200mm complete (A)	F3	4nr	8.00	76.00	630.36	105.95	812.31
25 x 225mm wrought softwood window board	F4	4.80m	1.44	13.68	29.86	6.53	50.07
Paint window board	F5	4.80m	0.76	7.22	4.61	1.77	13.60
Carried to summary			26.70	253.65	1,461.86	257.33	1,972.84
PART G INTERNAL PARTITIONS AND DOORS		N/A	N/A	N/A	N/A	N/A	N/A
PART H WALL FINISHES							
19 x 100mm wrought softwood skirting	H1	10.34m	1.76	16.72	14.17	4.63	35.52
12mm plasterboard fixed to walls with dabs	H2	20.04m2	7.82	74.29	39.88	17.13	131.30
12mm plasterboard fixed to walls less than 300mm wide with dabs	H3	18.83m	3.39	32.21	17.32	7.43	56.95
Two coats emulsion paint to walls	H4	22.86m2	5.94	56.43	13.49	10.49	80.41
Paint skirting	H5	10.34m	2.07	19.67	4.96	3.69	28.32
Carried to summary			20.98	199.31	89.82	43.37	332.50
PART J FLOOR FINISHES							
Cement and sand floor screed 40mm thick	J1	4.08m2	1.02	9.69	12.36	3.31	25.36
Vinyl floor tiles, size 300 x 300mm	J2	4.08m2	0.70	6.65	28.47	5.27	40.39
Carried forward			1.72	16.34	40.83	8.58	65.75

	Ref	Qty	Hours	Hours £	Mat'ls £	O & P £	Total £
Brought forward			1.72	16.34	40.83	8.58	65.75
25mm thick tongued and grooved boarding	J3	4.08m2	3.02	28.69	38.35	10.06	77.10
150 x 50mm sawn softwood joists	J4	9.50m	2.09	19.86	21.66	6.23	47.74
Cut and pin ends of joists to existing brick wall	J5	5nr	0.90	8.55	0.00	1.28	9.83
Build in ends of joists to blockwork	J6	5nr	0.50	4.75	0.00	0.71	5.46
Carried to summary			8.23	78.19	100.84	26.85	205.88
PART K CEILING FINISHES							
Plasterboard with taped butt joints fixed to joists	K1	8.16m2	2.94	2.08	15.10	2.58	19.76
5mm skim coat to plasterboard ceilings	K2	8.16m2	4.08	38.76	9.96	7.31	56.03
Two coats emulsion paint to ceilings	K3	8.16m2	2.12	20.14	4.81	3.74	28.69
Carried to summary			9.14	60.98	29.87	13.63	104.48
PART L ELECTRICAL WORK							
13 amp double switched socket outlet with neon	L1	4nr	1.60	22.40	33.04	8.32	63.76
Lighting point	L2	2nr	0.70	9.80	13.60	3.51	26.91
Lighting switch	L3	4nr	1.00	14.00	15.88	4.48	34.36
Lighting wiring	L4	10.00m	1.00	14.00	10.30	3.65	27.95
Power cable	L5	24.00m	3.60	50.40	26.40	11.52	88.32
Carried to summary			7.90	110.60	99.22	31.47	241.29

	Ref	Qty	Hours	Hours £	Mat'ls £	O & P £	Total £
PART M **HEATING WORK**							
15mm copper pipe	M1	9.00m	1.98	24.75	16.92	6.25	47.92
Elbow	M2	8nr	2.24	28.00	8.80	5.52	42.32
Tee	M3	2nr	0.44	5.50	3.90	1.41	10.81
Radiator, double convector size 1400 x 520mm	M4	4nr	5.20	65.00	462.72	79.16	606.88
Break into existing pipe and insert tee	M5	1nr	0.75	9.38	3.95	2.00	15.32
Carried to summary			10.61	132.63	496.29	94.34	723.25
PART N **ALTERATION WORK**							
Take out existing window size 1500 x 1000mm and lintel over, adapt opening to receive 1770 x 2000mm patio door and insert new lintel over (both measured separately) and make good	N1	1nr	20.00	190.00	20.00	31.50	241.50
Take out existing window size 1500 x 1000mm, enlarge opening to receive new PVC-U door (measured separately)	N2	1nr	25.00	237.50	45.00	42.38	324.88
Carried to summary			45.00	427.50	65.00	73.88	566.38

SUMMARY

	Hours	Hours £	Mat'ls £	O & P £	Total £
PART A PRELIMINARIES	0.00	0.00	0.00	0.00	1,689.00
PART B SUBSTRUCTURE TO DPC LEVEL	34.60	328.60	312.10	96.12	736.92
PART C EXTERNAL WALLS	82.98	788.31	1,689.44	371.66	2,849.41
PART D FLAT ROOF	0.00	0.00	0.00	0.00	0.00
PART E PITCHED ROOF	52.36	495.67	598.08	164.06	1,257.81
PART F WINDOWS AND EXTERNAL DOORS	26.70	253.65	1,461.86	257.33	1,972.84
PART G INTERNAL PARTITIONS AND DOORS	0.00	0.00	0.00	0.00	0.00
PART H WALL FINISHES	20.98	199.31	89.82	43.37	332.50
PART J FLOOR FINISHES	8.23	78.19	100.84	26.85	205.88
PART K CEILING FINISHES	9.14	60.98	29.87	13.63	104.48
PART L ELECTRICAL WORK	7.90	110.60	99.22	31.47	241.29
PART M HEATING WORK	10.61	132.63	496.29	94.34	723.25
PART N ALTERATION WORK	45.00	427.50	65.00	73.88	566.38
Final total	298.50	2,875.44	4,942.52	1,172.71	10,679.76

	Ref	Qty	Hours	Hours £	Mat'ls £	O & P £	Total £
PART A PRELIMINARIES							
Concrete mixer	A1	8 wks					360.00
Small tools	A2	9 wks					315.00
Scaffolding (m2/weeks)	A3	360.00					810.00
Skip	A4	5 wks					500.00
Clean up	A5	10 hrs					80.00
Carried to summary							2,065.00
PART B SUBSTRUCTURE TO DPC LEVEL							
Excavate topsoil 150mm thick by hand	B1	9.25m2	2.78	26.41	0.00	3.96	30.37
Excavate to reduce levels by hand	B2	2.77m3	6.93	65.84	0.00	9.88	75.71
Excavate for trench foundations by hand	B3	1.22m3	3.17	30.12	0.00	4.52	34.63
Earthwork support to sides of trenches	B4	10.80m2	4.32	41.04	12.96	8.10	62.10
Backfilling with excavated material	B5	0.47m3	0.28	2.66	0.00	0.40	3.06
Hardcore 225mm thick	B6	5.78m2	1.16	11.02	29.36	6.06	46.44
Hardcore filling to trench	B7	0.09m3	0.05	0.48	0.45	0.14	1.06
Concrete grade (1:3:6) in foundations	B8	1.00m3	1.16	11.02	67.94	11.84	90.80
Concrete grade (1:2:4) in bed 150mm thick	B9	5.78m2	1.22	11.59	62.87	11.17	85.63
Concrete (1:2:4) in cavity wall filling	B10	3.70m2	0.64	6.08	13.43	2.93	22.44
Carried forward			21.71	206.25	187.01	58.99	452.24

	Ref	Qty	Hours	Hours £	Mat'ls £	O & P £	Total £
Brought forward			21.71	206.25	187.01	58.99	452.24
Damp-proof membrane	B11	6.13m2	0.25	2.38	3.56	0.89	6.83
Reinforcement ref A193 in foundation	B12	3.70m2	0.44	4.18	4.81	1.35	10.34
Steel fabric reinforcement ref A193 in slab	B13	5.78m2	0.87	8.27	7.51	2.37	18.14
Solid blockwork 140mm thick in cavity wall	B14	4.81m2	6.25	59.38	61.47	18.13	138.97
Common bricks 112.5mm thick in cavity wall	B15	3.70m2	6.25	59.38	48.80	16.23	124.40
Facing bricks in 112.5mm thick in skin of cavity wall	B16	1.11m2	2.00	19.00	25.83	6.72	51.55
Form cavity 50mm wide in cavity wall	B17	4.81m2	0.14	1.33	2.89	0.63	4.85
DPC 112mm wide	B18	7.40m	0.37	3.52	5.99	1.43	10.93
DPC 140mm wide	B19	7.40m	0.44	4.18	7.18	1.70	13.06
Bond in block wall	B20	1.30m	0.58	5.51	2.83	1.25	9.59
Bond in half brick wall	B21	0.30m	0.46	4.37	0.61	0.75	5.73
50mm thick insulation board	B22	5.78m2	1.73	16.44	25.89	6.35	48.67
Carried to summary			41.49	394.16	384.38	116.78	895.32
PART C EXTERNAL WALLS							
Solid blockwork 140mm thick in cavity wall	C1	28.99m2	37.69	358.06	368.75	109.02	835.83
Facing brickwork 112.5mm thick in cavity wall	C2	28.99m2	52.18	495.71	668.51	174.63	1,338.85
75mm thick insulation in cavity wall	C3	28.99m2	6.38	60.61	167.72	34.25	262.58
Carried forward			96.25	914.38	1,204.98	317.90	2,437.26

	Ref	Qty	Hours	Hours £	Mat'ls £	O & P £	Total £
Brought forward			96.25	914.38	1,204.98	317.90	2,437.26
Steel lintel 2400mm long	C4	2nr	0.50	4.75	196.92	30.25	231.92
Steel lintel 1500mm long	C5	4nr	0.80	7.60	429.88	65.62	503.10
Close cavity wall at jambs	C7	13.76m	0.69	6.56	28.07	5.19	39.82
Close cavity wall at cills	C8	5.07m	0.25	2.38	10.34	1.91	14.62
Close cavity wall at top	C9	7.60m	0.37	3.52	15.10	2.79	21.41
DPC 112mm wide at jambs	C10	13.76m	0.69	6.56	11.15	2.66	20.36
DPC 112mm wide at cills	C11	5.07m	0.25	2.38	4.11	0.97	7.46
Carried to summary			99.80	948.10	1,900.55	427.30	3,275.95
PART D FLAT ROOF			N/A	N/A	N/A	N/A	N/A
PART E PITCHED ROOF							
100 x 75mm sawn softwood wall plate	E1	4.00m	1.40	13.30	8.88	3.33	25.51
150 x 50mm sawn softwood pole plate plugged to brickwork	E2	4.30m	1.29	12.26	13.85	3.92	30.02
100 x 50mm sawn softwood rafters	E3	17.50m	3.50	31.50	57.40	13.34	102.24
125 x 25mm sawn softwood purlin	E4	4.30m	0.86	8.17	10.36	2.78	21.31
150 x 50mm sawn softwood joists	E5	12.50m	2.75	26.13	28.50	8.19	62.82
150 x 50mm sawn softwood sprockets	E6	14nr	1.68	15.96	15.96	4.79	36.71
100mm layer of insulation quilt between joists fixed with chicken wire and 150mm layer over joists	E7	8.00m2	3.84	36.48	4.80	6.19	47.47
6mm softwood soffit 150mm wide	E8	9.30m	3.72	35.34	16.28	7.74	59.36
Carried forward			19.04	179.13	156.03	50.27	385.43

	Ref	Qty	Hours	Hours £	Mat'ls £	O & P £	Total £
Brought forward			19.04	179.13	156.03	50.27	385.43
19mm wrought softwood fascia/ barge board 200mm high	E9	4.30m	4.65	44.18	8.47	7.90	60.54
Marley Plain roof tiles on felt and battens	E10	15.05m2	28.60	271.70	389.34	99.16	760.20
Double eaves course	E11	4.30m	1.51	14.35	13.85	4.23	32.42
Verge with plain tile undercloak	E12	7.00m	1.08	10.26	22.54	4.92	37.72
Lead flashing code 5, 200mm girth	E13	4.30m	2.58	24.51	27.82	7.85	60.18
Rake out joint for flashing	E14	4.30m	1.51	14.35	0.65	2.25	17.24
112mm diameter PVC-U gutter	E15	4.30m	1.12	10.64	19.22	4.48	34.34
Stop end	E16	1nr	0.14	1.33	1.11	0.37	2.81
Stop end outlet	E17	1nr	0.25	2.38	1.11	0.52	4.01
68mm diameter PVC-U down pipe	E18	2.50m	0.63	5.99	8.48	2.17	16.63
Shoe	E19	1nr	0.30	2.85	2.09	0.74	5.68
Paint fascia and soffit	E20	2.79m2	1.81	17.20	2.68	2.98	22.86
Carried to summary			63.22	598.84	653.39	187.83	1,440.06
PART F WINDOWS AND EXTERNAL DOORS							
PVC-U door size 840 x 1980mm complete (B)	F1	1nr	2.50	23.75	209.99	35.06	268.80
PVC-U sliding patio door size 1700 x 2075mm (C)	F2	2nr	14.00	133.00	587.04	108.01	828.05
Carried forward			16.50	156.75	797.03	143.07	1,096.85

	Ref	Qty	Hours	Hours £	Mat'ls £	O & P £	Total £
Brought forward			16.50	156.75	797.03	143.07	1,096.85
PVC-U window size 1200 x 1200mm complete (A)	F3	4nr	8.00	76.00	630.36	105.95	812.31
25 x 225mm wrought softwood window board	F4	4.80m	1.44	13.68	29.86	6.53	50.07
Paint window board	F5	4.80m	0.76	7.22	4.61	1.77	13.60
Carried to summary			26.70	253.65	1,461.86	257.33	1,972.84
PART G INTERNAL PARTITIONS AND DOORS		N/A	N/A	N/A	N/A	N/A	N/A
PART H WALL FINISHES							
19 x 100mm wrought softwood skirting	H1	12.34m	2.07	19.67	16.91	5.49	42.06
12mm plasterboard fixed to walls with dabs	H2	24.94m2	9.73	92.44	49.63	21.31	163.37
12mm plasterboard fixed to walls less than 300mm wide with dabs	H3	18.33m	3.39	32.21	17.32	7.43	56.95
Two coats emulsion paint to walls	H4	26.76m2	5.49	52.16	15.79	10.19	78.14
Paint skirting	H5	12.34m	2.47	23.47	9.92	5.01	38.39
Carried to summary			23.15	219.93	109.57	49.42	378.92
PART J FLOOR FINISHES							
Cement and sand floor screed 40mm thick	J1	5.78m2	1.45	13.78	17.51	4.69	35.98
Vinyl floor tiles, size 300 x 300mm	J2	5.78m2	0.98	9.31	40.03	7.40	56.74
Carried forward			2.43	23.09	57.54	12.09	92.72

	Ref	Qty	Hours	Hours £	Mat'ls £	O & P £	Total £
Brought forward			2.43	23.09	57.54	12.09	92.72
25mm thick tongued and grooved boarding	J3	5.78m2	4.28	40.66	54.33	14.25	109.24
150 x 50mm sawn softwood joists	J4	13.30m	2.93	27.84	30.32	8.72	66.88
Cut and pin ends of joists to existing brick wall	J5	7nr	1.25	11.88	1.75	2.04	15.67
Build in ends of joists to blockwork	J6	7nr	0.70	6.65	1.05	1.16	8.86
Carried forward			11.59	110.11	144.99	38.26	293.36

PART K
CEILING FINISHES

	Ref	Qty	Hours	Hours £	Mat'ls £	O & P £	Total £
Plasterboard with taped butt joints fixed to joists	K1	11.56m2	4.16	39.52	21.39	9.14	70.05
5mm skim coat to plasterboard ceilings	K2	11.56m2	5.78	54.91	14.10	10.35	79.36
Two coats emulsion paint to ceilings	K3	11.56m2	3.00	28.50	4.81	5.00	38.31
Carried to summary			12.94	122.93	40.30	24.48	187.71

PART L
ELECTRICAL WORK

	Ref	Qty	Hours	Hours £	Mat'ls £	O & P £	Total £
13 amp double switched socket outlet with neon	L1	4nr	1.60	22.40	33.04	8.32	63.76
Lighting point	L2	4nr	0.70	9.80	27.20	5.55	42.55
Lighting switch	L3	4nr	1.00	14.00	15.88	4.48	34.36
Lighting wiring	L4	12.00m	1.00	14.00	12.36	3.95	30.31
Power cable	L5	28.00m	3.60	50.40	30.80	12.18	93.38
Carried to summary			7.90	110.60	119.28	34.48	264.36

	Ref	Qty	Hours	Hours £	Mat'ls £	O & P £	Total £
PART M **HEATING WORK**							
15mm copper pipe	M1	11.00m	1.98	24.75	20.68	6.81	52.24
Elbow	M2	8nr	2.24	28.00	8.80	5.52	42.32
Tee	M3	2nr	0.44	5.50	3.90	1.41	10.81
Radiator, double convector size 1400 x 520mm	M4	4nr	5.20	65.00	462.72	79.16	606.88
Break into existing pipe and insert tee	M5	1nr	0.75	9.38	3.95	2.00	15.32
Carried to summary			10.61	132.63	500.05	94.90	727.58
PART N **ALTERATION WORK**							
Take out existing window size 1500 x 1000mm and lintel over, adapt opening to receive 1770 x 2000mm patio door and insert new lintel over (both measured separately) and make good	N1	1nr	20.00	190.00	20.00	31.50	241.50
Take out existing window size 1500 x 1000mm, enlarge opening to receive new PVC-U door (measured separately)	N2	1nr	25.00	237.50	45.00	42.38	324.88
Carried to summary			45.00	427.50	65.00	73.88	566.38

SUMMARY

	Hours	Hours £	Mat'ls £	O & P £	Total £
PART A PRELIMINARIES	0.00	0.00	0.00	0.00	2,065.00
PART B SUBSTRUCTURE TO DPC LEVEL	41.49	394.16	384.38	116.78	895.32
PART C EXTERNAL WALLS	99.80	948.10	1,900.55	427.30	3,275.95
PART D FLAT ROOF	0.00	0.00	0.00	0.00	0.00
PART E PITCHED ROOF	63.22	598.84	653.39	187.83	1,440.06
PART F WINDOWS AND EXTERNAL DOORS	26.70	253.65	1,461.86	257.33	1,972.84
PART G INTERNAL PARTITIONS AND DOORS	0.00	0.00	0.00	0.00	0.00
PART H WALL FINISHES	23.15	219.93	109.57	49.42	378.92
PART J FLOOR FINISHES	11.59	110.11	144.99	38.26	293.36
PART K CEILING FINISHES	12.94	122.93	40.30	24.48	187.71
PART L ELECTRICAL WORK	7.90	110.60	119.28	34.48	264.36
PART M HEATING WORK	10.61	132.63	500.05	94.90	727.58
PART N ALTERATION WORK	45.00	427.50	65.00	73.88	566.38
Final total	342.40	3,318.45	5,379.37	1,304.66	12,067.48

	Ref	Qty	Hours	Hours £	Mat'ls £	O & P £	Total £
PART A PRELIMINARIES							
Concrete mixer	A1	9 wks					405.00
Small tools	A2	10 wks					350.00
Scaffolding (m2/weeks)	A3	450.00					1,012.50
Skip	A4	6 wks					600.00
Clean up	A5	12 hrs					96.00
Carried to summary							2,463.50
PART B SUBSTRUCTURE TO DPC LEVEL							
Excavate topsoil 150mm thick by hand	B1	11.4m2	3.42	32.49	0.00	4.87	37.36
Excavate to reduce levels by hand	B2	3.42m3	8.55	81.23	0.00	12.18	93.41
Excavate for trench foundations by hand	B3	1.39m3	3.61	34.30	0.00	5.14	39.44
Earthwork support to sides of trenches	B4	12.26m2	4.90	46.55	14.71	9.19	70.45
Backfilling with excavated material	B5	0.55m3	0.33	3.14	0.00	0.47	3.61
Hardcore 225mm thick	B6	7.48m2	1.50	14.25	38.00	7.84	60.09
Hardcore filling to trench	B7	0.12m3	0.06	0.57	0.60	0.18	1.35
Concrete grade (1:3:6) in foundations	B8	1.13m3	1.53	14.54	76.77	13.70	105.00
Concrete grade (1:2:4) in bed 150mm thick	B9	7.48m2	2.24	21.28	81.38	15.40	118.06
Concrete (1:2:4) in cavity wall filling	B10	4.20m2	0.84	7.98	15.25	3.48	26.71
Carried forward			26.98	256.31	226.71	72.45	555.47

	Ref	Qty	Hours	Hours £	Mat'ls £	O & P £	Total £
Brought forward			26.98	256.31	226.71	72.45	555.47
Damp-proof membrane	B11	7.88m2	0.32	3.04	4.57	1.14	8.75
Reinforcement ref A193 in foundation	B12	4.20m2	0.50	4.75	5.46	1.53	11.74
Steel fabric reinforcement ref A193 in slab	B13	7.48m2	1.12	10.64	9.72	3.05	23.41
Solid blockwork 140mm thick in cavity wall	B14	5.46m2	7.10	67.45	69.78	20.58	157.81
Common bricks 112.5mm thick in cavity wall	B15	4.20m2	7.10	67.45	55.40	18.43	141.28
Facing bricks in 112.5mm thick in skin of cavity wall	B16	1.26m2	2.27	21.57	29.32	7.63	58.52
Form cavity 50mm wide in cavity wall	B17	5.46m2	0.16	1.52	3.28	0.72	5.52
DPC 112mm wide	B18	8.40m	0.42	3.99	6.80	1.62	12.41
DPC 140mm wide	B19	8.40m	0.50	4.75	8.15	1.94	14.84
Bond in block wall	B20	1.30m	0.58	5.51	2.83	1.25	9.59
Bond in half brick wall	B21	0.30m	0.46	4.37	0.61	0.75	5.73
50mm thick insulation board	B22	7.48m2	2.24	21.28	33.51	8.22	63.01
Carried to summary			49.75	472.63	456.14	139.31	1,068.08
PART C EXTERNAL WALLS							
Solid blockwork 140mm thick in cavity wall	C1	34.04m2	44.25	420.38	432.99	128.00	981.37
Facing brickwork 112.5mm thick in cavity wall	C2	34.04m2	61.27	582.07	784.96	205.05	1,572.08
75mm thick insulation in cavity wall	C3	34.04m2	7.49	198.11	6.30	30.66	235.07
Carried forward			113.01	1,200.55	1,224.25	363.72	2,788.52

	Ref	Qty	Hours	Hours £	Mat'ls £	O & P £	Total £
Brought forward			113.01	1,200.55	1,224.25	363.72	2,788.52
Steel lintel 2400mm long	C4	2nr	0.50	4.75	196.92	30.25	231.92
Steel lintel 1500mm long	C5	4nr	0.80	7.60	429.88	65.62	503.10
Close cavity wall at jambs	C7	13.76m	0.69	6.56	28.07	5.19	39.82
Close cavity wall at cills	C8	5.07m	0.25	2.38	10.34	1.91	14.62
Close cavity wall at top	C9	8.40m	0.42	3.99	17.16	3.17	24.32
DPC 112mm wide at jambs	C10	13.76m	0.69	6.56	11.15	2.66	20.36
DPC 112mm wide at cills	C11	5.07m	0.25	2.38	4.11	0.97	7.46
Carried to summary			116.61	1,234.75	1,921.88	473.49	3,630.12
PART D FLAT ROOF			N/A	N/A	N/A	N/A	N/A
PART E PITCHED ROOF							
100 x 75mm sawn softwood wall plate	E1	5.00m	1.75	16.63	11.10	4.16	31.88
150 x 50mm sawn softwood pole plate plugged to brickwork	E2	5.30m	1.89	17.96	17.07	5.25	40.28
100 x 50mm sawn softwood rafters	E3	17.50m	3.50	31.50	57.40	13.34	102.24
125 x 25mm sawn softwood purlin	E4	5.30m	1.06	10.07	12.77	3.43	26.27
150 x 50mm sawn softwood joists	E5	12.50m	2.75	26.13	28.80	8.24	63.16
150 x 50mm sawn softwood sprockets	E6	14nr	1.68	15.96	15.96	4.79	36.71
100mm layer of insulation quilt between joists fixed with chicken wire and 150mm layer over joists	E7	10.00m2	4.80	45.60	109.10	23.21	177.91
6mm softwood soffit 150mm wide	E8	10.30m	4.12	39.14	18.03	8.58	65.75
Carried forward			21.55	202.98	270.23	70.98	544.19

	Ref	Qty	Hours	Hours £	Mat'ls £	O & P £	Total £
Brought forward			21.55	202.98	270.23	70.98	544.19
19mm wrought softwood fascia/ barge board 200mm high	E9	5.30m	5.15	48.93	10.44	8.90	68.27
Marley Plain roof tiles on felt and battens	E10	18.55m2	35.25	334.88	479.89	122.21	936.98
Double eaves course	E11	5.30m	1.86	17.67	17.07	5.21	39.95
Verge with plain tile undercloak	E12	7.00m	1.08	10.26	22.54	4.92	37.72
Lead flashing code 5, 200mm girth	E13	5.30m	3.18	30.21	35.40	9.84	75.45
Rake out joint for flashing	E14	5.30m	1.86	17.67	0.80	2.77	21.24
112mm diameter PVC-U gutter	E15	5.30m	1.38	13.11	23.69	5.52	42.32
Stop end	E16	1nr	0.14	1.33	1.11	0.37	2.81
Stop end outlet	E17	1nr	0.25	2.38	1.11	0.52	4.01
68mm diameter PVC-U down pipe	E18	2.50m	0.63	5.99	8.48	2.17	16.63
Shoe	E19	1nr	0.30	2.85	2.09	0.74	5.68
Paint fascia and soffit	E20	3.09m2	2.16	20.52	2.97	3.52	27.01
Carried to summary			74.79	708.76	875.82	237.69	1,822.26

PART F
WINDOWS AND
EXTERNAL DOORS

	Ref	Qty	Hours	Hours £	Mat'ls £	O & P £	Total £
PVC-U door size 840 x 1980mm complete (B)	F1	1nr	2.50	23.75	209.99	35.06	268.80
PVC-U sliding patio door size 1700 x 2075mm (C)	F2	2nr	14.00	133.00	587.04	108.01	828.05
Carried forward			16.50	156.75	797.03	143.07	1,096.85

	Ref	Qty	Hours	Hours £	Mat'ls £	O & P £	Total £
Brought forward			16.50	156.75	797.03	143.07	1,096.85
PVC-U window size 1200 x 1200mm complete (A)	F3	4nr	8.00	76.00	630.36	105.95	812.31
25 x 225mm wrought softwood window board	F4	4.80m	1.44	13.68	29.86	6.53	50.07
Paint window board	F5	4.80m	0.76	7.22	4.61	1.77	13.60
Carried to summary			26.70	253.65	1,461.86	257.33	1,972.84
PART G INTERNAL PARTITIONS AND DOORS		N/A	N/A	N/A	N/A	N/A	N/A
PART H WALL FINISHES							
19 x 100mm wrought softwood skirting	H1	14.34m	2.41	22.90	19.65	6.38	48.93
12mm plasterboard fixed to walls with dabs	H2	29.84m2	11.63	110.49	59.38	25.48	195.34
12mm plasterboard fixed to walls less than 300mm wide with dabs	H3	18.33m	3.39	32.21	17.32	7.43	56.95
Two coats emulsion paint to walls	H4	32.66m2	6.68	63.46	19.27	12.41	95.14
Paint skirting	H5	14.34m	1.80	17.10	6.88	3.60	27.58
Carried to summary			25.91	246.15	122.50	55.30	423.94
PART J FLOOR FINISHES							
Cement and sand floor screed 40mm thick	J1	7.48m2	1.87	17.77	22.66	6.06	46.49
Vinyl floor tiles, size 300 x 300mm	J2	7.48m2	1.80	17.10	52.22	10.40	79.72
Carried forward			3.67	34.87	74.88	16.46	126.21

	Ref	Qty	Hours	Hours £	Mat'ls £	O & P £	Total £
Brought forward			3.67	34.87	74.88	16.46	126.21
25mm thick tongued and grooved boarding	J3	7.48m2	5.54	52.63	70.30	18.44	141.37
150 x 50mm sawn softwood joists	J4	15.20m	3.34	31.73	34.66	9.96	76.35
Cut and pin ends of joists to existing brick wall	J5	8nr	1.49	14.16	2.00	2.42	18.58
Build in ends of joists to blockwork	J6	8nr	0.80	7.60	1.20	1.32	10.12
Carried to summary			14.84	140.98	183.04	48.60	372.62
PART K CEILING FINISHES							
Plasterboard with taped butt joints fixed to joists	K1	14.96m2	5.38	51.11	27.68	11.82	90.61
5mm skim coat to plasterboard ceilings	K2	14.96m2	7.48	71.06	18.25	13.40	102.71
Two coats emulsion paint to ceilings	K3	14.96m2	3.89	36.96	8.83	6.87	52.65
Carried to summary			16.75	159.13	54.76	32.08	245.97
PART L ELECTRICAL WORK							
13 amp double switched socket outlet with neon	L1	6nr	2.40	33.60	49.56	12.47	95.63
Lighting point	L2	4nr	1.40	19.60	27.20	7.02	53.82
Lighting switch	L3	4nr	1.00	14.00	15.88	4.48	34.36
Lighting wiring	L4	14.00m	1.40	19.60	14.42	5.10	39.12
Power cable	L5	32.00m	4.80	67.20	35.20	15.36	117.76
Carried to summary			11.00	154.00	142.26	44.44	340.70

	Ref	Qty	Hours	Hours £	Mat'ls £	O & P £	Total £
PART M HEATING WORK							
15mm copper pipe	M1	13.00m	2.86	35.75	24.44	9.03	69.22
Elbow	M2	8nr	2.24	28.00	8.80	5.52	42.32
Tee	M3	2nr	0.44	5.50	3.90	1.41	10.81
Radiator, double convector size 1400 x 520mm	M4	4nr	5.20	65.00	462.72	79.16	606.88
Break into existing pipe and insert tee	M5	1nr	0.75	9.38	3.95	2.00	15.32
Carried to summary			11.49	143.63	503.81	97.12	744.55
PART N ALTERATION WORK							
Take out existing window size 1500 x 1000mm and lintel over, adapt opening to receive 1770 x 2000mm patio door and insert new lintel over (both measured separately) and make good	N1	1nr	20.00	190.00	20.00	31.50	241.50
Take out existing window size 1500 x 1000mm, enlarge opening to receive new PVC-U door (measured separately)	N2	1nr	25.00	237.50	45.00	42.38	324.88
Carried to summary			45.00	427.50	65.00	73.88	566.38

SUMMARY

	Hours	Hours £	Mat'ls £	O & P £	Total £
PART A PRELIMINARIES	0.00	0.00	0.00	0.00	2,463.50
PART B SUBSTRUCTURE TO DPC LEVEL	49.75	472.63	456.14	139.31	1,068.08
PART C EXTERNAL WALLS	116.61	1,234.75	1,921.88	473.49	3,630.12
PART D FLAT ROOF	0.00	0.00	0.00	0.00	0.00
PART E PITCHED ROOF	74.79	708.76	875.82	237.69	1,822.26
PART F WINDOWS AND EXTERNAL DOORS	26.70	253.65	1,461.86	257.33	1,972.84
PART G INTERNAL PARTITIONS AND DOORS	0.00	0.00	0.00	0.00	0.00
PART H WALL FINISHES	25.91	246.15	122.50	55.30	423.94
PART J FLOOR FINISHES	14.84	140.98	183.04	48.60	372.62
PART K CEILING FINISHES	16.75	159.13	54.76	32.08	245.97
PART L ELECTRICAL WORK	11.00	154.00	142.26	44.44	340.70
PART M HEATING WORK	11.49	143.63	503.81	97.12	744.55
PART N ALTERATION WORK	45.00	427.50	65.00	73.88	566.38
Final total	392.84	3,941.18	5,787.07	1,459.24	13,650.96

	Ref	Qty	Hours	Hours £	Mat'ls £	O & P £	Total £
PART A PRELIMINARIES							
Concrete mixer	A1	9 wks					405.00
Small tools	A2	9 wks					315.00
Scaffolding (m2/weeks)	A3	360.00					810.00
Skip	A4	5 wks					500.00
Clean up	A5	10 hrs					80.00
Carried to summary							2,110.00
PART B SUBSTRUCTURE TO DPC LEVEL							
Excavate topsoil 150mm thick by hand	B1	10.40m2	3.12	29.64	0.00	4.45	34.09
Excavate to reduce levels by hand	B2	3.12m3	7.80	74.10	0.00	11.12	85.22
Excavate for trench foundations by hand	B3	1.39m3	3.61	34.30	0.00	5.14	39.44
Earthwork support to sides of trenches	B4	12.26m2	4.90	46.55	14.71	9.19	70.45
Backfilling with excavated material	B5	0.47m3	0.28	2.66	0.00	0.40	3.06
Hardcore 225mm thick	B6	6.48m2	1.30	12.35	32.92	6.79	52.06
Hardcore filling to trench	B7	0.09m3	0.05	0.48	0.45	0.14	1.06
Concrete grade (1:3:6) in foundations	B8	1.13m3	1.53	14.54	76.77	13.70	105.00
Concrete grade (1:2:4) in bed 150mm thick	B9	6.48m2	1.94	18.43	70.50	13.34	102.27
Concrete (1:2:4) in cavity wall filling	B10	4.20m2	0.84	7.98	15.25	3.48	26.71
Carried forward			25.37	241.02	210.60	67.74	519.36

	Ref	Qty	Hours	Hours £	Mat'ls £	O & P £	Total £
Brought forward			25.37	241.02	210.60	67.74	519.36
Damp-proof membrane	B11	6.88m2	0.28	2.66	3.99	1.00	7.65
Reinforcement ref A193 in foundation	B12	4.20m2	0.50	4.75	5.46	1.53	11.74
Steel fabric reinforcement ref A193 in slab	B13	6.48m2	0.97	9.22	8.42	2.65	20.28
Solid blockwork 140mm thick in cavity wall	B14	5.46m2	7.10	67.45	69.78	20.58	157.81
Common bricks 112.5mm thick in cavity wall	B15	4.20m2	7.10	67.45	55.40	18.43	141.28
Facing bricks in 112.5mm thick in skin of cavity wall	B16	1.26m2	2.27	21.57	29.32	7.63	58.52
Form cavity 50mm wide in cavity wall	B17	5.46m2	0.16	1.52	3.28	0.72	5.52
DPC 112mm wide	B18	8.40m	0.42	3.99	6.80	1.62	12.41
DPC 140mm wide	B19	8.40m	0.50	4.75	8.15	1.94	14.84
Bond in block wall	B20	1.30m	0.58	5.51	2.83	1.25	9.59
Bond in half brick wall	B21	0.30m	0.46	4.37	0.61	0.75	5.73
50mm thick insulation board	B22	6.48m2	1.94	18.43	29.03	7.12	54.58
Carried to summary			47.65	452.68	433.67	132.95	1,019.30
PART C EXTERNAL WALLS							
Solid blockwork 140mm thick in cavity wall	C1	34.04m2	44.25	420.38	432.99	128.00	981.37
Facing brickwork 112.5mm thick in cavity wall	C2	34.04m2	61.27	582.07	784.96	205.05	1,572.08
75mm thick insulation in cavity wall	C3	34.04m2	7.49	71.16	198.11	40.39	309.65
Carried forward			113.01	1,073.60	1,416.06	373.45	2,863.10

	Ref	Qty	Hours	Hours £	Mat'ls £	O & P £	Total £
Brought forward			113.01	1,073.60	1,416.06	373.45	2,863.10
Steel lintel 2400mm long	C4	2nr	0.50	4.75	196.92	30.25	231.92
Steel lintel 1500mm long	C5	4nr	0.80	7.60	429.88	65.62	503.10
Close cavity wall at jambs	C7	13.76m	0.69	6.56	28.07	5.19	39.82
Close cavity wall at cills	C8	5.07m	0.25	2.38	10.34	1.91	14.62
Close cavity wall at top	C9	8.40m	0.42	3.99	17.16	3.17	24.32
DPC 112mm wide at jambs	C10	13.76m	0.69	6.56	11.15	2.66	20.36
DPC 112mm wide at cills	C11	5.07m	0.25	2.38	4.11	0.97	7.46
Carried to summary			116.61	1,107.80	2,113.69	483.22	3,704.71
PART D FLAT ROOF			N/A	N/A	N/A	N/A	N/A
PART E PITCHED ROOF							
100 x 75mm sawn softwood wall plate	E1	3.00m	1.05	9.98	6.66	2.50	19.13
150 x 50mm sawn softwood pole plate plugged to brickwork	E2	3.30m	0.99	9.41	10.63	3.01	23.04
100 x 50mm sawn softwood rafters	E3	31.50	3.50	31.50	103.32	20.22	155.04
125 x 25mm sawn softwood purlin	E4	3.30m	0.66	6.27	7.91	2.13	16.31
150 x 50mm sawn softwood joists	E5	24.50m	2.75	26.13	55.86	12.30	94.28
150 x 50mm sawn softwood sprockets	E6	18nr	1.68	15.96	20.52	5.47	41.95
100mm layer of insulation quilt between joists fixed with chicken wire and 150mm layer over joists	E7	9.00m2	4.32	41.04	98.19	20.88	160.11
6mm softwood soffit 150mm wide	E8	10.30m	3.32	31.54	18.03	7.44	57.01
Carried forward			18.27	171.82	321.12	73.94	566.88

	Ref	Qty	Hours	Hours £	Mat'ls £	O & P £	Total £
Brought forward			18.27	171.82	321.12	73.94	566.88
19mm wrought softwood fascia/ barge board 200mm high	E9	3.30m	4.15	39.43	6.50	6.89	52.81
Marley Plain roof tiles on felt and battens	E10	14.85m2	21.95	208.53	384.17	88.90	681.60
Double eaves course	E11	3.30m	0.83	7.89	10.63	2.78	21.29
Verge with plain tile undercloak	E12	9.00m	1.75	16.63	28.98	6.84	52.45
Lead flashing code 5, 200mm girth	E13	3.30m	1.98	18.81	22.04	6.13	46.98
Rake out joint for flashing	E14	3.30m	1.16	11.02	0.50	1.73	13.25
112mm diameter PVC-U gutter	E15	3.30m	0.86	8.17	14.75	3.44	26.36
Stop end	E16	1nr	0.14	1.33	1.11	0.37	2.81
Stop end outlet	E17	1nr	0.25	2.38	1.11	0.52	4.01
68mm diameter PVC-U down pipe	E18	2.50m	0.63	5.99	8.48	2.17	16.63
Shoe	E19	1nr	0.30	2.85	2.09	0.74	5.68
Paint fascia and soffit	E20	3.09m2	1.53	14.54	2.97	2.63	20.13
Carried to summary			53.80	509.35	804.45	197.07	1,510.87
PART F WINDOWS AND EXTERNAL DOORS							
PVC-U door size 840 x 1980mm complete (B)	F1	1nr	2.50	23.75	209.99	35.06	268.80
PVC-U sliding patio door size 1700 x 2075mm (C)	F2	2nr	14.00	133.00	587.04	108.01	828.05
Carried forward			16.50	156.75	797.03	143.07	1,096.85

	Ref	Qty	Hours	Hours £	Mat'ls £	O & P £	Total £
Brought forward			16.50	156.75	797.03	143.07	1,096.85
PVC-U window size 1200 x 1200mm complete (A)	F3	4nr	8.00	76.00	630.36	105.95	812.31
25 x 225mm wrought softwood window board	F4	4.80m	1.44	13.68	29.86	6.53	50.07
Paint window board	F5	4.80m	0.76	7.22	4.61	1.77	13.60
Carried to summary			26.70	253.65	1,461.86	257.33	1,972.84
PART G INTERNAL PARTITIONS AND DOORS		N/A	N/A	N/A	N/A	N/A	N/A
PART H WALL FINISHES							
19 x 100mm wrought softwood skirting	H1	14.34m	2.41	22.90	19.65	6.38	48.93
12mm plasterboard fixed to walls with dabs	H2	29.84m2	11.63	110.49	59.38	25.48	195.34
12mm plasterboard fixed to walls less than 300mm wide with dabs	H3	18.33m	3.39	32.21	17.32	7.43	56.95
Two coats emulsion paint to walls	H4	32.66m2	6.68	63.46	19.27	12.41	95.14
Paint skirting	H5	14.34m	1.80	17.10	6.88	3.60	27.58
Carried to summary			25.91	246.15	122.50	55.30	423.94
PART J FLOOR FINISHES							
Cement and sand floor screed 40mm thick	J1	6.48m2	1.62	15.39	19.63	5.25	40.27
Vinyl floor tiles, size 300 x 300mm	J2	6.48m2	1.56	14.82	45.23	9.01	69.06
Carried forward			3.18	30.21	64.86	14.26	109.33

	Ref	Qty	Hours	Hours £	Mat'ls £	O & P £	Total £
Brought forward			3.18	30.21	64.86	14.26	109.33
25mm thick tongued and grooved boarding	J3	6.48m2	4.79	45.51	60.91	15.96	122.38
150 x 50mm sawn softwood joists	J4	14.50m	2.03	19.29	33.06	7.85	60.20
Cut and pin ends of joists to existing brick wall	J5	5nr	0.90	8.55	1.25	1.47	11.27
Build in ends of joists to blockwork	J6	5nr	0.50	4.75	0.75	0.83	6.33
Carried to summary			11.40	108.30	160.83	40.37	309.50
PART K CEILING FINISHES							
Plasterboard with taped butt joints fixed to joists	K1	12.96m2	4.67	44.37	23.98	10.25	78.60
5mm skim coat to plasterboard ceilings	K2	12.96m2	6.48	61.56	15.81	11.61	88.98
Two coats emulsion paint to ceilings	K3	12.96m2	3.37	32.02	7.65	5.95	45.61
Carried to summary			14.52	137.94	47.44	27.81	213.19
PART L ELECTRICAL WORK							
13 amp double switched socket outlet with neon	L1	4nr	1.60	22.40	33.04	8.32	63.76
Lighting point	L2	4nr	1.40	19.60	27.20	7.02	53.82
Lighting switch	L3	4nr	1.00	14.00	15.88	4.48	34.36
Lighting wiring	L4	12.00m	1.20	16.80	12.36	4.37	33.53
Power cable	L5	28.00m	4.20	58.80	30.80	13.44	103.04
Carried to summary			9.40	131.60	119.28	37.63	288.51

	Ref	Qty	Hours	Hours £	Mat'ls £	O & P £	Total £
PART M HEATING WORK							
15mm copper pipe	M1	11.00m	2.42	30.25	20.68	7.64	58.57
Elbow	M2	8nr	2.24	28.00	8.80	5.52	42.32
Tee	M3	2nr	0.68	8.50	3.90	1.86	14.26
Radiator, double convector size 1400 x 520mm	M4	4nr	5.20	65.00	462.72	79.16	606.88
Break into existing pipe and insert tee	M5	1nr	0.75	9.38	3.95	2.00	15.32
Carried to summary			11.29	141.13	500.05	96.18	737.35
PART N ALTERATION WORK							
Take out existing window size 1500 x 1000mm and lintel over, adapt opening to receive 1770 x 2000mm patio door and insert new lintel over (both measured separately) and make good	N1	1nr	20.00	190.00	20.00	31.50	241.50
Take out existing window size 1500 x 1000mm, enlarge opening to receive new PVC-U door (measured separately)	N2	1nr	25.00	237.50	45.00	42.38	324.88
Carried to summary			45.00	427.50	65.00	73.88	566.38

SUMMARY

	Hours	Hours £	Mat'ls £	O & P £	Total £
PART A PRELIMINARIES	0.00	0.00	0.00	0.00	2,110.00
PART B SUBSTRUCTURE TO DPC LEVEL	47.65	452.68	433.67	132.95	1,019.30
PART C EXTERNAL WALLS	116.61	1,107.80	2,113.69	483.22	3,704.71
PART D FLAT ROOF	0.00	0.00	0.00	0.00	0.00
PART E PITCHED ROOF	53.80	509.35	804.45	197.07	1,510.87
PART F WINDOWS AND EXTERNAL DOORS	26.70	253.65	1,461.86	257.33	1,972.84
PART G INTERNAL PARTITIONS AND DOORS	0.00	0.00	0.00	0.00	0.00
PART H WALL FINISHES	25.91	246.15	122.50	55.30	423.94
PART J FLOOR FINISHES	11.40	108.30	160.83	40.37	309.50
PART K CEILING FINISHES	14.52	137.94	47.44	27.81	213.19
PART L ELECTRICAL WORK	9.40	131.60	119.28	37.63	288.51
PART M HEATING WORK	11.29	141.13	500.05	96.18	737.35
PART N ALTERATION WORK	45.00	427.50	65.00	73.88	566.38
Final total	362.28	3,516.10	5,828.77	1,401.74	12,856.59

	Ref	Qty	Hours	Hours £	Mat'ls £	O & P £	Total £
PART A PRELIMINARIES							
Concrete mixer	A1	10 wks					450.00
Small tools	A2	10 wks					350.00
Scaffolding (m2/weeks)	A3	450.00					1,012.50
Skip	A4	6 wks					600.00
Clean up	A5	12 hrs					96.00
Carried to summary							2,508.50
PART B SUBSTRUCTURE TO DPC LEVEL							
Excavate topsoil 150mm thick by hand	B1	14.19m2	4.26	40.47	0.00	6.07	46.54
Excavate to reduce levels by hand	B2	4.06m3	10.15	96.43	0.00	14.46	110.89
Excavate for trench foundations by hand	B3	1.39m3	3.61	34.30	0.00	5.14	39.44
Earthwork support to sides of trenches	B4	13.72m2	5.49	52.16	16.46	10.29	78.91
Backfilling with excavated material	B5	0.55m3	0.22	2.09	0.00	0.31	2.40
Hardcore 225mm thick	B6	9.18m2	1.38	13.11	46.63	8.96	68.70
Hardcore filling to trench	B7	0.12m3	0.06	0.57	0.60	0.18	1.35
Concrete grade (1:3:6) in foundations	B8	1.27m3	1.71	16.25	86.28	15.38	117.90
Concrete grade (1:2:4) in bed 150mm thick	B9	9.18m2	2.75	26.13	99.88	18.90	144.91
Concrete (1:2:4) in cavity wall filling	B10	4.70m2	0.94	8.93	17.06	3.90	29.89
Carried forward			30.57	290.42	266.91	83.60	640.92

	Ref	Qty	Hours	Hours £	Mat'ls £	O & P £	Total £
Brought forward			30.57	290.42	266.91	83.60	640.92
Damp-proof membrane	B11	9.63m2	0.37	3.52	5.59	1.37	10.47
Reinforcement ref A193 in foundation	B12	4.70m2	0.44	4.18	6.11	1.54	11.83
Steel fabric reinforcement ref A193 in slab	B13	9.18m2	1.38	13.11	11.93	3.76	28.80
Solid blockwork 140mm thick in cavity wall	B14	6.11m2	7.94	75.43	78.08	23.03	176.54
Common bricks 112.5mm thick in cavity wall	B15	4.70m2	7.94	75.43	61.99	20.61	158.03
Facing bricks in 112.5mm thick in skin of cavity wall	B16	1.41m2	2.54	24.13	32.81	8.54	65.48
Form cavity 50mm wide in cavity wall	B17	1.41m2	0.18	1.71	3.67	0.81	6.19
DPC 112mm wide	B18	9.40m	0.47	4.47	7.61	1.81	13.89
DPC 140mm wide	B19	9.40m	0.56	5.32	9.12	2.17	16.61
Bond in block wall	B20	1.30m	0.58	5.51	2.83	1.25	9.59
Bond in half brick wall	B21	0.30m	0.46	4.37	0.61	0.75	5.73
50mm thick insulation board	B22	9.18m2	2.75	26.13	41.13	10.09	77.34
Carried to summary			56.18	533.71	528.39	159.32	1,221.42
PART C EXTERNAL WALLS							
Solid blockwork 140mm thick in cavity wall	C1	39.09m2	50.81	482.70	497.22	146.99	1,126.90
Facing brickwork 112.5mm thick in cavity wall	C2	39.09m2	70.36	668.42	901.42	235.48	1,805.32
75mm thick insulation in cavity wall	C3	39.09m2	8.60	81.70	271.50	52.98	406.18
Carried forward			129.77	1,232.82	1,670.14	435.44	3,338.40

	Ref	Qty	Hours	Hours £	Mat'ls £	O & P £	Total £
Brought forward			129.77	1,232.82	1,670.14	435.44	3,338.40
Steel lintel 2400mm long	C4	2nr	0.50	4.75	196.92	30.25	231.92
Steel lintel 1500mm long	C5	4nr	0.80	7.60	429.88	65.62	503.10
Close cavity wall at jambs	C7	13.76m	0.69	6.56	28.07	5.19	39.82
Close cavity wall at cills	C8	5.07m	0.25	2.38	10.34	1.91	14.62
Close cavity wall at top	C9	9.40m	0.47	4.47	19.18	3.55	27.19
DPC 112mm wide at jambs	C10	13.76m	0.69	6.56	11.15	2.66	20.36
DPC 112mm wide at cills	C11	5.07m	0.25	2.38	4.11	0.97	7.46
Carried to summary			133.42	1,267.49	2,369.79	545.59	4,182.87
PART D FLAT ROOF			N/A	N/A	N/A	N/A	N/A
PART E PITCHED ROOF							
100 x 75mm sawn softwood wall plate	E1	4.00m	1.40	13.30	8.85	3.32	25.47
150 x 50mm sawn softwood pole plate plugged to brickwork	E2	4.30m	1.29	12.26	13.85	3.92	30.02
100 x 50mm sawn softwood rafters	E3	31.50m	4.90	44.10	103.32	22.11	169.53
125 x 25mm sawn softwood purlin	E4	4.30m	0.86	8.17	10.30	2.77	21.24
150 x 50mm sawn softwood joists	E5	24.50m	5.39	51.21	55.80	16.05	123.06
150 x 50mm sawn softwood sprockets	E6	18nr	2.16	20.52	20.52	6.16	47.20
100mm layer of insulation quilt between joists fixed with chicken wire and 150mm layer over joists	E7	12.00m2	5.76	54.72	130.92	27.85	213.49
6mm softwood soffit 150mm wide	E8	11.30m	4.92	46.74	19.78	9.98	76.50
Carried forward			26.68	251.01	363.34	92.15	706.50

	Ref	Qty	Hours	Hours £	Mat'ls £	O & P £	Total £
Brought forward			26.68	251.01	363.34	92.15	706.50
19mm wrought softwood fascia/ barge board 200mm high	E9	4.30m	6.15	58.43	8.47	10.03	76.93
Marley Plain roof tiles on felt and battens	E10	19.55m2	36.77	349.32	500.59	127.49	977.39
Double eaves course	E11	4.30m	1.51	14.35	13.85	4.23	32.42
Verge with plain tile undercloak	E12	9.00m	2.25	21.38	28.98	7.55	57.91
Lead flashing code 5, 200mm girth	E13	4.30m	2.58	24.51	28.72	7.98	61.21
Rake out joint for flashing	E14	4.30m	1.51	14.35	0.65	2.25	17.24
112mm diameter PVC-U gutter	E15	4.30m	1.12	10.64	19.22	4.48	34.34
Stop end	E16	1nr	0.14	1.33	1.11	0.37	2.81
Stop end outlet	E17	1nr	0.25	2.38	1.11	0.52	4.01
68mm diameter PVC-U down pipe	E18	2.50m	0.63	5.99	8.48	2.17	16.63
Shoe	E19	1nr	0.30	2.85	2.09	0.74	5.68
Paint fascia and soffit	E20	3.39m2	2.37	22.52	3.25	3.86	29.63
Carried to summary			82.26	779.02	979.86	263.83	2,022.71
PART F WINDOWS AND EXTERNAL DOORS							
PVC-U door size 840 x 1980mm complete (B)	F1	1nr	2.50	23.75	209.99	35.06	268.80
PVC-U sliding patio door size 1700 x 2075mm (C)	F2	2nr	14.00	133.00	587.04	108.01	828.05
Carried forward			16.50	156.75	797.03	143.07	1,096.85

	Ref	Qty	Hours	Hours £	Mat'ls £	O & P £	Total £
Brought forward			16.50	156.75	797.03	143.07	1,096.85
PVC-U window size 1200 x 1200mm complete (A)	F3	4nr	8.00	76.00	630.36	105.95	812.31
25 x 225mm wrought softwood window board	F4	4.80m	1.44	13.68	29.86	6.53	50.07
Paint window board	F5	4.80m	0.76	7.22	4.61	1.77	13.60
Carried to summary			26.70	253.65	1,461.86	257.33	1,972.84
PART G INTERNAL PARTITIONS AND DOORS		N/A	N/A	N/A	N/A	N/A	N/A
PART H WALL FINISHES							
19 x 100mm wrought softwood skirting	H1	16.34m	2.75	26.13	22.39	7.28	55.79
12mm plasterboard fixed to walls with dabs	H2	34.74m2	13.55	128.73	69.13	29.68	227.53
12mm plasterboard fixed to walls less than 300mm wide with dabs	H3	18.33m	3.39	32.21	17.32	7.43	56.95
Two coats emulsion paint to walls	H4	38.56m2	7.89	74.96	22.75	14.66	112.36
Paint skirting	H5	16.34m	2.05	19.48	7.84	4.10	31.41
Carried to summary			29.63	281.49	139.43	63.14	484.05
PART J FLOOR FINISHES							
Cement and sand floor screed 40mm thick	J1	9.18m2	2.30	21.85	27.82	7.45	57.12
Vinyl floor tiles, size 300 x 300mm	J2	9.18m2	2.20	20.90	64.07	12.75	97.72
Carried forward			4.50	42.75	91.89	20.20	154.84

	Ref	Qty	Hours	Hours £	Mat'ls £	O & P £	Total £
Brought forward			4.50	42.75	91.89	20.20	154.84
25mm thick tongued and grooved boarding	J3	9.18m2	6.79	64.51	86.29	22.62	173.41
150 x 50mm sawn softwood joists	J4	20.30m	2.87	27.27	46.28	11.03	84.58
Cut and pin ends of joists to existing brick wall	J5	7nr	1.26	11.97	1.75	2.06	15.78
Build in ends of joists to blockwork	J6	7nr	0.70	6.65	1.05	1.16	8.86
Carried forward			16.12	153.14	227.26	57.06	437.46

PART K
CEILING FINISHES

	Ref	Qty	Hours	Hours £	Mat'ls £	O & P £	Total £
Plasterboard with taped butt joints fixed to joists	K1	18.36m2	6.60	62.70	33.97	14.50	111.17
5mm skim coat to plasterboard ceilings	K2	18.36m2	9.18	87.21	22.40	16.44	126.05
Two coats emulsion paint to ceilings	K3	18.36m2	4.77	45.32	10.83	8.42	64.57
Carried to summary			20.55	195.23	67.20	39.36	301.79

PART L
ELECTRICAL WORK

	Ref	Qty	Hours	Hours £	Mat'ls £	O & P £	Total £
13 amp double switched socket outlet with neon	L1	6nr	2.40	33.60	49.66	12.49	95.75
Lighting point	L2	4nr	1.40	19.60	27.20	7.02	53.82
Lighting switch	L3	4nr	1.00	14.00	15.88	4.48	34.36
Lighting wiring	L4	12.00m	1.20	16.80	12.36	4.37	33.53
Power cable	L5	32.00m	4.80	67.20	35.20	15.36	117.76
Carried to summary			10.80	151.20	140.30	43.73	335.23

	Ref	Qty	Hours	Hours £	Mat'ls £	O & P £	Total £
PART M HEATING WORK							
15mm copper pipe	M1	13.00m	2.86	35.75	24.46	9.03	69.24
Elbow	M2	8nr	2.24	28.00	8.80	5.52	42.32
Tee	M3	2nr	0.68	8.50	3.90	1.86	14.26
Radiator, double convector size 1400 x 520mm	M4	4nr	5.20	65.00	462.72	79.16	606.88
Break into existing pipe and insert tee	M5	1nr	0.75	9.38	3.95	2.00	15.32
Carried to summary			11.73	146.63	503.83	97.57	748.02
PART N ALTERATION WORK							
Take out existing window size 1500 x 1000mm and lintel over, adapt opening to receive 1770 x 2000mm patio door and insert new lintel over (both measured separately) and make good	N1	1nr	20.00	190.00	20.00	31.50	241.50
Take out existing window size 1500 x 1000mm, enlarge opening to receive new PVC-U door (measured separately)	N2	1nr	25.00	237.50	45.00	42.38	324.88
Carried to summary			45.00	427.50	65.00	73.88	566.38

SUMMARY

	Hours	Hours £	Mat'ls £	O & P £	Total £
PART A PRELIMINARIES	0.00	0.00	0.00	0.00	2,508.50
PART B SUBSTRUCTURE TO DPC LEVEL	56.18	533.71	528.39	159.32	1,221.42
PART C EXTERNAL WALLS	133.42	1,267.49	2,367.79	545.59	4,182.87
PART D FLAT ROOF	0.00	0.00	0.00	0.00	0.00
PART E PITCHED ROOF	82.26	779.02	979.86	263.83	2,022.71
PART F WINDOWS AND EXTERNAL DOORS	26.70	253.65	1,461.86	257.33	1,972.84
PART G INTERNAL PARTITIONS AND DOORS	0.00	0.00	0.00	0.00	0.00
PART H WALL FINISHES	29.63	281.49	139.43	63.14	484.05
PART J FLOOR FINISHES	16.12	153.14	227.26	57.06	437.46
PART K CEILING FINISHES	20.55	195.23	67.20	39.36	301.79
PART L ELECTRICAL WORK	10.80	151.20	140.30	43.73	335.23
PART M HEATING WORK	11.73	146.63	503.83	97.57	748.02
PART N ALTERATION WORK	45.00	427.50	65.00	73.88	566.38
Final total	432.39	4,189.06	6,480.92	1,600.81	14,781.27

	Ref	Qty	Hours	Hours £	Mat'ls £	O & P £	Total £
PART A PRELIMINARIES							
Concrete mixer	A1	11 wks					495.00
Small tools	A2	11 wks					385.00
Scaffolding (m2/weeks)	A3	550.00					1,237.50
Skip	A4	7 wks					700.00
Clean up	A5	12 hrs					96.00
Carried to summary							2,913.50
PART B SUBSTRUCTURE TO DPC LEVEL							
Excavate topsoil 150mm thick by hand	B1	17.50m2	5.25	49.88	0.00	7.48	57.36
Excavate to reduce levels by hand	B2	5.00m3	12.50	118.75	0.00	17.81	136.56
Excavate for trench foundations by hand	B3	1.72m3	4.47	42.47	0.00	6.37	48.83
Earthwork support to sides of trenches	B4	15.18m2	6.07	57.67	18.22	11.38	87.27
Backfilling with excavated material	B5	0.62m3	0.37	3.52	0.00	0.53	4.04
Hardcore 225mm thick	B6	11.88m2	2.38	22.61	60.35	12.44	95.40
Hardcore filling to trench	B7	0.14m3	0.07	0.67	0.70	0.20	1.57
Concrete grade (1:3:6) in foundations	B8	1.40m3	1.89	17.96	95.11	16.96	130.02
Concrete grade (1:2:4) in bed 150mm thick	B9	11.88m2	3.56	33.82	129.25	24.46	187.53
Concrete (1:2:4) in cavity wall filling	B10	5.20m2	1.04	9.88	18.88	4.31	33.07
Carried forward			37.60	357.20	322.51	101.96	781.67

	Ref	Qty	Hours	Hours £	Mat'ls £	O & P £	Total £
Brought forward			37.60	357.20	322.51	101.96	781.67
Damp-proof membrane	B11	12.38m2	0.52	4.94	7.18	1.82	13.94
Reinforcement ref A193 in foundation	B12	5.20m2	0.62	5.89	6.76	1.90	14.55
Steel fabric reinforcement ref A193 in slab	B13	11.38m2	1.78	16.91	15.46	4.86	37.23
Solid blockwork 140mm thick in cavity wall	B14	6.76m2	8.79	83.51	86.39	25.48	195.38
Common bricks 112.5mm thick in cavity wall	B15	5.20m2	8.79	83.51	68.59	22.81	174.91
Facing bricks in 112.5mm thick in skin of cavity wall	B16	1.56m2	2.81	26.70	36.30	9.45	72.44
Form cavity 50mm wide in cavity wall	B17	6.76m2	0.20	1.90	4.06	0.89	6.85
DPC 112mm wide	B18	10.40m	0.52	4.94	8.42	2.00	15.36
DPC 140mm wide	B19	10.40m	0.62	5.89	10.09	2.40	18.38
Bond in block wall	B20	1.30m	0.59	5.61	2.83	1.27	9.70
Bond in half brick wall	B21	0.30m	0.46	4.37	0.61	0.75	5.73
50mm thick insulation board	B22	11.88m2	3.56	33.82	53.22	13.06	100.10
Carried to summary			66.86	635.17	622.42	188.64	1,446.23
PART C EXTERNAL WALLS							
Solid blockwork 140mm thick in cavity wall	C1	44.14m2	57.38	545.11	561.46	165.99	1,272.56
Facing brickwork 112.5mm thick in cavity wall	C2	44.14m2	79.45	754.78	1,017.87	265.90	2,038.54
75mm thick insulation in cavity wall	C3	44.14m2	9.71	92.25	256.84	52.36	401.45
Carried forward			146.54	1,392.13	1,836.17	484.25	3,712.55

	Ref	Qty	Hours	Hours £	Mat'ls £	O & P £	Total £
Brought forward			146.54	1,392.13	1,836.17	484.25	3,712.55
Steel lintel 2400mm long	C4	2nr	0.50	4.75	196.92	30.25	231.92
Steel lintel 1500mm long	C5	4nr	0.80	7.60	429.88	65.62	503.10
Close cavity wall at jambs	C7	13.76m	0.69	6.56	28.07	5.19	39.82
Close cavity wall at cills	C8	5.07m	0.25	2.38	10.34	1.91	14.62
Close cavity wall at top	C9	10.40m	0.51	4.85	21.22	3.91	29.97
DPC 112mm wide at jambs	C10	13.76m	0.69	6.56	11.15	2.66	20.36
DPC 112mm wide at cills	C11	5.07m	0.25	2.38	4.11	0.97	7.46
Carried to summary			150.23	1,427.19	2,537.86	594.76	4,559.80
PART D FLAT ROOF			N/A	N/A	N/A	N/A	N/A
PART E PITCHED ROOF							
100 x 75mm sawn softwood wall plate	E1	5.00m	1.05	9.98	11.10	3.16	24.24
150 x 50mm sawn softwood pole plate plugged to brickwork	E2	5.30m	0.99	9.41	17.07	3.97	30.45
100 x 50mm sawn softwood rafters	E3	31.50m	3.50	31.50	103.32	20.22	155.04
125 x 25mm sawn softwood purlin	E4	5.30m	0.66	6.27	12.77	2.86	21.90
150 x 50mm sawn softwood joists	E5	24.50m	2.75	26.13	55.88	12.30	94.31
150 x 50mm sawn softwood sprockets	E6	18nr	1.68	15.96	20.52	5.47	41.95
100mm layer of insulation quilt between joists fixed with chicken wire and 150mm layer over joists	E7	15.00m2	7.20	68.40	163.65	34.81	266.86
6mm softwood soffit 150mm wide	E8	12.30m	3.32	31.54	21.53	7.96	61.03
Carried forward			21.15	199.18	405.84	90.75	695.77

	Ref	Qty	Hours	Hours £	Mat'ls £	O & P £	Total £
Brought forward			21.15	199.18	405.84	90.75	695.77
19mm wrought softwood fascia/ barge board 200mm high	E9	5.30m	4.15	39.43	10.44	7.48	57.34
Marley Plain roof tiles on felt and battens	E10	23.85m2	21.95	208.53	617.00	123.83	949.35
Double eaves course	E11	5.30m	0.83	7.89	17.07	3.74	28.70
Verge with plain tile undercloak	E12	9.00m	1.75	16.63	28.98	6.84	52.45
Lead flashing code 5, 200mm girth	E13	5.30m	1.98	18.81	35.40	8.13	62.34
Rake out joint for flashing	E14	5.30m	1.16	11.02	0.80	1.77	13.59
112mm diameter PVC-U gutter	E15	5.30m	0.86	8.17	23.69	4.78	36.64
Stop end	E16	1nr	0.14	1.33	1.11	0.37	2.81
Stop end outlet	E17	1nr	0.25	2.38	1.11	0.52	4.01
68mm diameter PVC-U down pipe	E18	2.50m	0.63	5.99	8.48	2.17	16.63
Shoe	E19	1nr	0.30	2.85	2.09	0.74	5.68
Paint fascia and soffit	E20	3.69m2	1.53	14.54	3.54	2.71	20.79
Carried to summary			56.68	536.71	1,155.55	253.84	1,946.10
PART F WINDOWS AND EXTERNAL DOORS							
PVC-U door size 840 x 1980mm complete (B)	F1	1nr	2.50	23.75	209.99	35.06	268.80
PVC-U sliding patio door size 1700 x 2075mm (C)	F2	2nr	14.00	133.00	587.04	108.01	828.05
Carried forward			16.50	156.75	797.03	143.07	1,096.85

	Ref	Qty	Hours	Hours £	Mat'ls £	O & P £	Total £
Brought forward			16.50	156.75	797.03	143.07	1,096.85
PVC-U window size 1200 x 1200mm complete (A)	F3	4nr	8.00	76.00	630.36	105.95	812.31
25 x 225mm wrought softwood window board	F4	4.80m	1.44	13.68	29.86	6.53	50.07
Paint window board	F5	4.80m	0.76	7.22	4.61	1.77	13.60
Carried to summary			26.70	253.65	1,461.86	257.33	1,972.84
PART G INTERNAL PARTITIONS AND DOORS		N/A	N/A	N/A	N/A	N/A	N/A
PART H WALL FINISHES							
19 x 100mm wrought softwood skirting	H1	18.34m	3.12	29.64	25.13	8.22	62.99
12mm plasterboard fixed to walls with dabs	H2	39.64m2	15.50	147.25	78.88	33.92	260.05
12mm plasterboard fixed to walls less than 300mm wide with dabs	H3	18.33m	3.39	32.21	17.32	7.43	56.95
Two coats emulsion paint to plastered walls	H4	42.46m2	8.49	80.66	25.05	15.86	121.56
Paint skirting	H5	18.34m	2.39	22.71	8.80	4.73	36.23
Carried to summary			32.89	312.46	155.18	70.15	537.78
PART J FLOOR FINISHES							
Cement and sand floor screed 40mm thick	J1	11.88m2	2.97	28.22	36.00	9.63	73.85
Vinyl floor tiles, size 300 x 300mm	J2	11.88m2	2.85	27.08	83.62	16.60	127.30
Carried forward			5.82	55.29	119.62	26.24	201.15

	Ref	Qty	Hours	Hours £	Mat'ls £	O & P £	Total £
Brought forward			5.82	55.29	119.62	26.24	201.15
25mm thick tongued and grooved boarding	J3	11.88m2	8.79	83.51	111.67	29.28	224.45
150 x 50mm sawn softwood joists	J4	23.20m	3.25	30.88	52.90	12.57	96.34
Cut and pin ends of joists to existing brick wall	J5	8nr	1.44	13.68	2.00	2.35	18.03
Build in ends of joists to blockwork	J6	8nr	0.80	7.60	1.20	1.32	10.12
Carried forward			20.10	190.95	287.39	71.75	550.09

PART K
CEILING FINISHES

	Ref	Qty	Hours	Hours £	Mat'ls £	O & P £	Total £
Plasterboard with taped butt joints fixed to joists	K1	23.76m2	6.60	62.70	43.96	16.00	122.66
5mm skim coat to plasterboard ceilings	K2	23.76m2	9.18	87.21	28.99	17.43	133.63
Two coats emulsion paint to ceilings	K3	23.76m2	4.77	45.32	14.02	8.90	68.24
Carried to summary			20.55	195.23	86.97	42.33	324.52

PART L
ELECTRICAL WORK

	Ref	Qty	Hours	Hours £	Mat'ls £	O & P £	Total £
13 amp double switched socket outlet with neon	L1	6nr	2.40	33.60	49.66	12.49	95.75
Lighting point	L2	4nr	1.40	19.60	27.20	7.02	53.82
Lighting switch	L3	4nr	1.00	14.00	15.88	4.48	34.36
Lighting wiring	L4	14.00m	1.40	19.60	14.42	5.10	39.12
Power cable	L5	36.00m	5.40	75.60	39.60	17.28	132.48
Carried to summary			11.60	162.40	146.76	46.37	355.53

	Ref	Qty	Hours	Hours £	Mat'ls £	O & P £	Total £
PART M **HEATING WORK**							
15mm copper pipe	M1	15.00m	3.30	41.25	28.20	10.42	79.87
Elbow	M2	8nr	2.24	28.00	8.80	5.52	42.32
Tee	M3	2nr	0.68	8.50	3.90	1.86	14.26
Radiator, double convector size 1400 x 520mm	M4	4nr	5.20	65.00	462.72	79.16	606.88
Break into existing pipe and insert tee	M5	1nr	0.75	9.38	3.95	2.00	15.32
Carried to summary			12.17	152.13	507.57	98.95	758.65
PART N **ALTERATION WORK**							
Take out existing window size 1500 x 1000mm and lintel over, adapt opening to receive 1770 x 2000mm patio door and insert new lintel over (both measured separately) and make good	N1	1nr	20.00	190.00	20.00	31.50	241.50
Take out existing window size 1500 x 1000mm, enlarge opening to receive new PVC-U door (measured separately)	N2	1nr	25.00	237.50	45.00	42.38	324.88
Carried to summary			45.00	427.50	65.00	73.88	566.38

SUMMARY

	Hours	Hours £	Mat'ls £	O & P £	Total £
PART A PRELIMINARIES	0.00	0.00	0.00	0.00	2,913.50
PART B SUBSTRUCTURE TO DPC LEVEL	66.86	635.17	622.42	188.64	1,446.23
PART C EXTERNAL WALLS	150.23	1,427.19	2,537.86	594.76	4,559.80
PART D FLAT ROOF	0.00	0.00	0.00	188.81	0.00
PART E PITCHED ROOF	56.68	536.71	1,155.55	253.84	1,946.10
PART F WINDOWS AND EXTERNAL DOORS	26.70	253.65	1,461.86	257.33	1,972.84
PART G INTERNAL PARTITIONS AND DOORS	0.00	0.00	0.00	0.00	0.00
PART H WALL FINISHES	32.89	312.46	155.18	70.15	537.78
PART J FLOOR FINISHES	20.10	190.95	287.39	71.75	550.09
PART K CEILING FINISHES	20.55	195.23	86.97	42.33	324.52
PART L ELECTRICAL WORK	11.60	162.40	146.76	46.37	355.53
PART M HEATING WORK	12.17	152.13	507.57	98.95	758.65
PART N ALTERATION WORK	45.00	427.50	65.00	73.88	566.38
Final total	442.78	4,293.39	7,026.56	1,886.81	15,931.42

	Ref	Qty	Hours	Hours £	Mat'ls £	O & P £	Total £
PART A PRELIMINARIES							
Concrete mixer	A1	12 wks					540.00
Small tools	A2	12 wks					420.00
Scaffolding (m2/weeks)	A3	720.00					1,620.00
Skip	A4	8 wks					800.00
Clean up	A5	12 hrs					96.00
Carried to summary							3,476.00
PART B SUBSTRUCTURE TO DPC LEVEL							
Excavate topsoil 150mm thick by hand	B1	19.85m2	5.96	56.62	0.00	8.49	65.11
Excavate to reduce levels by hand	B2	5.95m3	14.88	141.36	0.00	21.20	162.56
Excavate for trench foundations by hand	B3	1.88m3	4.89	46.46	0.00	6.97	53.42
Earthwork support to sides of trenches	B4	16.64m2	6.66	63.27	19.97	12.49	95.73
Backfilling with excavated material	B5	0.70m3	0.42	3.99	0.00	0.60	4.59
Hardcore 225mm thick	B6	14.58m2	2.92	27.74	74.07	15.27	117.08
Hardcore filling to trench	B7	0.15m3	0.08	0.76	0.76	0.23	1.75
Concrete grade (1:3:6) in foundations	B8	1.54m3	2.08	19.76	104.63	18.66	143.05
Concrete grade (1:2:4) in bed 150mm thick	B9	14.58m2	4.37	41.52	158.94	30.07	230.52
Concrete (1:2:4) in cavity wall filling	B10	5.70m2	1.14	10.83	20.69	4.73	36.25
Carried forward			43.40	412.30	379.06	118.70	910.06

	Ref	Qty	Hours	Hours £	Mat'ls £	O & P £	Total £
Brought forward			43.40	412.30	379.06	118.70	910.06
Damp-proof membrane	B11	15.13m2	0.61	5.80	8.88	2.20	16.88
Reinforcement ref A193 in foundation	B12	5.70m2	0.68	6.46	7.41	2.08	15.95
Steel fabric reinforcement ref A193 in slab	B13	14.58m2	2.19	20.81	18.95	5.96	45.72
Solid blockwork 140mm thick in cavity wall	B14	7.41m2	9.63	91.49	94.70	27.93	214.11
Common bricks 112.5mm thick in cavity wall	B15	5.70m2	9.63	91.49	75.18	25.00	191.66
Facing bricks in 112.5mm thick in skin of cavity wall	B16	1.71m2	3.08	29.26	39.79	10.36	79.41
Form cavity 50mm wide in cavity wall	B17	7.41m2	0.22	2.09	4.45	0.98	7.52
DPC 112mm wide	B18	11.40m	0.57	5.42	9.23	2.20	16.84
DPC 140mm wide	B19	11.40m	0.68	6.46	11.06	2.63	20.15
Bond in block wall	B20	1.30m	0.58	5.51	2.83	1.25	9.59
Bond in half brick wall	B21	0.30m	0.46	4.37	0.61	0.75	5.73
50mm thick insulation board	B22	14.58m2	4.37	41.52	65.32	16.03	122.86
Carried to summary			76.10	722.95	717.47	216.06	1,656.48
PART C EXTERNAL WALLS							
Solid blockwork 140mm thick in cavity wall	C1	46.65m2	60.65	576.18	593.39	175.43	1,345.00
Facing brickwork 112.5mm thick in cavity wall	C2	46.65m2	83.97	797.72	1,075.75	281.02	2,154.48
75mm thick insulation in cavity wall	C3	46.65m2	9.82	93.29	271.50	54.72	419.51
Carried forward			154.44	1,467.18	1,940.64	511.17	3,918.99

	Ref	Qty	Hours	Hours £	Mat'ls £	O & P £	Total £
Brought forward			154.44	1,467.18	1,940.64	511.17	3,918.99
Steel lintel 2400mm long	C4	2nr	0.50	4.75	196.92	30.25	231.92
Steel lintel 1500mm long	C5	6nr	1.20	11.40	644.82	98.43	754.65
Steel lintel 1150mm long	C6	1nr	0.15	1.43	47.18	7.29	55.90
Close cavity wall at jambs	C7	24.52m	1.23	11.69	50.02	9.26	70.96
Close cavity wall at cills	C8	9.31m	0.47	4.47	18.99	3.52	26.97
Close cavity wall at top	C9	11.40m	0.57	5.42	23.26	4.30	32.98
DPC 112mm wide at jambs	C10	24.52m	1.23	11.69	19.86	4.73	36.28
DPC 112mm wide at cills	C11	9.31m	0.47	4.47	7.54	1.80	13.81
Carried to summary			160.11	1,521.05	2,902.05	663.46	5,086.56
PART D FLAT ROOF			N/A	N/A	N/A	N/A	N/A
PART E PITCHED ROOF							
100 x 75mm sawn softwood wall plate	E1	6.00m	1.80	17.10	13.32	4.56	34.98
150 x 50mm sawn softwood pole plate plugged to brickwork	E2	6.30m	1.89	17.96	20.29	5.74	43.98
100 x 50mm sawn softwood rafters	E3	31.50m	6.30	56.70	103.32	24.00	184.02
125 x 25mm sawn softwood purlin	E4	6.30m	0.86	8.17	15.18	3.50	26.85
150 x 50mm sawn softwood joists	E5	6.30m	0.86	8.17	55.86	9.60	73.63
150 x 50mm sawn softwood sprockets	E6	18nr	2.16	20.52	20.52	6.16	47.20
100mm layer of insulation quilt between joists fixed with chicken wire and 150mm layer over joists	E7	18.00m2	8.64	82.08	196.38	41.77	320.23
6mm softwood soffit 150mm wide	E8	13.30m	5.32	50.54	23.28	11.07	84.89
Carried forward			27.83	261.24	448.15	106.41	815.79

	Ref	Qty	Hours	Hours £	Mat'ls £	O & P £	Total £
Brought forward			27.83	261.24	448.15	106.41	815.79
19mm wrought softwood fascia/ barge board 200mm high	E9	6.30m	6.65	63.18	12.41	11.34	86.92
Marley Plain roof tiles on felt and battens	E10	28.35m2	33.87	321.77	733.42	158.28	1,213.46
Double eaves course	E11	6.30m	2.21	21.00	20.29	6.19	47.48
Verge with plain tile undercloak	E12	9.00m	2.25	21.38	28.98	7.55	57.91
Lead flashing code 5, 200mm girth	E13	6.30m	3.78	35.91	42.80	11.81	90.52
Rake out joint for flashing	E14	6.30m	2.21	21.00	0.95	3.29	25.24
112mm diameter PVC-U gutter	E15	6.30m	1.64	15.58	28.16	6.56	50.30
Stop end	E16	1nr	0.14	1.33	1.11	0.37	2.81
Stop end outlet	E17	1nr	0.25	2.38	1.11	0.52	4.01
68mm diameter PVC-U down pipe	E18	2.50m	0.63	5.99	8.48	2.17	16.63
Shoe	E19	1nr	0.30	2.85	2.09	0.74	5.68
Paint fascia and soffit	E20	3.99m2	2.79	26.51	3.83	4.55	34.89
Carried to summary			84.55	800.08	1,331.78	319.78	2,451.63

PART F
WINDOWS AND EXTERNAL DOORS

	Ref	Qty	Hours	Hours £	Mat'ls £	O & P £	Total £
PVC-U door size 840 x 1980mm complete (B)	F1	2nr	5.00	47.50	419.98	70.12	537.60
PVC-U sliding patio door size 1700 x 2075mm (C)	F2	2nr	14.00	133.00	587.04	108.01	828.05
Carried forward			19.00	180.50	1,007.02	178.13	1,365.65

	Ref	Qty	Hours	Hours £	Mat'ls £	O & P £	Total £
Brought forward			19.00	180.50	1,007.02	178.13	1,365.65
PVC-U window size 1200 x 1200mm complete (A)	F3	6nr	12.00	114.00	945.54	158.93	1,218.47
25 x 225mm wrought softwood window board	F4	7.20m	2.16	20.52	44.79	9.80	75.11
Paint window board	F5	7.20m	1.22	11.59	6.91	2.78	21.28
Carried to summary			34.38	326.61	2,004.26	349.63	2,680.50
PART G INTERNAL PARTITIONS AND DOORS		N/A	N/A	N/A	N/A	N/A	N/A
PART H WALL FINISHES							
19 x 100mm wrought softwood skirting	H1	19.56m	3.33	31.64	26.80	8.77	67.20
12mm plasterboard fixed to walls with dabs	H2	40.00m2	15.60	148.20	79.60	34.17	261.97
12mm plasterboard fixed to walls less than 300mm wide with dabs	H3	33.83m	6.08	57.76	31.12	13.33	102.21
Two coats emulsion paint to plastered walls	H4	25.07m2	9.01	85.60	26.59	16.83	129.01
Paint skirting	H5	19.56m	2.54	24.13	9.39	5.03	38.55
Carried to summary			36.56	347.32	173.50	78.12	598.94
PART J FLOOR FINISHES							
Cement and sand floor screed 40mm thick	J1	14.58m2	3.65	34.68	44.84	11.93	91.44
Vinyl floor tiles, size 300 x 300mm	J2	14.58m2	3.50	33.25	101.77	20.25	155.27
Carried forward			7.15	67.93	146.61	32.18	246.72

	Ref	Qty	Hours	Hours £	Mat'ls £	O & P £	Total £
Brought forward			7.15	67.93	146.61	32.18	246.72
25mm thick tongued and grooved boarding	J3	14.58m2	10.79	102.51	137.05	35.93	275.49
150 x 50mm sawn softwood joists	J4	29.00m	4.06	38.57	66.12	15.70	120.39
Cut and pin ends of joists to existing brick wall	J5	10nr	1.80	17.10	2.50	2.94	22.54
Build in ends of joists to blockwork	J6	10nr	1.00	9.50	1.50	1.65	12.65
Carried to summary			24.80	235.60	353.78	88.41	677.79
PART K **CEILING FINISHES**							
Plasterboard with taped butt joints fixed to joists	K1	29.18m2	10.51	99.85	53.98	23.07	176.90
5mm skim coat to plasterboard ceilings	K2	29.18m2	14.59	138.61	35.60	26.13	200.34
Two coats emulsion paint to ceilings	K3	29.18m2	7.59	72.11	17.22	13.40	102.72
Carried to summary			32.69	310.56	106.80	62.60	479.96
PART L **ELECTRICAL WORK**							
13 amp double switched socket outlet with neon	L1	8nr	3.20	44.80	66.08	16.63	127.51
Lighting point	L2	4nr	1.40	19.60	27.20	7.02	53.82
Lighting switch	L3	5nr	1.25	17.50	19.85	5.60	42.95
Lighting wiring	L4	16.00m	1.60	22.40	16.48	5.83	44.71
Power cable	L5	40.00m	6.00	84.00	44.00	19.20	147.20
Carried to summary			13.45	188.30	173.61	54.29	416.20

	Ref	Qty	Hours	Hours £	Mat'ls £	O & P £	Total £
PART M **HEATING WORK**							
15mm copper pipe	M1	18.00m	3.96	49.50	33.84	12.50	95.84
Elbow	M2	8nr	2.24	28.00	8.80	5.52	42.32
Tee	M3	2nr	0.68	8.50	3.90	1.86	14.26
Radiator, double convector size 1400 x 520mm	M4	4nr	5.20	65.00	462.72	79.16	606.88
Break into existing pipe and insert tee	M5	1nr	0.75	9.38	3.95	2.00	15.32
Carried to summary			12.83	160.38	513.21	101.04	774.62
PART N **ALTERATION WORK**							
Take out existing window size 1500 x 1000mm and lintel over, adapt opening to receive 1770 x 2000mm patio door and insert new lintel over (both measured separately) and make good	N1	1nr	20.00	190.00	20.00	31.50	241.50
Take out existing window size 1500 x 1000mm, enlarge opening to receive new PVC-U door (measured separately)	N2	1nr	25.00	237.50	45.00	42.38	324.88
Carried to summary			45.00	427.50	65.00	73.88	566.38

SUMMARY

	Hours	Hours £	Mat'ls £	O & P £	Total £
PART A PRELIMINARIES	0.00	0.00	0.00	0.00	3,476.00
PART B SUBSTRUCTURE TO DPC LEVEL	76.10	722.95	717.47	216.06	1,656.48
PART C EXTERNAL WALLS	160.11	1,521.05	2,902.05	663.46	5,086.56
PART D FLAT ROOF	0.00	0.00	0.00	0.00	0.00
PART E PITCHED ROOF	84.55	800.08	1,331.78	319.78	2,451.63
PART F WINDOWS AND EXTERNAL DOORS	34.38	326.61	2,004.26	349.63	2,680.50
PART G INTERNAL PARTITIONS AND DOORS	0.00	0.00	0.00	0.00	0.00
PART H WALL FINISHES	36.56	347.32	173.50	78.12	598.94
PART J FLOOR FINISHES	24.80	235.60	353.78	88.41	677.79
PART K CEILING FINISHES	32.69	310.56	106.80	62.60	479.96
PART L ELECTRICAL WORK	13.45	188.30	173.61	54.29	416.20
PART M HEATING WORK	12.83	160.38	513.21	101.04	774.62
PART N ALTERATION WORK	45.00	427.50	65.00	73.88	566.38
Final total	520.47	5,040.35	8,341.46	2,007.27	18,865.06

	Ref	Qty	Hours	Hours £	Mat'ls £	O & P £	Total £
PART A PRELIMINARIES							
Concrete mixer	A1	11 wks					495.00
Small tools	A2	11 wks					385.00
Scaffolding (m2/weeks)	A3	720.00					1,620.00
Skip	A4	7 wks					700.00
Clean up	A5	12 hrs					96.00
Carried to summary							3,296.00
PART B SUBSTRUCTURE TO DPC LEVEL							
Excavate topsoil 150mm thick by hand	B1	17.85m2	5.36	50.92	0.00	7.64	58.56
Excavate to reduce levels by hand	B2	5.35m3	13.38	127.11	0.00	19.07	146.18
Excavate for trench foundations by hand	B3	1.88m3	4.89	46.46	0.00	6.97	53.42
Earthwork support to sides of trenches	B4	16.64m2	6.66	63.27	19.97	12.49	95.73
Backfilling with excavated material	B5	0.62m3	0.37	3.52	0.00	0.53	4.04
Hardcore 225mm thick	B6	13.69m2	2.74	26.03	69.54	14.34	109.91
Hardcore filling to trench	B7	0.14m3	0.07	0.67	0.70	0.20	1.57
Concrete grade (1:3:6) in foundations	B8	1.54m3	2.08	19.76	104.63	18.66	143.05
Concrete grade (1:2:4) in bed 150mm thick	B9	14.58m2	4.11	39.05	148.94	28.20	216.18
Concrete (1:2:4) in cavity wall filling	B10	5.70m2	1.14	10.83	20.69	4.73	36.25
Carried forward			40.80	387.60	364.47	112.81	864.88

	Ref	Qty	Hours	Hours £	Mat'ls £	O & P £	Total £
Brought forward			40.80	387.60	364.47	112.81	864.88
Damp-proof membrane	B11	13.13m2	0.53	5.04	7.62	1.90	14.55
Reinforcement ref A193 in foundation	B12	5.70m2	0.68	6.46	7.41	2.08	15.95
Steel fabric reinforcement ref A193 in slab	B13	13.69m2	2.05	19.48	17.80	5.59	42.87
Solid blockwork 140mm thick in cavity wall	B14	7.41m2	9.63	91.49	94.70	27.93	214.11
Common bricks 112.5mm thick in cavity wall	B15	5.70m2	9.63	91.49	75.18	25.00	191.66
Facing bricks in 112.5mm thick in skin of cavity wall	B16	1.71m2	2.27	21.57	39.79	9.20	70.56
Form cavity 50mm wide in cavity wall	B17	7.41m2	0.22	2.09	4.45	0.98	7.52
DPC 112mm wide	B18	11.40m	0.57	5.42	9.23	2.20	16.84
DPC 140mm wide	B19	11.40m	0.68	6.46	11.06	2.63	20.15
Bond in block wall	B20	1.30m	0.58	5.51	2.83	1.25	9.59
Bond in half brick wall	B21	0.30m	0.46	4.37	0.61	0.75	5.73
50mm thick insulation board	B22	13.69m2	4.11	39.05	61.33	15.06	115.43
Carried to summary			72.21	686.00	696.48	207.37	1,589.85
PART C EXTERNAL WALLS							
Solid blockwork 140mm thick in cavity wall	C1	44.65m2	58.05	551.48	567.95	167.91	1,287.34
Facing brickwork 112.5mm thick in cavity wall	C2	44.65m2	80.37	763.52	1,029.63	268.97	2,062.12
75mm thick insulation in cavity wall	C3	44.65m2	9.82	93.29	271.50	54.72	419.51
Carried forward			148.24	1,408.28	1,869.08	491.60	3,768.96

	Ref	Qty	Hours	Hours £	Mat'ls £	O & P £	Total £
Brought forward			148.24	1,408.28	1,869.08	491.60	3,768.96
Steel lintel 2400mm long	C4	2nr	0.50	4.75	196.92	30.25	231.92
Steel lintel 1500mm long	C5	6nr	1.20	11.40	644.82	98.43	754.65
Steel lintel 1150mm long	C6	1nr	0.15	1.43	47.18	7.29	55.90
Close cavity wall at jambs	C7	24.52m	1.23	11.69	50.02	9.26	70.96
Close cavity wall at cills	C8	9.31m	0.47	4.47	18.99	3.52	26.97
Close cavity wall at top	C9	11.40m	0.57	5.42	23.26	4.30	32.98
DPC 112mm wide at jambs	C10	24.52m	1.23	11.69	19.86	4.73	36.28
DPC 112mm wide at cills	C11	9.31m	0.47	4.47	7.54	1.80	13.81
Carried to summary			153.91	1,462.15	2,830.49	643.90	4,936.53
PART D FLAT ROOF			N/A	N/A	N/A	N/A	N/A
PART E PITCHED ROOF							
100 x 75mm sawn softwood wall plate	E1	4.00m	1.40	13.30	8.88	3.33	25.51
150 x 50mm sawn softwood pole plate plugged to brickwork	E2	4.30m	1.29	12.26	13.85	3.92	30.02
100 x 50mm sawn softwood rafters	E3	49.50m	9.90	89.10	162.36	37.72	289.18
125 x 25mm sawn softwood purlin	E4	4.30m	0.86	8.17	10.36	2.78	21.31
150 x 50mm sawn softwood joists	E5	40.50	8.10	76.95	92.34	25.39	194.68
150 x 50mm sawn softwood sprockets	E6	20nr	2.40	22.80	22.80	6.84	52.44
100mm layer of insulation quilt between joists fixed with chicken wire and 150mm layer over joists	E7	16.00m2	7.68	72.96	174.56	37.13	284.65
6mm softwood soffit 150mm wide	E8	13.30m	5.32	50.54	23.28	11.07	84.89
Carried forward			36.95	346.08	508.43	128.18	982.68

	Ref	Qty	Hours	Hours £	Mat'ls £	O & P £	Total £
Brought forward			36.95	346.08	508.43	128.18	982.68
19mm wrought softwood fascia/ barge board 200mm high	E9	4.30m	6.65	63.18	8.47	10.75	82.39
Marley Plain roof tiles on felt and battens	E10	23.65m2	44.94	426.93	611.83	155.81	1,194.57
Double eaves course	E11	4.30m	1.51	14.35	13.85	4.23	32.42
Verge with plain tile undercloak	E12	11.00m	2.75	26.13	35.42	9.23	70.78
Lead flashing code 5, 200mm girth	E13	4.30m	2.58	24.51	28.72	7.98	61.21
Rake out joint for flashing	E14	4.30m	1.51	14.35	0.65	2.25	17.24
112mm diameter PVC-U gutter	E15	4.30m	1.12	10.64	19.26	4.49	34.39
Stop end	E16	1nr	0.14	1.33	1.11	0.37	2.81
Stop end outlet	E17	1nr	0.25	2.38	1.11	0.52	4.01
68mm diameter PVC-U down pipe	E18	2.50m	0.63	5.99	8.48	2.17	16.63
Shoe	E19	1nr	0.30	2.85	2.09	0.74	5.68
Paint fascia and soffit	E20	3.99m2	2.79	26.51	3.83	4.55	34.89
Carried to summary			102.12	965.19	1,243.25	331.27	2,539.71

PART F
WINDOWS AND
EXTERNAL DOORS

	Ref	Qty	Hours	Hours £	Mat'ls £	O & P £	Total £
PVC-U door size 840 x 1980mm complete (B)	F1	2nr	5.00	47.50	419.98	70.12	537.60
PVC-U sliding patio door size 1700 x 2075mm (C)	F2	2nr	14.00	133.00	587.04	108.01	828.05
Carried forward			19.00	180.50	1,007.02	178.13	1,365.65

	Ref	Qty	Hours	Hours £	Mat'ls £	O & P £	Total £
Brought forward			19.00	180.50	1,007.02	178.13	1,365.65
PVC-U window size 1200 x 1200mm complete (A)	F3	6nr	12.00	114.00	945.54	158.93	1,218.47
25 x 225mm wrought softwood window board	F4	7.20m	2.16	20.52	44.78	9.80	75.10
Paint window board	F5	7.20m	1.22	11.59	6.91	2.78	21.28
Carried to summary			34.38	326.61	2,004.25	349.63	2,680.49
PART G INTERNAL PARTITIONS AND DOORS		N/A	N/A	N/A	N/A	N/A	N/A
PART H WALL FINISHES							
19 x 100mm wrought softwood skirting	H1	19.56m	3.33	31.64	26.80	8.77	67.20
12mm plasterboard fixed to walls with dabs	H2	40.00m2	15.60	148.20	79.60	34.17	261.97
12mm plasterboard fixed to walls less than 300mm wide with dabs	H3	33.83m	6.08	57.76	31.12	13.33	102.21
Two coats emulsion paint to walls	H4	45.07m2	9.01	85.60	26.59	16.83	129.01
Paint skirting	H5	19.56m	2.54	24.13	9.39	5.03	38.55
Carried to summary			36.56	347.32	173.50	78.12	598.94
PART J FLOOR FINISHES							
Cement and sand floor screed 40mm thick	J1	12.58m2	3.15	29.93	38.12	10.21	78.25
Vinyl floor tiles, size 300 x 300mm	J2	12.58m2	3.02	28.69	87.81	17.48	133.98
Carried forward			6.17	58.62	125.93	27.68	212.23

	Ref	Qty	Hours	Hours £	Mat'ls £	O & P £	Total £
Brought forward			6.17	58.62	125.93	27.68	212.23
25mm thick tongued and grooved boarding	J3	12.58m2	9.31	88.45	118.25	31.00	237.70
150 x 50mm sawn softwood joists	J4	27.30m	3.82	36.29	62.24	14.78	113.31
Cut and pin ends of joists to existing brick wall	J5	7nr	1.26	11.97	1.75	2.06	15.78
Build in ends of joists to blockwork	J6	7nr	0.70	6.65	1.05	1.16	8.86
Carried to summary			21.26	201.97	309.22	76.68	587.87
PART K CEILING FINISHES							
Plasterboard with taped butt joints fixed to joists	K1	25.16m2	9.06	86.07	46.55	19.89	152.51
5mm skim coat to plasterboard ceilings	K2	25.16m2	12.58	119.51	30.70	22.53	172.74
Two coats emulsion paint to ceilings	K3	25.16m2	6.54	62.13	14.84	11.55	88.52
Carried to summary			28.18	267.71	92.09	53.97	413.77
PART L ELECTRICAL WORK							
13 amp double switched socket outlet with neon	L1	6nr	2.40	33.60	49.36	12.44	95.40
Lighting point	L2	4nr	1.40	19.60	27.20	7.02	53.82
Lighting switch	L3	5nr	1.25	17.50	19.85	5.60	42.95
Lighting wiring	L4	16.00m	1.60	22.40	16.48	5.83	44.71
Power cable	L5	36.00m	5.40	75.60	39.60	17.28	132.48
Carried to summary			12.05	168.70	152.49	48.18	369.37

	Ref	Qty	Hours	Hours £	Mat'ls £	O & P £	Total £
PART M HEATING WORK							
15mm copper pipe	M1	19.00m	4.18	52.25	35.72	13.20	101.17
Elbow	M2	8nr	2.24	28.00	8.80	5.52	42.32
Tee	M3	2nr	0.68	8.50	3.90	1.86	14.26
Radiator, double convector size 1400 x 520mm	M4	4nr	5.20	65.00	462.72	79.16	606.88
Break into existing pipe and insert tee	M5	1nr	0.75	9.38	3.95	2.00	15.32
Carried to summary			13.05	163.13	515.09	101.73	779.95
PART N ALTERATION WORK							
Take out existing window size 1500 x 1000mm and lintel over, adapt opening to receive 1770 x 2000mm patio door and insert new lintel over (both measured separately) and make good	N1	1nr	20.00	190.00	20.00	31.50	241.50
Take out existing window size 1500 x 1000mm, enlarge opening to receive new PVC-U door (measured separately)	N2	1nr	25.00	237.50	45.00	42.38	324.88
Carried to summary			45.00	427.50	65.00	73.88	566.38

SUMMARY

	Hours	Hours £	Mat'ls £	O & P £	Total £
PART A PRELIMINARIES	0.00	0.00	0.00	0.00	3,296.00
PART B SUBSTRUCTURE TO DPC LEVEL	72.21	686.00	696.48	207.37	1,589.85
PART C EXTERNAL WALLS	153.91	1,426.15	2,830.49	643.90	4,936.53
PART D FLAT ROOF	0.00	0.00	0.00	0.00	0.00
PART E PITCHED ROOF	102.12	965.19	1,243.25	331.27	2,539.71
PART F WINDOWS AND EXTERNAL DOORS	34.88	326.61	2,004.25	349.63	2,680.49
PART G INTERNAL PARTITIONS AND DOORS	0.00	0.00	0.00	0.00	0.00
PART H WALL FINISHES	36.56	347.42	173.50	78.12	598.94
PART J FLOOR FINISHES	21.26	201.97	309.22	76.68	587.87
PART K CEILING FINISHES	28.18	267.71	92.09	53.97	413.77
PART L ELECTRICAL WORK	12.05	168.70	152.49	48.18	369.37
PART M HEATING WORK	13.05	163.13	515.09	101.73	779.95
PART N ALTERATION WORK	45.00	427.50	65.00	73.88	566.38
Final total	519.22	4,980.38	8,081.86	1,964.73	18,358.86

	Ref	Qty	Hours	Hours £	Mat'ls £	O & P £	Total £
PART A PRELIMINARIES							
Concrete mixer	A1	12 wks					540.00
Small tools	A2	12 wks					420.00
Scaffolding (m2/weeks)	A3	780.00					1,755.00
Skip	A4	8 wks					800.00
Clean up	A5	12 hrs					96.00
Carried to summary							3,611.00
PART B SUBSTRUCTURE TO DPC LEVEL							
Excavate topsoil 150mm thick by hand	B1	22.00m2	6.60	62.70	0.00	9.41	72.11
Excavate to reduce levels by hand	B2	6.60m3	16.60	157.70	0.00	23.66	181.36
Excavate for trench foundations by hand	B3	2.05m3	5.33	50.64	0.00	7.60	58.23
Earthwork support to sides of trenches	B4	18.10m2	7.24	68.78	21.72	13.58	104.08
Backfilling with excavated material	B5	0.70m3	0.42	3.99	0.00	0.60	4.59
Hardcore 225mm thick	B6	16.28m2	2.44	23.18	82.70	15.88	121.76
Hardcore filling to trench	B7	0.15m3	0.08	0.76	0.76	0.23	1.75
Concrete grade (1:3:6) in foundations	B8	1.67m3	2.25	21.38	113.46	20.23	155.06
Concrete grade (1:2:4) in bed 150mm thick	B9	16.28m2	4.88	46.36	177.13	33.52	257.01
Concrete (1:2:4) in cavity wall filling	B10	6.20m2	1.24	11.78	22.51	5.14	39.43
Carried forward			47.08	447.26	418.28	129.83	995.37

	Ref	Qty	Hours	Hours £	Mat'ls £	O & P £	Total £
Brought forward			47.08	447.26	418.28	129.83	995.37
Damp-proof membrane	B11	16.88m2	0.68	6.46	9.79	2.44	18.69
Reinforcement ref A193 in foundation	B12	6.20m2	0.74	7.03	8.06	2.26	17.35
Steel fabric reinforcement ref A193 in slab	B13	16.28m2	2.44	23.18	21.16	6.65	50.99
Solid blockwork 140mm thick in cavity wall	B14	8.06m2	10.48	99.56	103.00	30.38	232.94
Common bricks 112.5mm thick in cavity wall	B15	6.20m2	10.48	99.56	81.78	27.20	208.54
Facing bricks in 112.5mm thick in skin of cavity wall	B16	1.86m2	3.35	31.83	43.28	11.27	86.37
Form cavity 50mm wide in cavity wall	B17	8.06m2	0.24	2.28	4.84	1.07	8.19
DPC 112mm wide	B18	12.40m	0.62	5.89	10.04	2.39	18.32
DPC 140mm wide	B19	12.40m	0.74	7.03	12.03	2.86	21.92
Bond in block wall	B20	1.30m	0.58	5.51	2.83	1.25	9.59
Bond in half brick wall	B21	0.30m	0.46	4.37	0.61	0.75	5.73
50mm thick insulation board	B22	16.28m2	4.88	46.36	72.93	17.89	137.18
Carried to summary			82.77	786.32	788.63	236.24	1,811.19
PART C EXTERNAL WALLS							
Solid blockwork 140mm thick in cavity wall	C1	46.82m2	60.87	578.27	595.55	176.07	1,349.89
Facing brickwork 112.5mm thick in cavity wall	C2	46.82m2	84.28	800.66	1,079.67	282.05	2,162.38
75mm thick insulation in cavity wall	C3	46.82m2	10.30	97.85	272.49	55.55	425.89
Carried forward			155.45	1,476.78	1,947.71	513.67	3,938.16

	Ref	Qty	Hours	Hours £	Mat'ls £	O & P £	Total £
Brought forward			155.45	1,476.78	1,947.71	513.67	3,938.16
Steel lintel 2400mm long	C4	2nr	0.50	4.75	196.92	30.25	231.92
Steel lintel 1500mm long	C5	8nr	1.60	15.20	859.76	131.24	1,006.20
Steel lintel 1150mm long	C6	1nr	0.15	1.43	47.18	7.29	55.90
Close cavity wall at jambs	C7	29.32m	1.47	13.97	59.81	11.07	84.84
Close cavity wall at cills	C8	11.71m	0.59	5.61	23.89	4.42	33.92
Close cavity wall at top	C9	12.40m	0.62	5.89	25.30	4.68	35.87
DPC 112mm wide at jambs	C10	29.32m	1.47	13.97	23.75	5.66	43.37
DPC 112mm wide at cills	C11	11.71m	0.59	5.61	9.48	2.26	17.35
Carried to summary			162.29	1,541.76	3,146.62	703.26	5,391.63
PART D FLAT ROOF			N/A	N/A	N/A	N/A	N/A
PART E PITCHED ROOF							
100 x 75mm sawn softwood wall plate	E1	5.00m	1.75	16.63	11.10	4.16	31.88
150 x 50mm sawn softwood pole plate plugged to brickwork	E2	5.30m	1.59	15.11	17.07	4.83	37.00
100 x 50mm sawn softwood rafters	E3	49.50m	9.90	89.10	162.36	37.72	289.18
125 x 25mm sawn softwood purlin	E4	5.30m	1.06	10.07	12.77	3.43	26.27
150 x 50mm sawn softwood joists	E5	40.50m	8.91	84.65	92.34	26.55	203.53
150 x 50mm sawn softwood sprockets	E6	20nr	2.40	22.80	22.80	6.84	52.44
100mm layer of insulation quilt between joists fixed with chicken wire and 150mm layer over joists	E7	20.00m2	9.60	91.20	218.20	46.41	355.81
6mm softwood soffit 150mm wide	E8	16.30m	5.72	54.34	25.03	11.91	91.28
Carried forward			40.93	383.89	561.67	141.83	1,087.39

	Ref	Qty	Hours	Hours £	Mat'ls £	O & P £	Total £
Brought forward			40.93	383.89	561.67	141.83	1,087.39
19mm wrought softwood fascia/ barge board 200mm high	E9	5.30m	7.15	67.93	10.46	11.76	90.14
Marley Plain roof tiles on felt and battens	E10	29.15m2	55.39	526.21	754.11	192.05	1,472.36
Double eaves course	E11	5.30m	1.86	17.67	17.07	5.21	39.95
Verge with plain tile undercloak	E12	11.00m	2.75	26.13	35.47	9.24	70.83
Lead flashing code 5, 200mm girth	E13	5.30m	3.18	30.21	35.40	9.84	75.45
Rake out joint for flashing	E14	5.30m	1.86	17.67	0.80	2.77	21.24
112mm diameter PVC-U gutter	E15	5.30m	1.38	13.11	23.69	5.52	42.32
Stop end	E16	1nr	0.14	1.33	1.11	0.37	2.81
Stop end outlet	E17	1nr	0.25	2.38	1.11	0.52	4.01
68mm diameter PVC-U down pipe	E18	2.50m	0.63	5.99	8.48	2.17	16.63
Shoe	E19	1nr	0.30	2.85	2.09	0.74	5.68
Paint fascia and soffit	E20	4.29m2	3.00	28.50	4.12	4.89	37.51
Carried to summary			118.82	1,123.84	1,455.58	386.91	2,966.33
PART F WINDOWS AND EXTERNAL DOORS							
PVC-U door size 840 x 1980mm complete (B)	F1	2nr	5.00	47.50	419.98	70.12	537.60
PVC-U sliding patio door size 1700 x 2075mm (C)	F2	2nr	14.00	133.00	587.04	108.01	828.05
Carried forward			19.00	180.50	1,007.02	178.13	1,365.65

	Ref	Qty	Hours	Hours £	Mat'ls £	O & P £	Total £
Brought forward			19.00	180.50	1,007.02	178.13	1,365.65
PVC-U window size 1200 x 1200mm complete (A)	F3	8nr	16.00	152.00	1,260.72	211.91	1,624.63
25 x 225mm wrought softwood window board	F4	9.60m	2.88	27.36	59.71	13.06	100.13
Paint window board	F5	9.60m	1.52	14.44	9.22	3.55	27.21
Carried to summary			39.40	374.30	2,336.67	406.65	3,117.62
PART G INTERNAL PARTITIONS AND DOORS		N/A	N/A	N/A	N/A	N/A	N/A
PART H WALL FINISHES							
19 x 100mm wrought softwood skirting	H1	21.50m	3.66	34.77	29.46	9.63	73.86
12mm plasterboard fixed to walls with dabs	H2	42.02m2	16.39	155.71	83.62	35.90	275.22
12mm plasterboard fixed to walls less than 300mm wide with dabs	H3	41.03m	7.38	70.11	37.75	16.18	124.04
Two coats emulsion paint to walls	H4	48.17m2	9.63	91.49	28.42	17.99	137.89
Paint skirting	H5	21.50m	2.80	26.60	10.32	5.54	42.46
Carried to summary			39.86	378.67	189.57	85.24	653.48
PART J FLOOR FINISHES							
Cement and sand floor screed 40mm thick	J1	16.28m2	4.07	38.67	49.33	13.20	101.19
Vinyl floor tiles, size 300 x 300mm	J2	16.28m2	3.90	37.05	113.63	22.60	173.28
Carried forward			7.97	75.72	162.96	35.80	274.48

	Ref	Qty	Hours	Hours £	Mat'ls £	O & P £	Total £
Brought forward			7.97	75.72	162.96	35.80	274.48
25mm thick tongued and grooved boarding	J3	16.28m^2	12.04	114.38	153.03	40.11	307.52
150 x 50mm sawn softwood joists	J4	31.20m	4.37	41.52	71.14	16.90	129.55
Cut and pin ends of joists to existing brick wall	J5	8nr	1.44	13.68	2.04	2.36	18.08
Build in ends of joists to blockwork	J6	8nr	8.00	76.00	1.20	11.58	88.78
Carried to summary			33.82	321.30	390.37	106.75	818.41
PART K **CEILING FINISHES**							
Plasterboard with taped butt joints fixed to joists	K1	33.56m^2	12.08	114.76	62.09	26.53	203.38
5mm skim coat to plasterboard ceilings	K2	33.56m^2	16.78	159.41	40.96	30.06	230.43
Two coats emulsion paint to ceilings	K3	33.56m^2	8.73	82.94	19.80	15.41	118.15
Carried to summary			37.59	357.11	122.85	71.99	551.95
PART L **ELECTRICAL WORK**							
13 amp double switched socket outlet with neon	L1	8nr	3.20	44.80	66.08	16.63	127.51
Lighting point	L2	6nr	2.10	29.40	40.80	10.53	80.73
Lighting switch	L3	5nr	1.25	17.50	19.85	5.60	42.95
Lighting wiring	L4	18.00m	1.80	25.20	18.54	6.56	50.30
Power cable	L5	40.00m	6.00	84.00	44.00	19.20	147.20
Carried to summary			14.35	200.90	189.27	58.53	448.70

	Ref	Qty	Hours	Hours £	Mat'ls £	O & P £	Total £
PART M **HEATING WORK**							
15mm copper pipe	M1	21.00m	4.62	57.75	39.48	14.58	111.81
Elbow	M2	8nr	2.24	28.00	8.80	5.52	42.32
Tee	M3	2nr	0.68	8.50	3.90	1.86	14.26
Radiator, double convector size 1400 x 520mm	M4	6nr	7.80	97.50	674.08	115.74	887.32
Break into existing pipe and insert tee	M5	1nr	0.75	9.38	3.95	2.00	15.32
Carried to summary			16.09	201.13	730.21	139.70	1,071.04
PART N **ALTERATION WORK**							
Take out existing window size 1500 x 1000mm and lintel over, adapt opening to receive 1770 x 2000mm patio door and insert new lintel over (both measured separately) and make good	N1	1nr	20.00	190.00	20.00	31.50	241.50
Take out existing window size 1500 x 1000mm, enlarge opening to receive new PVC-U door (measured separately)	N2	1nr	25.00	237.50	45.00	42.38	324.88
Carried to summary			45.00	427.50	65.00	73.88	566.38

SUMMARY

	Hours	Hours £	Mat'ls £	O & P £	Total £
PART A PRELIMINARIES	0.00	0.00	0.00	0.00	3,611.00
PART B SUBSTRUCTURE TO DPC LEVEL	82.77	786.32	788.63	236.24	1,811.19
PART C EXTERNAL WALLS	162.29	1,541.76	3,146.62	703.26	5,391.63
PART D FLAT ROOF	0.00	0.00	0.00	0.00	0.00
PART E PITCHED ROOF	118.82	1,123.84	1,455.58	386.91	2,966.33
PART F WINDOWS AND EXTERNAL DOORS	39.40	374.30	2,336.67	406.65	3,117.62
PART G INTERNAL PARTITIONS AND DOORS	0.00	0.00	0.00	0.00	0.00
PART H WALL FINISHES	39.86	378.67	189.57	85.24	653.48
PART J FLOOR FINISHES	33.82	321.29	390.37	106.75	818.41
PART K CEILING FINISHES	37.59	357.11	122.85	71.99	551.95
PART L ELECTRICAL WORK	14.35	200.90	189.27	58.53	448.70
PART M HEATING WORK	16.09	201.13	730.21	139.70	1,071.04
PART N ALTERATION WORK	45.00	427.50	65.00	73.88	566.38
Final total	589.99	5,712.82	9,414.77	2,269.15	21,007.73

	Ref	Qty	Hours	Hours £	Mat'ls £	O & P £	Total £
PART A PRELIMINARIES							
Concrete mixer	A1	12 wks					540.00
Small tools	A2	12 wks					420.00
Scaffolding (m2/weeks)	A3	840.00					1,890.00
Skip	A4	9 wks					900.00
Clean up	A5	12 hrs					96.00
Carried to summary							3,846.00
PART B SUBSTRUCTURE TO DPC LEVEL							
Excavate topsoil 150mm thick by hand	B1	26.15m2	7.85	74.58	0.00	11.19	85.76
Excavate to reduce levels by hand	B2	7.85m3	19.63	186.49	0.00	27.97	214.46
Excavate for trench foundations by hand	B3	2.21m3	5.75	54.63	0.00	8.19	62.82
Earthwork support to sides of trenches	B4	19.56m2	7.81	74.20	23.47	14.65	112.31
Backfilling with excavated material	B5	0.77m3	0.31	2.95	0.00	0.44	3.39
Hardcore 225mm thick	B6	19.98m2	3.00	28.50	101.50	19.50	149.50
Hardcore filling to trench	B7	0.17m3	0.09	0.86	0.86	0.26	1.97
Concrete grade (1:3:6) in foundations	B8	1.81m3	2.44	23.18	122.97	21.92	168.07
Concrete grade (1:2:4) in bed 150mm thick	B9	19.98m2	5.99	56.91	211.94	40.33	309.17
Concrete (1:2:4) in cavity wall filling	B10	6.70m2	1.34	12.73	24.32	5.56	42.61
Carried forward			54.21	515.00	485.06	150.01	1,150.06

	Ref	Qty	Hours	Hours £	Mat'ls £	O & P £	Total £
Brought forward			54.21	515.00	485.06	150.01	1,150.06
Damp-proof membrane	B11	20.63m2	0.83	7.89	11.97	2.98	22.83
Reinforcement ref A193 in foundation	B12	6.70m2	0.80	7.60	8.71	2.45	18.76
Steel fabric reinforcement ref A193 in slab	B13	19.98m2	3.00	28.50	25.97	8.17	62.64
Solid blockwork 140mm thick in cavity wall	B14	8.71m2	11.32	107.54	111.31	32.83	251.68
Common bricks 112.5mm thick in cavity wall	B15	6.70m2	11.32	107.54	88.37	29.39	225.30
Facing bricks in 112.5mm thick in skin of cavity wall	B16	2.01m2	3.62	34.39	46.77	12.17	93.33
Form cavity 50mm wide in cavity wall	B17	8.71m2	0.26	2.47	5.23	1.16	8.86
DPC 112mm wide	B18	13.40m	0.67	6.37	10.85	2.58	19.80
DPC 140mm wide	B19	13.40m	0.80	7.60	13.00	3.09	23.69
Bond in block wall	B20	1.30m	0.58	5.51	2.83	1.25	9.59
Bond in half brick wall	B21	0.30m	0.46	4.37	0.61	0.75	5.73
50mm thick insulation board	B22	19.98m2	5.99	56.91	89.51	21.96	168.38
Carried to summary			93.86	891.67	900.19	268.78	2,060.64
PART C EXTERNAL WALLS							
Solid blockwork 140mm thick in cavity wall	C1	51.87m2	67.43	640.59	659.79	195.06	1,495.43
Facing brickwork 112.5mm thick in cavity wall	C2	51.87m2	93.34	886.73	1,196.12	312.43	2,395.28
75mm thick insulation in cavity wall	C3	51.87m2	11.41	108.40	301.88	61.54	471.82
Carried forward			172.18	1,635.71	2,157.79	569.03	4,362.53

	Ref	Qty	Hours	Hours £	Mat'ls £	O & P £	Total £
Brought forward			172.18	1,635.71	2,157.79	569.03	4,362.53
Steel lintel 2400mm long	C4	2nr	0.50	4.75	196.92	30.25	231.92
Steel lintel 1500mm long	C5	8nr	1.60	15.20	859.76	131.24	1,006.20
Steel lintel 1150mm long	C6	1nr	0.15	1.43	47.18	7.29	55.90
Close cavity wall at jambs	C7	29.32m	1.47	13.97	59.81	11.07	84.84
Close cavity wall at cills	C8	11.71m	0.59	5.61	23.89	4.42	33.92
Close cavity of hollow wall at top	C9	13.40m	0.67	6.37	27.34	5.06	38.76
DPC 112mm wide at jambs	C10	29.32m	1.47	13.97	23.75	5.66	43.37
DPC 112mm wide at cills	C11	11.71m	0.59	5.61	9.48	2.26	17.35
Carried to summary			179.07	1,701.17	3,358.74	758.99	5,818.89
PART D FLAT ROOF			N/A	N/A	N/A	N/A	N/A
PART E PITCHED ROOF							
100 x 75mm sawn softwood wall plate	E1	6.00m	1.80	17.10	13.32	4.56	34.98
150 x 50mm sawn softwood pole plate plugged to brickwork	E2	6.30m	1.89	17.96	20.29	5.74	43.98
100 x 50mm sawn softwood rafters	E3	49.50m	9.90	89.10	162.36	37.72	289.18
125 x 25mm sawn softwood purlin	E4	6.30m	0.86	8.17	15.18	3.50	26.85
125 x 25mm sawn softwood joists	E5	40.50m	8.91	84.65	92.34	26.55	203.53
150 x 50mm sawn softwood sprockets	E6	20nr	2.40	22.80	22.80	6.84	52.44
100mm layer of insulation quilt between joists fixed with chicken wire and 150mm layer over joists	E7	24.00m2	11.52	109.44	261.84	55.69	426.97
6mm softwood soffit 150mm wide	E8	15.30m	6.12	58.14	26.78	12.74	97.66
Carried forward			43.40	407.35	614.91	153.34	1,175.60

	Ref	Qty	Hours	Hours £	Mat'ls £	O & P £	Total £
Brought forward			43.40	407.35	614.91	153.34	1,175.60
19mm wrought softwood fascia/ barge board 200mm high	E9	6.30m	7.65	72.68	12.41	12.76	97.85
Marley Plain roof tiles on felt and battens	E10	36.65m2	65.84	625.48	896.40	228.28	1,750.16
Double eaves course	E11	6.30m	2.21	21.00	20.29	6.19	47.48
Verge with plain tile undercloak	E12	11.00m	2.75	26.13	35.42	9.23	70.78
Lead flashing code 5, 200mm girth	E13	6.30m	3.78	35.91	42.08	11.70	89.69
Rake out joint for flashing	E14	6.30m	2.21	21.00	0.95	3.29	25.24
112mm diameter PVC-U gutter	E15	6.30m	1.64	15.58	28.16	6.56	50.30
Stop end	E16	1nr	0.14	1.33	1.11	0.37	2.81
Stop end outlet	E17	1nr	0.25	2.38	1.11	0.52	4.01
68mm diameter PVC-U down pipe	E18	2.50m	0.63	5.99	8.48	2.17	16.63
Shoe	E19	1nr	0.30	2.85	2.09	0.74	5.68
Paint fascia and soffit	E20	4.59m2	3.21	30.50	4.41	5.24	40.14
Carried to summary			134.01	1,268.15	1,667.82	440.39	3,376.36

PART F
WINDOWS AND
EXTERNAL DOORS

	Ref	Qty	Hours	Hours £	Mat'ls £	O & P £	Total £
PVC-U door size 840 x 1980mm complete (B)	F1	2nr	5.00	47.50	419.98	70.12	537.60
PVC-U sliding patio door size 1700 x 2075mm (C)	F2	2nr	14.00	133.00	587.04	108.01	828.05
Carried forward			19.00	180.50	1,007.02	178.13	1,365.65

	Ref	Qty	Hours	Hours £	Mat'ls £	O & P £	Total £
Brought forward			19.00	180.50	1,007.02	178.13	1,365.65
PVC-U window size 1200 x 1200mm complete (A)	F3	8nr	16.00	152.00	1,260.72	211.91	1,624.63
25 x 225mm wrought softwood window board	F4	9.60m	2.88	27.36	59.71	13.06	100.13
Paint window board	F5	9.60m	1.52	14.44	9.22	3.55	27.21
Carried to summary			39.40	374.30	2,336.67	406.65	3,117.62
PART G INTERNAL PARTITIONS AND DOORS							
50 x 75mm sawn softwood sole plate	G1	3.70m	0.81	7.70	4.14	1.24	9.52
50 x 75mm sawn softwood head	G2	3.70m	0.81	7.70	4.14	1.24	9.52
50 x 75mm sawn softwood studs	G3	17.15m	4.80	45.60	19.21	5.76	44.18
50 x 75mm sawn softwood noggings	G4	3.70m	1.04	9.88	4.14	1.24	9.52
Plasterboard	G5	14.80m2	5.33	50.64	27.38	8.21	62.97
35mm thick veneered internal door size 762 x 1981mm	G6	1nr	1.25	11.88	55.32	16.60	127.24
38 x 150mm wrought softwood lining	G7	4.87m	1.07	10.17	42.81	12.84	98.46
13 x 38mm wrought softwood stop	G8	4.87m	0.73	6.94	2.39	0.72	5.50
19 x 50mm wrought softwood architrave	G9	9.74m	1.46	13.87	2.24	0.67	5.15
19 x 100mm wrought softwood skirting	G10	5.88m	1.00	9.50	8.06	2.42	18.54
100mm rising steel butts	G11	1pr	0.30	2.85	3.95	1.19	9.09
Carried forward			18.60	176.70	173.78	52.13	399.69

	Ref	Qty	Hours	Hours £	Mat'ls £	O & P £	Total £
Brought forward			18.60	176.70	173.78	52.13	399.69
SAA mortice latch with lever furniture	G12	1nr	0.80	7.60	14.10	4.23	32.43
Two coats emulsion on plasterboard walls	G13	14.80m2	3.85	36.58	8.76	2.63	20.15
Paint general surfaces	G14	4.70m2	3.29	31.26	1.40	0.42	3.22
Carried to summary			26.54	252.13	198.04	59.41	455.49
PART H WALL FINISHES							
19 x 100mm wrought softwood skirting	H1	28.90m2	4.91	46.65	35.59	12.34	94.57
12mm plasterboard fixed to walls with dabs	H2	46.22m2	18.03	171.29	92.77	39.61	303.66
12mm plasterboard fixed to walls less than 300mm wide with dabs	H3	41.03m	7.38	70.11	37.75	16.18	124.04
Two coats emulsion paint to walls	H4	67.91m2	13.58	129.01	40.07	25.36	194.44
Paint skirting	H5	28.90m	3.76	35.72	13.87	7.44	57.03
Carried to summary			47.66	452.77	220.05	100.92	773.74
PART J FLOOR FINISHES							
Cement and sand floor screed 40mm thick	J1	19.98m2	5.00	47.50	60.54	16.21	124.25
Vinyl floor tiles, size 300 x 300mm	J2	19.98m2	4.79	45.51	139.46	27.74	212.71
25mm thick tongued and grooved boarding	J3	19.98m2	14.79	140.51	187.81	49.25	377.56
Carried forward			24.58	233.51	387.81	93.20	714.52

	Ref	Qty	Hours	Hours £	Mat'ls £	O & P £	Total £
Brought forward			24.58	233.51	387.81	93.20	714.52
150 x 50mm sawn softwood joists	J4	39.00m	8.46	80.37	88.92	25.39	194.68
Cut and pin ends of joists to existing brick wall	J5	10nr	1.80	17.10	2.50	2.94	22.54
Build in ends of joists to blockwork	J6	10nr	1.00	9.50	1.50	1.65	12.65
Carried to summary			35.84	340.48	480.73	123.18	944.39
PART K CEILING FINISHES							
Plasterboard with taped butt joints fixed to joists	K1	39.96m2	14.39	136.71	73.93	31.60	242.23
5mm skim coat to plasterboard ceilings	K2	39.36m2	19.98	189.81	48.75	35.78	274.34
Two coats emulsion paint to ceilings	K3	39.96m2	10.39	98.71	23.58	18.34	140.63
Carried to summary			44.76	425.22	146.26	85.72	657.20
PART L ELECTRICAL WORK							
13 amp double switched socket outlet with neon	L1	10nr	4.00	56.00	82.60	20.79	159.39
Lighting point	L2	6nr	2.10	29.40	40.80	10.53	80.73
Lighting switch	L3	5nr	1.25	17.50	19.85	5.60	42.95
Lighting wiring	L4	22.00m	2.20	30.80	22.66	8.02	61.48
Power cable	L5	44.00m	6.60	92.40	48.40	21.12	161.92
Carried to summary			16.15	226.10	214.31	66.06	506.47

	Ref	Qty	Hours	Hours £	Mat'ls £	O & P £	Total £
PART M **HEATING WORK**							
15mm copper pipe	M1	23.00m	5.06	63.25	43.24	15.97	122.46
Elbow	M2	8nr	2.24	28.00	8.80	5.52	42.32
Tee	M3	2nr	0.68	8.50	3.90	1.86	14.26
Radiator, double convector size 1400 x 520mm	M4	6nr	7.80	97.50	694.08	118.74	910.32
Break into existing pipe and insert tee	M5	1nr	0.75	9.38	3.95	2.00	15.32
Carried to summary			16.53	206.63	753.97	144.09	1,104.68
PART N **ALTERATION WORK**							
Take out existing window size 1500 x 1000mm and lintel over, adapt opening to receive 1770 x 2000mm patio door and insert new lintel over (both measured separately) and make good	N1	1nr	20.00	190.00	20.00	31.50	241.50
Take out existing window size 1500 x 1000mm, enlarge opening to receive new PVC-U door (measured separately)	N2	1nr	25.00	237.50	45.00	42.38	324.88
Carried to summary			45.00	427.50	65.00	73.88	566.38

SUMMARY

	Hours	Hours £	Mat'ls £	O & P £	Total £
PART A PRELIMINARIES	0.00	0.00	0.00	0.00	3,846.00
PART B SUBSTRUCTURE TO DPC LEVEL	93.86	891.67	900.19	268.78	2,060.64
PART C EXTERNAL WALLS	179.07	1,701.17	3,358.74	758.99	5,818.89
PART D FLAT ROOF	0.00	0.00	0.00	0.00	0.00
PART E PITCHED ROOF	134.01	1,268.15	1,667.82	440.39	3,376.36
PART F WINDOWS AND EXTERNAL DOORS	39.40	374.30	2,336.67	406.65	3,117.62
PART G INTERNAL PARTITIONS AND DOORS	26.54	252.13	198.04	59.41	455.49
PART H WALL FINISHES	47.66	452.77	220.05	100.92	773.74
PART J FLOOR FINISHES	35.84	340.48	480.73	123.18	944.39
PART K CEILING FINISHES	44.76	452.22	146.26	85.72	657.20
PART L ELECTRICAL WORK	16.15	226.10	214.31	66.06	506.47
PART M HEATING WORK	16.53	206.63	753.97	144.09	1,104.68
PART N ALTERATION WORK	45.00	427.50	65.00	73.88	566.38
Final total	678.82	6,593.12	10,341.78	2,528.07	23,227.86

SUMMARY OF EXTENSION COSTS

One storey flat roof

	2 x 3m £	%	2 x 4m £	%	2 x 5m £	%	3 x 3m £	%	3 x 4m £	%
PART A PRELIMINARIES	896	15	1043	15	1318	17	1082	15	1324	16
PART B SUBSTRUCT-URE TO DPC LEVEL	739	12	898	13	1070	13	1006	13	1211	14
PART C EXTERNAL WALLS	1440	23	1653	24	1866	23	1866	25	1927	23
PART D FLAT ROOF	812	14	1003	14	1208	15	1093	14	1374	16
PART E PITCHED ROOF	0.00	0	0.00	0	0.00	0	0.00	0	0.00	0
PART F WINDOWS & EXTERNAL DOORS	1264	21	1264	19	1260	16	1264	17	1264	16
PART G INTERNAL PARTITIONS, DOORS	0.00	0	0.00	0	0.00	0	0.00	0	0.00	0
PART H WALL FINISHES	160	3	191	3	222	3	218	3	249	3
PART J FLOOR FINISHES	69	1	97	1	120	2	109	2	155	2
PART K CEILING FINISHES	53	1	75	1	96	1	83	1	117	1
PART L ELECTRICAL WORK	121	2	144	2	171	2	136	2	168	2
PART M HEATING WORK	215	4	372	5	377	5	372	5	377	4
PART N ALTERATION WORK	241	4	241	3	241	3	241	3	241	3
Final total £	6010	100	6981	100	7949	100	7470	100	8407	100
Floor area m2	4.08		6.29		7.48		6.48		9.18	
£ per m2 £	1,473		1,110		1,063		1,153		916	

SUMMARY OF EXTENSION COSTS

One storey flat roof

	3 x 5m		3 x 6m		4 x 4m		4 x 5m		4 x 6m	
	£	%	£	%	£	%	£	%	£	%
PART A PRELIMINARIES	1610	17	1908	17	1863	17	2166	18	2481	18
PART B SUBSTRUCT-URE TO DPC LEVEL	1457	15	1589	14	1572	15	1811	14	2061	15
PART C EXTERNAL WALLS	2293	23	2620	23	2417	22	2695	22	2779	20
PART D FLAT ROOF	1659	17	1908	17	1720	16	2066	16	2420	18
PART E PITCHED ROOF	0.00		0.00		0.00		0.00		0.00	
PART F WINDOWS & EXTERNAL DOORS	1263	13	1754	15	1754	16	1973	16	1973	14
PART G INTERNAL PARTITIONS, DOORS	0.00	0	0.00	0	0.00	0	0.00	0	0.00	0
PART H WALL FINISHES	282	3	316	3	316	3	344	3	374	3
PART J FLOOR FINISHES	201	2	247	2	212	2	274	2	337	2
PART K CEILING FINISHES	151	2	185	2	160	1	181	1	252	2
PART L ELECTRICAL WORK	178	2	212	2	185	2	246	2	258	2
PART M HEATING WORK	383	4	388	3	393	4	550	4	563	4
PART N ALTERATION WORK	241	2	241	2	241	2	241	2	241	2
Final total £	9718	100	11368	100	10833	100	12547	100	13739	100
Floor area m2	11.88		14.58		12.58		16.28		19.98	
£ per m2 £	818		780		861		771		688	

SUMMARY OF EXTENSION COSTS

One storey pitched roof

		2 x 3m		2 x 4m		2 x 5m		3 x 3m		3 x 4m	
		£	%	£	%	£	%	£	%	£	%
PART A PRELIMINARIES		896	13	1043	13	1318	15	1082	13	1324	14
PART B SUBSTRUCT-URE TO DPC LEVEL		739	11	898	12	1070	12	1006	12	1211	12
PART C EXTERNAL WALLS		1693	25	1906	25	2118	25	2371	28	2595	26
PART D FLAT ROOF		0.00	0	0.00	0	0.00	0	0.00	0	0.00	0
PART E PITCHED ROOF		1258	19	1514	20	1608	19	1511	18	2023	21
PART F WINDOWS & EXTERNAL DOORS		1264	19	1264	16	1260	15	1264	15	1264	13
PART G INTERNAL PARTITIONS, DOORS		0.00	0	0.00	0	0.00	0	0.00	0	0.00	0
PART H WALL FINISHES		160	2	191	2	222	3	218	3	249	3
PART J FLOOR FINISHES		69	1	97	1	120	1	109	1	155	2
PART K CEILING FINISHES		53	1	75	1	96	1	84	1	117	1
PART L ELECTRICAL WORK		121	2	144	2	171	2	136	2	168	2
PART M HEATING WORK		215	3	372	5	377	4	372	4	377	4
PART N ALTERATION WORK		241	4	241	3	241	3	241	3	241	2
Final total	**£**	6709	100	7745	100	8601	100	8394	100	9724	100
Floor area	**m2**	4.08		6.29		7.48		6.48		9.18	
£ per m2	**£**	1,644		1,231		1,150		1,295		1,059	

SUMMARY OF EXTENSION COSTS

One storey pitched roof

	3 x 5m		3 x 6m		4 x 4m		4 x 5m		4 x 6m	
	£	%	£	%	£	%	£	%	£	%
PART A PRELIMINARIES	1610	15	1908	16	1863	15	2166	15	2481	15
PART B SUBSTRUCT-URE TO DPC LEVEL	1444	14	1589	13	1589	13	1811	13	2061	13
PART C EXTERNAL WALLS	2803	27	2922	24	3447	27	3659	24	3788	24
PART D FLAT ROOF	0.00	0	0.00	0	0.00	0	0.00	0	0.00	0
PART E PITCHED ROOF	1946	19	2496	19	2540	20	2966	21	3376	21
PART F WINDOWS & EXTERNAL DOORS	1264	12	1754	14	1754	14	1973	14	1973	13
PART G INTERNAL PARTITIONS, DOORS	0.00	0	0.00	0	0.00	0	0.00	0	0.00	0
PART H WALL FINISHES	282	3	314	3	314	2	344	2	374	2
PART J FLOOR FINISHES	201	2	247	2	212	2	275	2	337	2
PART K CEILING FINISHES	151	1	185	2	160	1	181	1	252	2
PART L ELECTRICAL WORK	178	2	212	2	185	1	246	2	258	2
PART M HEATING WORK	383	3	388	3	393	3	550	4	563	4
PART N ALTERATION WORK	241	2	241	2	241	2	241	2	241	2
Final total £	10503	100	12256	100	12698	100	14412	100	15704	100
Floor area m2	11.88		14.58		12.58		16.28		19.98	
£ per m2 £	884		841		1,009		885		786	

SUMMARY OF EXTENSION COSTS

Two storey flat roof

	2 x 3m		2 x 4m		2 x 5m		3 x 3m		3 x 4m	
	£	%	£	%	£	%	£	%	£	%
PART A PRELIMINARIES	1689	17	2065	17	2464	18	2110	17	2509	18
PART B SUBSTRUCT-URE TO DPC LEVEL	737	7	895	8	1068	8	1019	8	1221	9
PART C EXTERNAL WALLS	2849	28	3277	28	3705	29	3705	29	4132	29
PART D FLAT ROOF	826	8	1017	9	1223	9	1120	9	1385	10
PART E PITCHED ROOF	0.00	0	0.00	0	0.00	0	0.00	0	0.00	0
PART F WINDOWS & EXTERNAL DOORS	1973	19	1973	17	1973	15	1973	16	1973	14
PART G INTERNAL PARTITIONS, DOORS	0.00	0	0.00	0	0.00	0	0.00	0	0.00	0
PART H WALL FINISHES	314	3	396	3	441	3	442	4	501	4
PART J FLOOR FINISHES	206	2	293	3	373	3	310	2	437	3
PART K CEILING FINISHES	104	1	188	2	246	2	213	2	302	2
PART L ELECTRICAL WORK	241	2	264	2	341	3	289	2	335	2
PART M HEATING WORK	723	7	728	6	745	6	737	6	748	5
PART N ALTERATION WORK	566	6	566	5	566	4	566	5	566	4
Final total £	10228	100	11662	100	13145	100	12484	100	14109	100
Floor area m2	8.16		12.58		14.96		12.96		18.36	
£ per m2 £	1,253		927		879		963		768	

SUMMARY OF EXTENSION COSTS

Two storey flat roof

	3 x 5m		3 x 6m		4 x 4m		4 x 5m		4 x 6m	
	£	%	£	%	£	%	£	%	£	%
PART A PRELIMINARIES	2914	19	3476	19	3296	20	3611	19	3711	17
PART B SUBSTRUCTURE TO DPC LEVEL	1368	9	1656	9	1590	10	1811	9	2061	10
PART C EXTERNAL WALLS	4560	28	5091	28	4941	29	5392	27	5612	26
PART D FLAT ROOF	1651	10	1943	10	1747	10	2074	10	2179	10
PART E PITCHED ROOF	0.00	0	0.00	0	0.00	0	0.00	0	0.00	0
PART F WINDOWS & EXTERNAL DOORS	1973	13	2680	15	2680	15	3118	15	3118	15
PART G INTERNAL PARTITIONS, DOORS	0.00	0	0.00	0	0.00	0	0.00	0	455	2
PART H WALL FINISHES	555	4	629	3	630	4	691	3	740	3
PART J FLOOR FINISHES	550	4	678	4	588	3	818	4	818	4
PART K CEILING FINISHES	325	2	480	3	414	2	552	3	552	3
PART L ELECTRICAL WORK	356	2	416	2	369	0	449	2	449	2
PART M HEATING WORK	759	5	775	4	780	4	1071	5	1071	5
PART N ALTERATION WORK	566	4	566	3	566	3	566	3	566	3
Final total £	15577	100	18390	100	17601	100	20153	100	21332	100
Floor area m2	23.76		29.16		25.16		32.56		39.96	
£ per m2 £	654		631		699		619		534	

SUMMARY OF EXTENSION COSTS

Two storey pitched roof

	2 x 3m		2 x 4m		2 x 5m		3 x 3m		3 x 4m	
	£	%	£	%	£	%	£	%	£	%
PART A PRELIMINARIES	1689	16	2065	17	2464	18	2110	17	2509	17
PART B SUBSTRUCTURE TO DPC LEVEL	737	7	895	7	1068	8	1019	8	1221	8
PART C EXTERNAL WALLS	2849	27	3276	27	3630	27	3705	29	4183	28
PART D FLAT ROOF	0.00	0	0.00	0	0.00	0	0.00	0	0.00	0
PART E PITCHED ROOF	1258	12	1440	12	1822	13	1510	12	2023	14
PART F WINDOWS & EXTERNAL DOORS	1973	18	1973	17	1973	15	1973	15	1973	13
PART G INTERNAL PARTITIONS, DOORS	0.00	0	0.00	0	0.00	0	0.00	0	0.00	0
PART H WALL FINISHES	333	3	378	3	424	3	424	3	484	3
PART J FLOOR FINISHES	206	2	293	2	373	3	310	2	437	3
PART K CEILING FINISHES	104	1	188	2	246	2	213	2	302	2
PART L ELECTRICAL WORK	241	2	264	2	341	2	289	2	335	2
PART M HEATING WORK	723	7	728	6	745	5	737	6	748	5
PART N ALTERATION WORK	566	5	566	5	566	4	566	4	566	4
Final total £	10679	100	12066	100	13652	100	12856	101	14781	100
Floor area m2	8.16		12.58		14.96		12.96		18.36	
£ per m2 £	1,309		959		913		992		805	

SUMMARY OF EXTENSION COSTS

Two storey pitched roof

	3 x 5m		3 x 6m		4 x 4m		4 x 5m		4 x 6m	
	£	%	£	%	£	%	£	%	£	%
PART A PRELIMINARIES	2914	18	3476	18	3296	18	3611	17	3846	17
PART B SUBSTRUCT-URE TO DPC LEVEL	1446	9	1656	9	1590	9	1811	9	2061	9
PART C EXTERNAL WALLS	4560	29	5087	27	4937	27	5392	26	5819	25
PART D FLAT ROOF	0.00	0	0.00	0	0.00	0	0.00	0	0.00	0
PART E PITCHED ROOF	1946	12	2452	13	2540	14	2966	13	3376	15
PART F WINDOWS & EXTERNAL DOORS	1973	13	2681	14	2680	15	3117	15	3117	13
PART G INTERNAL PARTITIONS, DOORS	0.00	0	0.00	0	0.00	0	0.00	0	455	2
PART H WALL FINISHES	538	3	599	3	599	3	654	3	774	3
PART J FLOOR FINISHES	550	3	678	4	588	3	818	4	944	4
PART K CEILING FINISHES	325	2	480	3	414	2	552	3	657	3
PART L ELECTRICAL WORK	356	2	416	2	369	2	449	2	506	2
PART M HEATING WORK	759	5	775	4	780	4	1071	5	1105	5
PART N ALTERATION WORK	566	4	566	3	566	3	566	3	566	2
Final total £	15933	100	18866	100	18359	100	21007	100	23226	100
Floor area m2	23.76		29.16		25.16		32.56		39.56	
£ per m2 £	671		647		729		645		581	

	Qty	Hours	Hours £	Mat'ls £	O & P £	Total £
ALTERNATIVE ITEMS						
The following items are alternatives to those listed in the previous extension take-offs and can be substituted to suit the needs of any particular project.						
Excavate topsoil 150mm thick by machine and deposit on site	m2	0.00	0.00	0.90	0.14	1.04
Excavate to reduce levels by machine and deposit on site	m3	0.00	0.00	1.62	0.24	1.86
Load excavated material by machine into skips and remove to tip	m3	0.00	0.00	15.56	2.33	17.89
Common bricks (£175 per 1,000) in leaf of cavity wall	m2	1.70	16.15	12.40	4.28	32.83
Common bricks (£200 per 1,000) in leaf of cavity wall	m2	1.70	16.15	13.91	4.51	34.57
Common bricks (£225 per 1,000) in leaf of cavity wall	m2	1.70	16.15	15.41	4.73	36.29
Facing bricks (£300 per 1,000) in leaf of cavity wall	m2	1.85	17.58	20.36	5.69	43.63
Facing bricks (£350 per 1,000) in leaf of cavity wall	m2	1.85	17.58	23.36	6.14	47.08
Facing bricks (£400 per 1,000) in leaf of cavity wall	m2	1.85	17.58	27.86	6.82	52.25
Marley Ludlow Plus roofing tiles smooth finish, size 387 x 229mm 75mm lap, type 1F reinforced roofing felt and 38 x 19mm sawn softwood battens	m2	1.10	10.45	11.84	3.34	25.63
Marley Ludlow Major roofing tiles granule finish, size 420 x 330mm 75mm lap, type 1F reinforced roofing felt and 38 x 19mm sawn softwood battens	m2	1.10	10.45	11.20	3.25	24.90

	Qty	Hours	Hours £	Mat'ls £	O & P £	Total £
Brought forward		1.10	10.45	11.20	3.25	24.90
Marley Modern roofing tiles smooth finish, size 420 x 330mm 75mm lap, type 1F reinforced roofing felt and 38 x 19mm sawn softwood battens	m2	1.10	10.45	11.62	3.31	25.38
Marley Mendip roofing tiles smooth finish, size 420 x 330mm 75mm lap, type 1F reinforced roofing felt and 25 x 19mm sawn softwood battens	m2	1.10	10.45	11.98	3.36	25.79
100mm diameter cast iron gutter	m	0.40	3.80	12.14	2.39	18.33
Stop end	nr	0.14	1.33	3.88	0.78	5.99
Stop end outlet	nr	0.16	1.52	4.55	0.91	6.98
65mm diameter cast iron down pipe	m	0.38	3.61	15.50	2.87	21.98
Shoe	nr	0.32	3.04	8.86	1.79	13.69
Quarry floor tiles, size 150 x 150 x 12.5mm	m2	1.00	9.50	23.87	5.01	38.38
Ceramic floor tiles, size 150 x 150 x 12.5mm	m2	1.00	9.50	24.58	5.11	39.19
Vinyl floor floor sheeting 2.5mm thick with welded joints	m2	0.38	3.61	9.63	1.99	15.23
Acrylic reinforced bath1700mm long complete with chromium plated grip handles, pair of taps, waste fitting, overflow, chain, plug, trap and bath panels	nr	3.50	43.75	207.50	37.69	288.94
Vitreous china wash basin size 540 x 430mm, complete with pair of taps, waste fitting, overflow, chain, plug, trap and pedestal	nr	2.00	25.00	118.56	21.53	165.09
Vitreous china bidet complete	nr	1.50	18.75	136.70	23.32	178.77

	Qty	Hours	Hours £	Mat'ls £	O & P £	Total £
Vitreous china WC suite comprising pan, plastic seat and cover, 9 litre cistern and brackets, ball valve and plastic connecting pipe, complete	nr	1.50	18.75	136.70	23.32	178.77
Shower cubicle with plastic tray, aluminium and acrylic sides with opening door and surface-mounted thermostatically-controlled shower fitting	nr	1.50	18.75	136.70	23.32	178.77
Prepare, size and hang wallpaper to walls with adhesive						
lining paper, £1.00 per roll	m2	0.25	2.38	0.24	0.39	3.01
lining paper, £1.10 per roll	m2	0.25	2.38	0.26	0.40	3.03
lining paper, £1.20 per roll	m2	0.25	2.38	0.28	0.40	3.05
lining paper, £1.30 per roll	m2	0.25	2.38	0.30	0.40	3.08
washable paper, £3.50 per roll	m2	0.30	2.85	0.82	0.55	4.22
washable paper, £4.50 per roll	m2	0.30	2.85	1.06	0.59	4.50
washable paper, £5.50 per roll	m2	0.30	2.85	1.30	0.62	4.77
embossed paper, £4.00 per roll	m2	0.32	3.04	1.06	0.62	4.72
embossed paper, £5.00 per roll	m2	0.32	3.04	1.18	0.63	4.85
embossed paper, £6.00 per roll	m2	0.32	3.04	1.30	0.65	4.99
embossed paper, £7.00 per roll	m2	0.32	3.04	1.42	0.67	5.13
Prepare, size and hang wallpaper to ceilings with adhesive						
lining paper, £1.00 per roll	m2	0.30	2.85	0.24	0.46	3.55
lining paper, £1.10 per roll	m2	0.30	2.85	0.26	0.47	3.58
lining paper, £1.20 per roll	m2	0.30	2.85	0.28	0.47	3.60
lining paper, £1.30 per roll	m2	0.30	2.85	0.30	0.47	3.62
washable paper, £3.50 per roll	m2	0.35	3.33	0.82	0.62	4.77
washable paper, £4.50 per roll	m2	0.35	3.33	1.06	0.66	5.04
washable paper, £5.50 per roll	m2	0.35	3.33	1.30	0.69	5.32
embossed paper, £4.00 per roll	m2	0.37	3.52	1.06	0.69	5.26
embossed paper, £5.00 per roll	m2	0.37	3.52	1.18	0.70	5.40
embossed paper, £6.00 per roll	m2	0.37	3.52	1.30	0.72	5.54
embossed paper, £7.00 per roll	m2	0.37	3.52	1.42	0.74	5.68

Part Two

LOFT CONVERSIONS

Standard items

Drawings

Loft conversions

4.5 x 4.5m

4.5 x 5.5m

4.5 x 6.5m

Summary of loft conversion costs

	Ref	Unit	Hours	Hours £	Mat'ls £	O & P £	Total £
PART A PRELIMINARIES							
Small tools	A1	wk					35.00
Scaffolding (m2/weeks)	A3	m2/wk					2.25
Skip	A4	wk					100.00
Clean up	A5	hour					8.00
PART B PREPARATION							
Clear out roof space	B1	item	4.00	38.00	0.00	5.70	43.70
Remove carpet from bedroom	B2	item	2.00	19.00	0.00	2.85	21.85
Erect temporary screen (10m2) consisting of 75 x 50mm covered both sides with polythene sheeting	B3	item	6.00	57.00	118.40	26.31	201.71
Disconnect pipes from cold water storage tank, move tank to new position and reconnect pipes	B4	item	1.90	18.05	3.50	3.23	24.78
PART C DORMER WINDOW							
Remove clay tiles or slates, felt and battens and remove	C1	m2	0.75	7.13	0.00	1.07	8.19
Cut into rafters and purlin to form new opening size 2000 x 3000mm, trim with 150 x 50mm sawn softwood bearers	C2	nr	14.00	133.00	29.56	24.38	186.94
Cut into rafters and purlin to form new opening size 2600 x 3000mm, trim with 150 x 50mm sawn softwood bearers	C3	nr	16.00	152.00	34.87	28.03	214.90
200 x 75mm sawn softwood purlin	C4	m	0.25	2.38	3.78	0.92	7.08
Cut and pin end of purlin to existing brickwork	C5	nr	1.00	9.50	0.40	1.49	11.39

	Ref	Unit	Hours	Hours £	Mat'ls £	O & P £	Total £
50 x 75mm sawn softwood sole plate	C6	m	0.22	2.09	1.22	0.50	3.81
50 x 75mm sawn softwood head	C7	m	0.22	2.09	1.22	0.50	3.81
50 x 75mm sawn softwood studs	C8	m	0.28	2.66	1.22	0.58	4.46
50 x 75mm sawn joists	C9	m	0.20	1.90	2.28	0.63	4.81
Mild steel bolt, M10 x 100mm	C10	nr	0.15	1.43	1.25	0.40	3.08
18mm thick WPB grade plywood roof decking fixed to roof joists	C11	m2	0.90	8.55	7.86	2.46	18.87
50mm wide sawn softwood tapered firring pieces average depth 50mm	C12	m	0.18	1.71	2.26	0.60	4.57
Crown Wool insulation or similar 100mm thick fixed between joists with chicken wire and 150mm thick layer laid over joists	C13	m2	0.90	8.55	10.91	2.92	22.38
19mm thick wrought softwood fascia board 200mm high	C14	m	0.50	4.75	1.97	1.01	7.73
6mm thick asbestos-free insulation board soffit 150mm wide	C15	m	0.40	3.80	1.75	0.83	6.38
Three-layer polyester-base mineral-surfaced roofing felt	C16	m2	0.55	5.23	8.41	2.05	15.68
Turn down to edge of roof	C17	m	0.10	0.95	0.84	0.27	2.06
Lead flashing code 5, 200mm girth dressing under existing tiles and over new dormer roof	C18	m	0.60	5.70	6.68	6.68	19.06
Lead stepped flashing code 5, 200mm girth dressing under vertical tiles and over existing roof	C19	m	0.90	8.55	7.42	6.68	22.65
Marley Plain roofing tiles size 278 x 165mm, 65mm lap, type 1F reinforced fel, 38 x 19mm battens hung, vertically	C20	m2	1.90	18.05	25.87	6.59	50.51
Raking cutting on tiling	C21	m	0.15	1.43	0.00	0.21	1.64

	Ref	Unit	Hours	Hours £	Mat'ls £	O & P £	Total £
Make good existing tiling up to new dormer	C22	m	0.18	1.71	0.00	0.26	1.97
PVC-U window size 1800 x 1200mm complete	C23	nr	2.00	19.00	252.16	40.67	311.83
PVC-U window size 2400 x 1200mm complete	C24	nr	2.30	21.85	391.82	62.05	475.72
25 x 225mm wrought softwood window board	C25	m	0.30	2.85	6.22	1.36	10.43
25 x 150mm wrought softwood lining	C26	m	0.24	2.28	2.48	0.71	5.47
PVC-U gutter, 112mm half round with gutter union joints, fixed to wrought softwood fascia board with support brackets at 1m maximum centres	C27	m	0.26	2.47	5.50	1.20	9.17
Extra for stop end	C28	nr	0.14	1.33	1.11	0.37	2.81
Extra for stop end outlet	C29	nr	0.14	1.33	1.11	0.37	2.81
PVC-U down pipe 68mm diameter, loose spigot and socket joints, plugged to faced brickwork with pipe clips at 2m centres	C30	m	0.25	2.38	3.39	0.86	6.63
Extra for shoe	C31	nr	0.30	2.85	2.09	0.74	5.68
Plasterboard 9.5mm thick fixed to softwood studding, joints filled with filler and taped to receive decoration	C32	m2	0.36	3.42	2.35	0.87	6.64
Plasterboard 9.5mm thick fixed to softwood joists, joints filled with filler and taped to receive decoration	C33	m2	0.36	3.42	2.35	0.87	6.64
One coat skim plaster to plaster-board ceiling including scrimming joints	C34	m2	0.50	4.75	1.22	0.90	6.87

	Ref	Unit	Hours	Hours £	Mat'ls £	O & P £	Total £
Two coats emulsion paint to plasterboard walls and ceilings	C35	m2	0.36	3.42	2.35	0.87	6.64
Apply one coat primer, one oil-based undercoat and one coat gloss paint on surfaces not exceeding 300mm girth	C36	m	0.20	1.90	0.30	0.33	2.53
PART D ROOF WINDOW							
Remove clay tiles or slates, felt and battens and remove	D1	m2	0.75	7.13	0.00	1.07	8.19
Cut into rafters and purlin to form new opening size 880 x 1080mm, trim with 150 x 50mm sawn softwood bearers	D2	nr	3.00	28.50	15.40	6.59	50.49
Make good existing tiling up to new dormer	D3	m	0.18	1.71	0.00	0.26	1.97
Velux window size 780 x 980mm complete with flashings	D4	nr	3.25	30.88	232.00	39.43	302.31
25 x 150mm wrought softwood lining	D5	m	0.24	2.28	2.48	0.71	5.47
Apply one coat primer, one oil-based undercoat and one coat gloss paint on surfaces not exceeding 300mm girth	D6	m	0.20	1.90	0.30	0.33	2.53
PART E STAIRS							
Break into existing ceiling joists and plasterboard ceiling, trim with 150 x 50mm trimmer including temporary supports, to form new opening size 2750 x 1200mm	E1	nr	3.00	28.50	12.00	6.08	46.58
Make good plasterboard ceiling up to new opening	E2	m	0.20	1.90	0.50	0.36	2.76

	Ref	Unit	Hours	Hours £	Mat'ls £	O & P £	Total £
Wrought softwood straight-flight staircase with 13 close treads, 2700mm going, 2600mm rise complete with 38 x 200mm strings, 66 x 66 x 1350mm newel post, 63 x 44mm handrail, balusters 848mm high and spacers	E3	nr	32.00	304.00	565.00	130.35	999.35
25 x 150mm wrought softwood lining	E4	m	0.24	2.28	2.48	0.71	5.47
Apply one coat primer, one oil-based undercoat and one coat gloss paint on surfaces exceeding 300mm girth	E5	m2	0.70	6.65	0.96	1.14	8.75
PART F FLOORING							
25mm thick tongued and grooved wrought softwood flooring	F1	m2	0.74	7.03	9.40	2.46	18.89
19 x 100mm wrought softwood skirting	F2	m	0.17	1.62	1.37	0.45	3.43
Apply one coat primer, one oil-based undercoat and one coat gloss paint on skirting not exceeding 300mm girth	F3	m	0.20	1.90	0.30	0.33	2.53
PART G INTERNAL PARTITIONS AND DOORS							
50 x 75mm sawn softwood sole plate	G1	m	0.22	2.09	1.12	0.48	3.69
50 x 75mm sawn softwood head	G2	m	0.22	2.09	1.12	0.48	3.69
50 x 75mm sawn softwood studs	G3	m	0.28	2.66	1.12	0.57	4.35
50 x 75mm sawn softwood noggings	G4	m	0.28	2.66	1.12	0.57	4.35

	Ref	Unit	Hours	Hours £	Mat'ls £	O & P £	Total £
Plasterboard 9.5mm thick fixed to softwood studding, filled joints and taped to receive decoration	G5	m2	0.36	3.42	1.85	0.79	6.06
Flush door 35mm thick, size 762 x 1981mm, internal quality, half hour fire check, veneered finish both sides	G6	nr	1.25	11.88	55.32	10.08	77.27
38 x 150mm wrought softwood lining	G7	m	0.22	2.09	0.88	0.45	3.42
13 x 38mm wrought softwood door stop	G8	m	0.20	1.90	0.49	0.36	2.75
19 x 50mm wrought softwood chamfered architrave	G9	m	0.15	1.43	0.46	0.28	2.17
19 x 100mm wrought softwood chamfered skirting	G10	m	0.17	1.62	1.37	0.45	3.43
100mm rising steel butts	G11	pair	0.30	2.85	3.95	1.02	7.82
Silver anodised aluminium mortice latch with lever furniture	G12	nr	0.80	7.60	14.10	3.26	24.96
Two coats emulsion paint to plasterboard walls	G13	m2	0.26	2.47	0.59	0.46	3.52
Apply one coat primer, one oil-based undercoat and one coat gloss paint to general surfaces exceeding 300mm girth	G14	m2	0.70	6.65	0.96	1.14	8.75
PART H **CEILINGS AND SOFFITS**							
Plasterboard 9.5mm thick fixed to ceiling joists, joints filled with filler and taped to receive decoration	H1	m2	0.36	3.42	2.35	0.87	6.64
Plasterboard 9.5mm thick fixed to sloping ceiling softwood joists, joints filled with filler and taped to receive decoration	H2	m2	0.36	3.42	2.35	0.87	6.64

	Ref	Unit	Hours	Hours £	Mat'ls £	O & P £	Total £
One coat skim plaster to plaster-board ceiling and scrim joints	H3	m2	0.50	4.75	1.22	0.90	6.87
Two coats emulsion paint to plaster-board walls and ceilings	H4	m2	0.26	2.47	0.59	0.46	3.52
PART J WALL FINISHES							
Plaster, first coat 11mm bonding, second coat 2mm finish to walls	J1	m2	0.39	3.71	1.99	0.85	6.55
Two coats emulsion paint to plaster-board walls and ceilings	J2	m2	0.26	2.47	0.59	0.46	3.52
PART K ELECTRICAL WORK							
13 amp double switched socket outlet with neon	K1	nr	0.40	5.60	8.26	2.08	15.94
Lighting point	K2	nr	0.35	4.90	6.80	1.76	13.46
Lighting switch 1 way	K3	nr	0.35	4.90	3.97	1.33	10.20
Lighting switch 2 way	K4	nr	0.40	5.60	7.84	2.02	15.46
Lighting wiring	K5	m	0.40	5.60	7.84	2.02	15.46
Power cable	K6	m	0.10	1.40	1.03	0.36	2.79
Three-floor linked smoke alarm system	K7	nr	1.00	14.00	24.50	5.78	44.28
PART L HEATING WORK							
15mm copper pipe	L1	m	0.22	2.75	1.88	0.69	5.32
Elbow	L2	nr	0.28	3.50	1.10	0.69	5.29
Tee	L3	nr	0.34	4.25	1.95	0.93	7.13
Radiator, double convector size 1400 x 520mm	L4	nr	1.30	16.25	115.68	19.79	151.72
Break into existing pipe, insert tee	L5	nr	0.75	9.38	3.95	2.00	15.32

Loft conversion size 4.5 × 4.5m (not to scale)

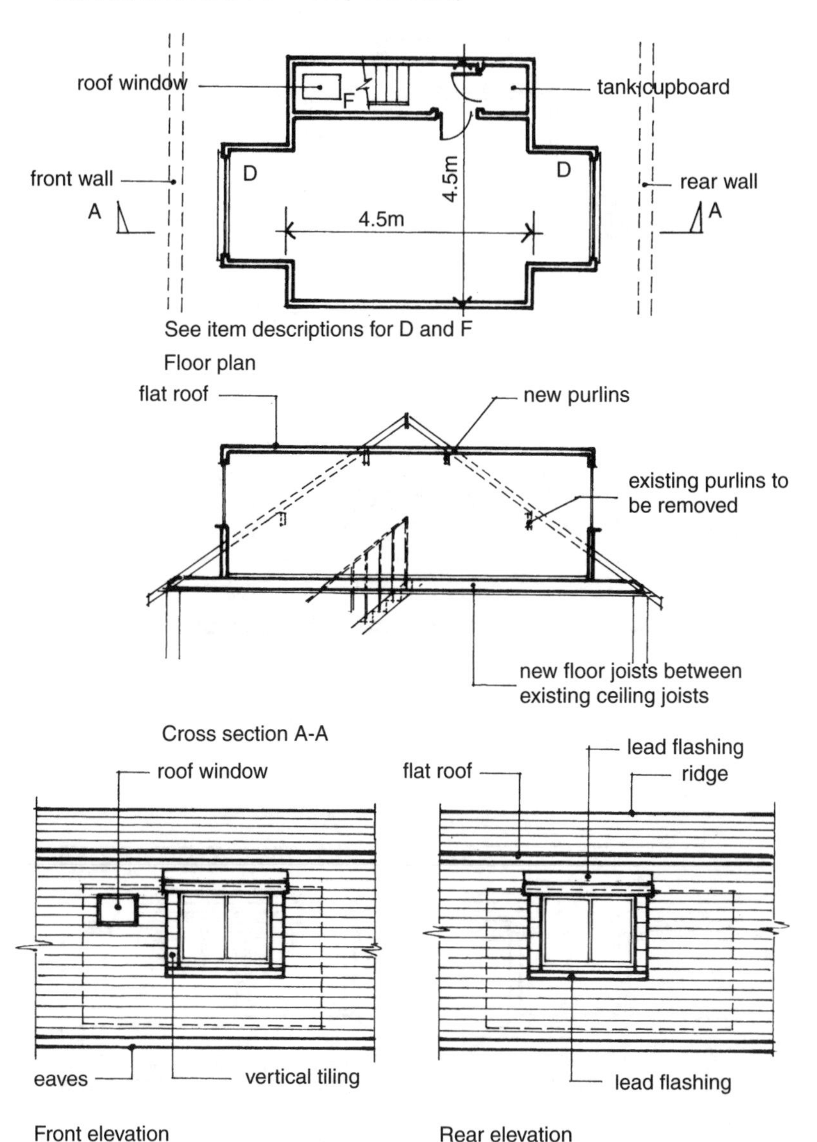

Loft conversion size 4.5×5.5m (not to scale)

roof window
front wall
A
D
4.5m
5.5m
F
F
tank cupboard
A
roof window
D
rear wall

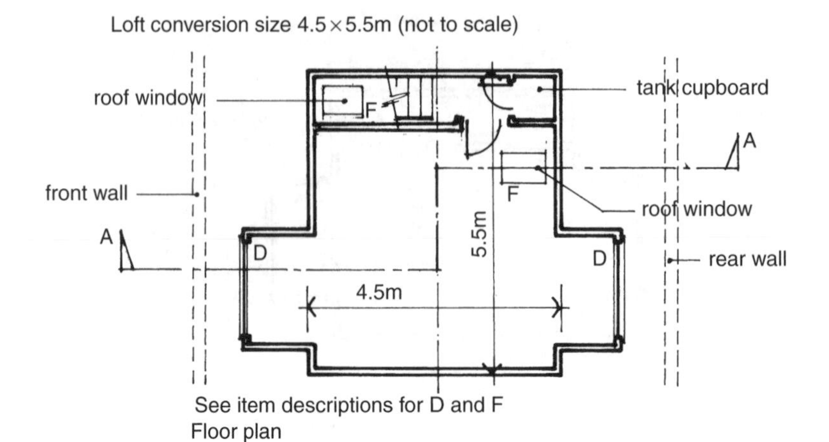

See item descriptions for D and F

Floor plan

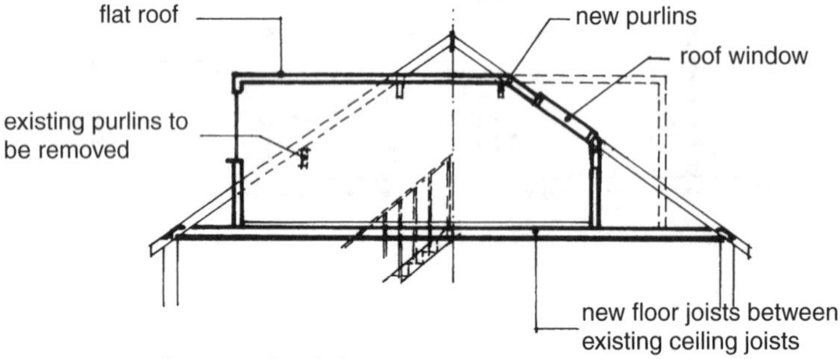

flat roof
existing purlins to be removed
new purlins
roof window
new floor joists between existing ceiling joists

Cross section A-A

ridge
flat roof
lead flashing
eaves
vertical tile hanging

Front elevation

roof window
lead flashing

Rear elevation

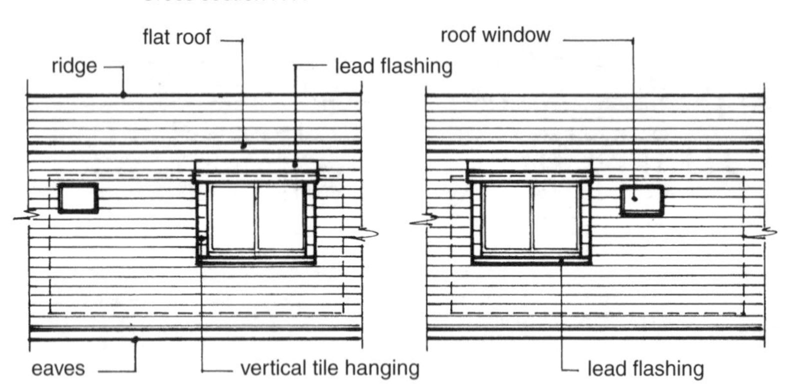

Loft conversion size 4.5 × 6.5m (not to scale)

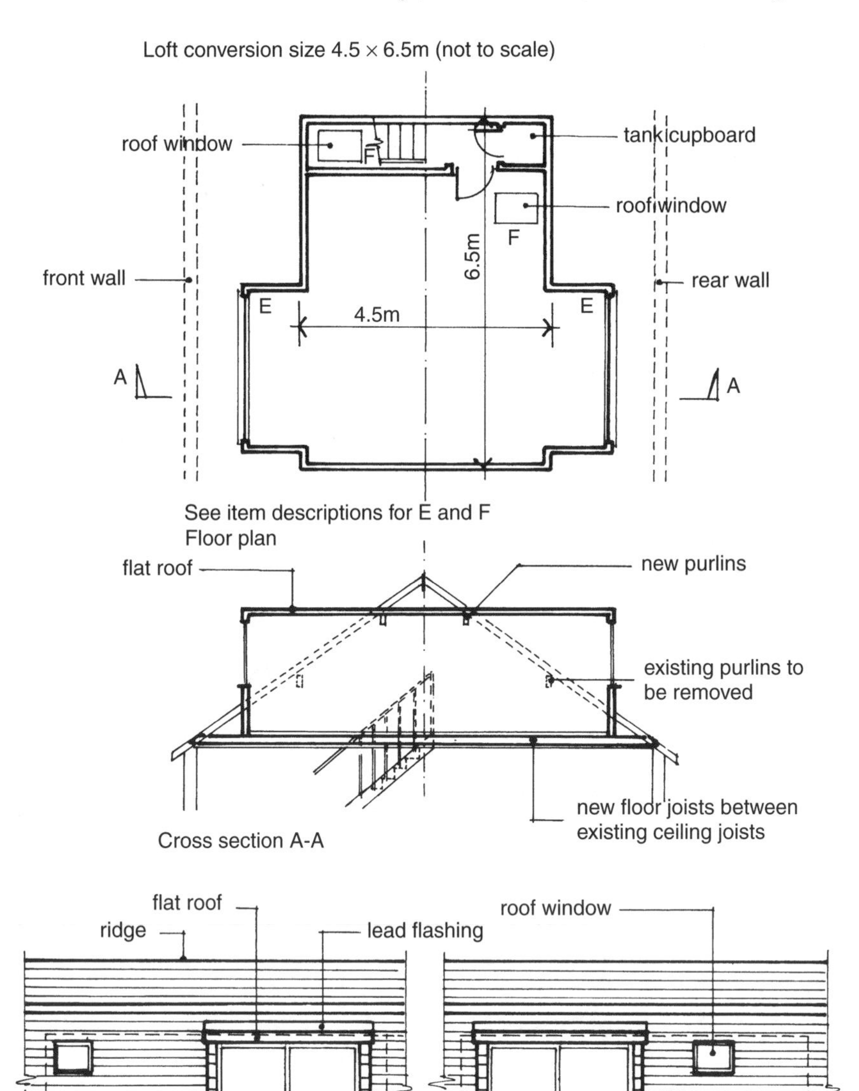

	Ref	Qty	Hours	Hours £	Mat'ls £	O & P £	Total £
PART A PRELIMINARIES							
Small tools	A1	6 wks					210.00
Scaffolding (m2/weeks)	A3	432.00					972.00
Skip	A4	6 wks					600.00
Clean up	A5	8 hrs					64.00
Carried to summary							1,846.00
PART B PREPARATION							
Clear out roof space	B1	item	4.00	38.00	0.00	5.70	43.70
Remove carpet from bedroom	B2	item	2.00	19.00	0.00	2.85	21.85
Erect temporary screen in bedroom consisting of 75 x 50mm covered both sides with polythene sheeting	B3	10.00m2	6.00	57.00	11.84	10.33	79.17
Disconnect pipes from cold water storage tank, move tank to new position and reconnect pipes	B4	item	1.90	18.05	3.50	3.23	24.78
Carried to summary			13.90	132.05	15.34	22.11	169.50
PART C DORMER WINDOW							
Remove clay tiles or slates, felt and battens and remove	C1	17.50m2	12.79	121.51	0.00	18.23	139.73
Cut into rafters and purlin to form new opening size 2000 x 3000mm, trim with 150 x 50mm sawn softwood bearers	C2	1nr	14.00	133.00	29.56	24.38	186.94
Cut into rafters and purlin to form new opening size 2600 x 3000mm, trim with 150 x 50mm sawn softwood bearers	C3	1nr	0.00	0.00	0.00	0.00	0.00
Carried forward			26.79	254.51	29.56	42.61	326.67

	Ref	Qty	Hours	Hours £	Mat'ls £	O & P £	Total £
Brought forward			26.79	254.51	29.56	42.61	326.67
200 x 75mm sawn softwood purlin	C4	9.60m	2.40	22.80	30.53	8.00	61.33
Cut and pin end of purlin to existing brickwork	C5	4nr	4.00	38.00	1.60	5.94	45.54
50 x 75mm sawn softwood sole plate	C6	9.40m	2.08	19.76	11.47	4.68	35.91
50 x 75mm sawn softwood head	C7	9.40m	2.08	19.76	11.47	4.68	35.91
50 x 75mm sawn softwood studs	C8	36.00m	2.82	26.79	43.92	10.61	81.32
50 x 75mm sawn joists	C9	21.00m	4.20	39.90	47.88	13.17	100.95
Mild steel bolt, M10 x 100mm	C10	14nr	2.10	19.95	17.50	5.62	43.07
18mm thick WPB grade plywood roof decking fixed to roof joists	C11	13.00m2	11.70	111.15	102.18	32.00	245.33
50mm wide sawn softwood tapered firring pieces average depth 50mm	C12	21.00m	3.78	35.91	47.46	12.51	95.88
Crown Wool insulation or similar 100mm thick fixed between joists with chicken wire and 150mm thick layer laid over joists	C13	23.00m2	20.70	196.65	250.93	67.14	514.72
19mm thick wrought softwood fascia board 200mm high	C14	15.20m	7.60	72.20	29.94	15.32	117.46
6mm thick asbestos-free insulation board soffit 150mm wide	C15	15.20m	6.08	57.76	26.60	12.65	97.01
Three-layer polyester-base mineral-surfaced roofing felt	C16	13.00m2	7.15	67.93	109.33	26.59	203.84
Turn down to edge of roof	C17	15.20m	1.52	14.44	12.77	4.08	31.29
Lead flashing code 5, 200mm girth dressing under existing tiles and over new dormer roof	C18	5.00m	3.00	28.50	33.40	9.29	71.19
Carried forward			108.00	1,026.00	806.54	274.88	2,107.42

	Ref	Qty	Hours	Hours £	Mat'ls £	O & P £	Total £
Brought forward			108.00	1,026.00	806.54	274.88	2,107.42
Lead stepped flashing code 5, 200mm girth dressing under vertical tiles and over existing roof	C19	12.80m	11.52	109.44	94.98	30.66	235.08
Marley Plain roofing tiles, hung vertically, size 278 x 165mm, 65mm lap, type 1F reinforced roofing felt and 38 x 19mm softwood battens	C20	10.00m	19.00	180.50	258.70	65.88	505.08
Raking cutting on tiling	C21	12.80m	1.92	18.24	0.00	2.74	20.98
Make good existing tiling up to new dormer	C22	15.30m	1.92	18.24	0.00	2.74	20.98
PVC-U window size 1800 x 1200mm complete	C23	2nr	4.00	38.00	504.32	81.35	623.67
PVC-U window size 2400 x 1200mm complete	C24	0.00	0.00	0.00	0.00	0.00	0.00
25 x 225mm wrought softwood window board	C25	5.00m	1.50	14.25	31.10	6.80	52.15
25 x 150mm wrought softwood lining	C26	20.00m	4.80	45.60	49.60	14.28	109.48
PVC-U gutter, 112mm half round with gutter union joints, fixed to wrought softwood fascia board with support brackets at 1m maximum centres	C27	5.00m	1.30	12.35	27.50	5.98	45.83
Extra for stop end	C28	4nr	0.56	5.32	4.44	1.46	11.22
Extra for stop end outlet	C29	2nr	0.28	2.66	2.22	0.73	5.61
PVC-U down pipe 68mm diameter, loose spigot and socket joints, plugged to faced brickwork with pipe clips at 2m centres	C30	5.00m	1.25	11.88	16.95	4.32	33.15
Extra for shoe	C31	2nr	0.60	5.70	4.18	1.48	11.36
Carried forward			156.65	1,488.18	1,800.53	493.31	3,782.01

	Ref	Qty	Hours	Hours £	Mat'ls £	O & P £	Total £
Brought forward			156.65	1,488.18	1,800.53	493.31	3,782.01
Plasterboard 9.5mm thick fixed to softwood studding, joints filled with filler and taped to receive decoration	C32	15.40m2	5.54	52.63	28.49	12.17	93.29
Plasterboard 9.5mm thick fixed to softwood joists, joints filled with filler and taped to receive decoration	C33	13.00m2	4.68	44.46	24.05	10.28	78.79
One coat skim plaster to plaster-board ceiling including scrimming joints	C34	28.00m2	14.00	133.00	34.16	25.07	192.23
Two coats emulsion paint to plasterboard walls and ceilings	C35	28.00m2	7.28	69.16	16.52	12.85	98.53
Apply one coat primer, one oil-based undercoat and one coat gloss paint on surfaces not exceeding 300mm girth	C36	20.00m	4.00	38.00	6.00	6.60	50.60
Carried to summary			192.15	1,825.43	1,909.75	560.28	4,295.45

PART D
ROOF WINDOW

	Ref	Qty	Hours	Hours £	Mat'ls £	O & P £	Total £
Remove clay tiles or slates, felt and battens and remove	D1	1.16m2	0.87	8.27	0.00	1.24	9.50
Cut into rafters and purlin to form new opening size 880 x 1080mm, trim with 150 x 50mm sawn softwood bearers	D2	2nr	6.00	57.00	30.80	13.17	100.97
Make good existing tiling up to new dormer	D3	2.16m	0.39	3.71	0.00	0.56	4.26
Velux window size 780 x 980mm complete with flashings	D4	2nr	6.50	61.75	464.00	78.86	604.61
25 x 150mm wrought softwood lining	D5	3.52m	0.84	7.98	8.73	2.51	19.22
Carried forward			14.60	138.70	503.53	96.33	738.56

	Ref	Qty	Hours	Hours £	Mat'ls £	O & P £	Total £
Brought forward			14.60	138.70	503.53	96.33	738.56
Apply one coat primer, one oil-based undercoat and one coat gloss paint on surfaces not exceeding 300mm girth	D6	3.52m	0.70	6.65	1.06	1.16	8.87
Carried to summary			15.30	145.35	504.59	97.49	747.43
PART E STAIRS							
Break into existing ceiling joists and plasterboard ceiling, trim with 150 x 50mm trimmer including temporary supports, to form new opening size 2750 x 1200mm	E1	1nr	3.00	28.50	12.00	6.08	46.58
Make good plasterboard ceiling up to new opening	E2	7.90m	1.58	15.01	3.95	2.84	21.80
Wrought softwood straight-flight staircase with 13 close treads, 2700mm going, 2600mm rise complete with 38 x 200mm strings, 66 x 66 x 1350mm newel post, 63 x 44mm handrail, balusters 848mm high and spacers	E3	1nr	32.00	304.00	565.00	130.35	999.35
25 x 150mm wrought softwood lining	E4	7.90m	1.90	18.05	19.59	5.65	43.29
Apply one coat primer, one oil-based undercoat and one coat gloss paint on surfaces exceeding 300mm girth	E5	23.00m2	16.10	152.95	23.96	26.54	203.45
Carried to summary			54.58	518.51	624.50	171.45	1,314.46
PART F FLOORING							
25mm thick tongued and grooved wrought softwood flooring	F1	21.00m2	15.49	147.16	197.40	51.68	396.24
Carried forward			15.49	147.16	197.40	51.68	396.24

	Ref	Qty	Hours	Hours £	Mat'ls £	O & P £	Total £
Brought forward			15.49	147.16	197.40	51.68	396.24
19 x 100mm wrought softwood skirting	F2	22.00m	3.74	35.53	30.14	9.85	75.52
Apply one coat primer, one oil-based undercoat and one coat gloss paint on skirting not exceeding 300mm girth	F3	22.00m	4.40	41.80	6.60	7.26	55.66
Carried to summary			23.63	224.49	234.14	68.79	527.42
PART G INTERNAL PARTITIONS AND DOORS							
50 x 75mm sawn softwood sole plate	G1	9.20m	2.02	19.19	10.30	4.42	33.91
50 x 75mm sawn softwood head	G2	10.10m	2.22	21.09	11.31	4.86	37.26
50 x 75mm sawn softwood studs	G3	29.80m	8.34	79.23	33.38	16.89	129.50
50 x 75mm sawn softwood noggings	G4	9.20m	2.58	24.51	10.30	5.22	40.03
Plasterboard 9.5mm thick fixed to softwood studding, filled joints and taped to receive decoration	G5	17.56m2	6.32	60.04	32.49	13.88	106.41
Flush door 35mm thick, size 762 x 1981mm, internal quality, half hour fire check, veneered finish both sides	G6	2nr	2.50	23.75	110.64	20.16	154.55
38 x 150mm wrought softwood lining	G7	9.52m	2.09	19.86	8.38	4.24	32.47
13 x 38mm wrought softwood door stop	G8	9.52m	1.43	13.59	4.38	2.69	20.66
19 x 50mm wrought softwood chamfered architrave	G9	9.52m	0.15	1.43	0.46	0.28	2.17
Carried forward			27.65	262.68	221.64	72.65	556.96

	Ref	Qty	Hours	Hours £	Mat'ls £	O & P £	Total £
Brought forward			27.65	262.68	221.64	72.65	556.96
19 x 50mm wrought softwood chamfered skirting	G10	8.20m	1.62	15.39	13.04	4.26	32.69
100mm rising steel butts	G11	2 pair	0.60	5.70	7.90	2.04	15.64
Silver anodised aluminium mortice latch with lever furniture	G12	2nr	1.60	15.20	28.20	6.51	49.91
Two coats emulsion paint to plasterboard walls	G13	6.04m2	4.23	40.19	5.80	6.90	52.88
Apply one coat primer, one oil-based undercoat and one coat gloss paint to general surfaces exceeding 300mm girth	G14	6.04m2	0.70	6.65	0.96	1.14	8.75
Carried to summary			34.78	330.41	264.50	89.24	684.15
PART H CEILINGS AND SOFFITS							
Plasterboard 9.5mm thick fixed to ceiling joists, joints filled with filler and taped to receive decoration	H1	7.65m2	2.75	26.13	17.98	6.62	50.72
Plasterboard 9.5mm thick fixed to sloping ceiling softwood joists, joints filled with filler and taped to receive decoration	H2	30.60m2	12.24	116.28	71.91	28.23	216.42
One coat skim plaster to plasterboard ceiling and scrim joints	H3	38.25m2	19.13	181.74	46.67	34.26	262.67
Two coats emulsion paint to plasterboard walls and ceilings	H4	38.25m2	9.95	94.53	22.57	17.56	134.66
Carried to summary			44.07	418.67	159.13	86.67	664.46
PART J WALL FINISHES							
Plaster, first coat 11mm bonding, and 2mm finish coat to walls	J1	19.80m2	7.73	73.44	39.44	16.93	129.81
Carried forward			7.73	73.44	39.44	16.93	129.81

	Ref	Qty	Hours	Hours £	Mat'ls £	O & P £	Total £
Brought forward			7.73	73.44	39.44	16.93	129.81
Two coats emulsion paint to plasterboard walls and ceilings	J2	19.80m2	5.15	48.93	11.69	9.09	69.71
Carried to summary			12.88	122.36	51.13	26.02	199.51
PART K ELECTRICAL WORK							
13 amp double switched socket outlet with neon	K1	3nr	1.20	16.80	24.78	6.24	47.82
Lighting point	K2	4nr	1.20	16.80	27.20	6.60	50.60
Lighting switch 1 way	K3	2nr	0.70	9.80	7.94	2.66	20.40
Lighting switch 2 way	K4	2nr	0.80	11.20	15.48	4.00	30.68
Lighting wiring	K5	16.00m	1.60	22.40	16.48	5.83	44.71
Power cable	K6	12.00m	1.80	25.20	13.20	5.76	44.16
Three-floor linked smoke alarm system	K7	1nr	1.00	14.00	24.50	5.78	44.28
Carried to summary			8.30	116.20	129.58	36.87	282.65
PART L HEATING WORK							
15mm copper pipe	L1	9.00m	1.98	24.75	16.92	6.25	47.92
Elbow	L2	5nr	1.40	17.50	5.50	3.45	26.45
Tee	L3	1nr	0.34	4.25	1.95	0.93	7.13
Radiator, double convector size 1400 x 520mm	L4	2nr	2.60	32.50	231.36	39.58	303.44
Break into existing pipe, insert tee	L5	1nr	0.75	9.38	3.95	2.00	15.32
Carried to summary			7.07	88.38	259.68	52.21	400.26

SUMMARY

	Hours	Hours £	Mat'ls £	O & P £	Total £
PART A PRELIMINARIES	0.00	0.00	0.00	0.00	1,846.00
PART B PREPARATION	13.90	132.05	15.34	22.11	169.50
PART C DORMER WINDOW	192.15	1,825.43	1,909.75	560.28	4,295.45
PART D ROOF WINDOW	15.30	145.35	504.59	97.49	747.43
PART E STAIRS	54.58	518.51	624.50	171.45	1,314.46
PART F FLOORING	23.53	224.49	234.14	68.79	527.42
PART G INTERNAL PARTITIONS AND DOORS	34.78	330.41	264.50	89.24	684.15
PART H CEILINGS AND SOFFITS	44.07	418.67	159.13	86.67	664.46
PART J WALL FINISHES	12.88	122.36	51.13	26.02	199.51
PART K ELECTRICAL WORK	8.30	116.20	129.58	36.87	282.65
PART L HEATING WORK	7.07	88.38	259.68	52.21	400.26
Final total	406.56	3,921.85	4,152.34	1,211.13	11,131.29

	Ref	Qty	Hours	Hours £	Mat'ls £	O & P £	Total £
PART A PRELIMINARIES							
Small tools	A1	7 wks					245.00
Scaffolding (m2/weeks)	A3	504.00					1,134.00
Skip	A4	7 wks					700.00
Clean up	A5	8 hrs					64.00
Carried to summary							2,143.00
PART B PREPARATION							
Clear out roof space	B1	item	4.00	38.00	0.00	5.70	43.70
Remove carpet from bedroom	B2	item	2.00	19.00	0.00	2.85	21.85
Erect temporary screen in bedroom consisting of 75 x 50mm covered both sides with polythene sheeting	B3	10.00m2	6.00	57.00	11.84	10.33	79.17
Disconnect pipes from cold water storage tank, move tank to new position and reconnect pipes	B4	item	1.90	18.05	3.50	3.23	24.78
Carried to summary			13.90	132.05	15.34	22.11	169.50
PART C DORMER WINDOW							
Remove clay tiles or slates, felt and battens and remove	C1	17.50m2	12.79	121.51	0.00	18.23	139.73
Cut into rafters and purlin to form new opening size 2000 x 3000mm, trim with 150 x 50mm sawn softwood bearers	C2	1nr	14.00	133.00	29.56	24.38	186.94
Cut into rafters and purlin to form new opening size 2600 x 3000mm, trim with 150 x 50mm sawn softwood bearers	C3	1nr	0.00	0.00	0.00	0.00	0.00
Carried forward			26.79	254.51	29.56	42.61	326.67

390 Loft, size 4.5 x 5.5m

	Ref	Qty	Hours	Hours £	Mat'ls £	O & P £	Total £
Brought forward			26.79	254.51	29.56	42.61	326.67
200 x 75mm sawn softwood purlin	C4	9.60m	2.40	22.80	30.53	8.00	61.33
Cut and pin end of purlin to existing brickwork	C5	4nr	4.00	38.00	1.60	5.94	45.54
50 x 75mm sawn softwood sole plate	C6	9.40m	2.08	19.76	11.47	4.68	35.91
50 x 75mm sawn softwood head	C7	9.40m	2.08	19.76	11.47	4.68	35.91
50 x 75mm sawn softwood studs	C8	36.00m	2.82	26.79	43.92	10.61	81.32
50 x 75mm sawn joists	C9	21.00m	4.20	39.90	47.88	13.17	100.95
Mild steel bolt, M10 x 100mm	C10	14nr	2.10	19.95	17.50	5.62	43.07
18mm thick WPB grade plywood roof decking fixed to roof joists	C11	13.00m2	11.70	111.15	102.18	32.00	245.33
50mm wide sawn softwood tapered firring pieces average depth 50mm	C12	23.00m	4.14	39.33	51.98	13.70	105.01
Crown Wool insulation or similar 100mm thick fixed between joists with chicken wire and 150mm thick layer laid over joists	C13	23.00m2	20.70	196.65	250.93	67.14	514.72
19mm thick wrought softwood fascia board 200mm high	C14	15.20m	7.60	72.20	29.94	15.32	117.46
6mm thick asbestos-free insulation board soffit 150mm wide	C15	15.20m	6.08	57.76	26.60	12.65	97.01
Three-layer polyester-base mineral-surfaced roofing felt	C16	13.00m2	7.15	67.93	109.33	26.59	203.84
Turn down to edge of roof	C17	15.20m	1.52	14.44	12.77	4.08	31.29
Lead flashing code 5, 200mm girth dressing under existing tiles and over new dormer roof	C18	5.00m	3.00	28.50	33.40	9.29	71.19
Carried forward			108.36	1,029.42	811.06	276.07	2,116.55

	Ref	Qty	Hours	Hours £	Mat'ls £	O & P £	Total £
Brought forward			108.36	1,029.42	811.06	276.07	2,116.55
Lead stepped flashing code 5, 200mm girth dressing under vertical tiles and over existing roof	C19	12.80m	11.52	109.44	94.98	30.66	235.08
Marley Plain roofing tiles, hung vertically, size 278 x 165mm, 65mm lap, type 1F reinforced roofing felt and 38 x 19mm softwood battens	C20	10.00m	19.00	180.50	258.70	65.88	505.08
Raking cutting on tiling	C21	12.80m	1.92	18.24	0.00	2.74	20.98
Make good existing tiling up to new dormer	C22	15.30m	1.92	18.24	0.00	2.74	20.98
PVC-U window size 1800 x 1200mm complete	C23	2nr	4.00	38.00	504.32	81.35	623.67
PVC-U window size 2400 x 1200mm complete	C24	0.00	0.00	0.00	0.00	0.00	0.00
25 x 225mm wrought softwood window board	C25	5.00m	1.50	14.25	31.10	6.80	52.15
25 x 150mm wrought softwood lining	C26	20.00m	4.80	45.60	49.60	14.28	109.48
PVC-U gutter, 112mm half round with gutter union joints, fixed to wrought softwood fascia board with support brackets at 1m maximum centres	C27	5.00m	1.30	12.35	27.50	5.98	45.83
Extra for stop end	C28	4nr	0.56	5.32	4.44	1.46	11.22
Extra for stop end outlet	C29	2nr	0.28	2.66	2.22	0.73	5.61
PVC-U down pipe 68mm diameter, loose spigot and socket joints, plugged to faced brickwork with pipe clips at 2m centres	C30	5.00m	1.25	11.88	16.95	4.32	33.15
Extra for shoe	C31	2nr	0.60	5.70	4.18	1.48	11.36
Carried forward			157.01	1,491.60	1,805.05	494.50	3,791.14

	Ref	Qty	Hours	Hours £	Mat'ls £	O & P £	Total £
Brought forward			157.01	1,491.60	1,805.05	494.50	3,791.14
Plasterboard 9.5mm thick fixed to softwood studding, joints filled with filler and taped to receive decoration	C32	15.40m2	5.54	52.63	28.49	12.17	93.29
Plasterboard 9.5mm thick fixed to softwood joists, joints filled with filler and taped to receive decoration	C33	13.00m2	4.68	44.46	24.05	10.28	78.79
One coat skim plaster to plaster-board ceiling including scrimming joints	C34	28.00m2	14.00	133.00	34.16	25.07	192.23
Two coats emulsion paint to plasterboard walls and ceilings	C35	28.00m2	7.28	69.16	16.52	12.85	98.53
Apply one coat primer, one oil-based undercoat and one coat gloss paint on surfaces not exceeding 300mm girth	C36	20.00m	4.00	38.00	6.00	6.60	50.60
Carried to summary			192.51	1,828.85	1,914.27	561.47	4,304.58

PART D
ROOF WINDOW

	Ref	Qty	Hours	Hours £	Mat'ls £	O & P £	Total £
Remove clay tiles or slates, felt and battens and remove	D1	1.16m2	0.87	8.27	0.00	1.24	9.50
Cut into rafters and purlin to form new opening size 880 x 1080mm, trim with 150 x 50mm sawn softwood bearers	D2	2nr	6.00	57.00	30.80	13.17	100.97
Make good existing tiling up to new dormer	D3	2.16m	0.39	3.71	0.00	0.56	4.26
Velux window size 780 x 980mm complete with flashings	D4	2nr	6.50	61.75	464.00	78.86	604.61
25 x 150mm wrought softwood lining	D5	3.52m	0.84	7.98	8.73	2.51	19.22
Carried forward			14.60	138.70	503.53	96.33	738.56

	Ref	Qty	Hours	Hours £	Mat'ls £	O & P £	Total £
Brought forward			14.60	138.70	503.53	96.33	738.56
Apply one coat primer, one oil-based undercoat and one coat gloss paint on surfaces not exceeding 300mm girth	D6	3.52m	0.70	6.65	1.06	1.16	8.87
Carried to summary			15.30	145.35	504.59	97.49	747.43
PART E STAIRS							
Break into existing ceiling joists and plasterboard ceiling, trim with 150 x 50mm trimmer including temporary supports, to form new opening size 2750 x 1200mm	E1	1nr	3.00	28.50	12.00	6.08	46.58
Make good plasterboard ceiling up to new opening	E2	7.90m	1.58	15.01	3.95	2.84	21.80
Wrought softwood straight-flight staircase with 13 close treads, 2700mm going, 2600mm rise complete with 38 x 200mm strings, 66 x 66 x 1350mm newel post, 63 x 44mm handrail, balusters 848mm high and spacers	E3	1nr	32.00	304.00	565.00	130.35	999.35
25 x 150mm wrought softwood lining	E4	7.90m	1.90	18.05	19.59	5.65	43.29
Apply one coat primer, one oil-based undercoat and one coat gloss paint on surfaces exceeding 300mm girth	E5	23.00m2	16.10	152.95	23.96	26.54	203.45
Carried to summary			54.58	518.51	624.50	171.45	1,314.46
PART F FLOORING							
25mm thick tongued and grooved wrought softwood flooring	F1	25.40m2	18.80	178.60	238.76	62.60	479.96
Carried forward			18.80	178.60	238.76	62.60	479.96

	Ref	Qty	Hours	Hours £	Mat'ls £	O & P £	Total £
Brought forward			18.80	178.60	238.76	62.60	479.96
19 x 100mm wrought softwood skirting	F2	24.00m	4.32	41.04	32.88	11.09	85.01
Apply one coat primer, one oil-based undercoat and one coat gloss paint on skirting not exceeding 300mm girth	F3	24.00m	4.80	45.60	7.20	7.92	60.72
Carried to summary			27.92	265.24	278.84	81.61	625.69
PART G INTERNAL PARTITIONS AND DOORS							
50 x 75mm sawn softwood sole plate	G1	11.20m	2.46	23.37	12.54	5.39	41.30
50 x 75mm sawn softwood head	G2	12.10m	2.66	25.27	13.55	5.82	44.64
50 x 75mm sawn softwood studs	G3	32.80m	9.18	87.21	36.74	18.59	142.54
50 x 75mm sawn softwood noggings	G4	11.20m	3.14	29.83	12.54	6.36	48.73
Plasterboard 9.5mm thick fixed to softwood studding, filled joints and taped to receive decoration	G5	18.76m2	6.74	64.03	34.71	14.81	113.55
Flush door 35mm thick, size 762 x 1981mm, internal quality, half hour fire check, veneered finish both sides	G6	2nr	2.50	23.75	110.64	20.16	154.55
38 x 150mm wrought softwood lining	G7	9.52m	2.09	19.86	8.38	4.24	32.47
13 x 38mm wrought softwood door stop	G8	9.52m	1.43	13.59	4.38	2.69	20.66
19 x 50mm wrought softwood chamfered architrave	G9	9.52m	0.15	1.43	0.46	0.28	2.17
Carried forward			30.35	288.33	233.94	78.34	600.60

	Ref	Qty	Hours	Hours £	Mat'ls £	O & P £	Total £
Brought forward			30.35	288.33	233.94	78.34	600.60
19 x 50mm wrought softwood chamfered skirting	G10	8.20m	1.62	15.39	13.04	4.26	32.69
100mm rising steel butts	G11	2 pair	0.60	5.70	7.90	2.04	15.64
Silver anodised aluminium mortice latch with lever furniture	G12	2nr	1.60	15.20	28.20	6.51	49.91
Two coats emulsion paint to plasterboard walls	G13	9.00m2	2.34	22.23	5.31	4.13	31.67
Apply one coat primer, one oil-based undercoat and one coat gloss paint to general surfaces exceeding 300mm girth	G14	6.04m2	4.23	40.19	5.80	6.90	52.88
Carried to summary			39.12	371.64	281.15	97.92	750.71

PART H
CEILINGS AND SOFFITS

	Ref	Qty	Hours	Hours £	Mat'ls £	O & P £	Total £
Plasterboard 9.5mm thick fixed to ceiling joists, joints filled with filler and taped to receive decoration	H1	9.35m2	3.37	32.02	21.97	8.10	62.08
Plasterboard 9.5mm thick fixed to sloping ceiling softwood joists, joints filled with filler and taped to receive decoration	H2	37.40m2	14.96	142.12	87.89	34.50	264.51
One coat skim plaster to plasterboard ceiling and scrim joints	H3	47.75m2	23.88	226.86	58.26	42.77	327.89
Two coats emulsion paint to plasterboard walls and ceilings	H4	47.75m2	12.42	117.99	28.17	21.92	168.08
Carried to summary			54.63	518.99	196.29	107.29	822.57

PART J
WALL FINISHES

	Ref	Qty	Hours	Hours £	Mat'ls £	O & P £	Total £
Plaster, first coat 11mm bonding, and 2mm finish coat to walls	J1	19.80m2	7.73	73.44	39.44	16.93	129.81
Carried forward			7.73	73.44	39.44	16.93	129.81

	Ref	Qty	Hours	Hours £	Mat'ls £	O & P £	Total £
Brought forward			7.73	73.44	39.44	16.93	129.81
Two coats emulsion paint to plaster-board walls and ceilings	J2	19.80m2	5.15	48.93	11.69	9.09	69.71
Carried to summary			12.88	122.36	51.13	26.02	199.51
PART K ELECTRICAL WORK							
13 amp double switched socket outlet with neon	K1	3nr	1.20	16.80	24.78	6.24	47.82
Lighting point	K2	4nr	1.20	16.80	27.20	6.60	50.60
Lighting switch 1 way	K3	2nr	0.70	9.80	7.94	2.66	20.40
Lighting switch 2 way	K4	2nr	0.80	11.20	15.48	4.00	30.68
Lighting wiring	K5	16.00m	1.60	22.40	16.48	5.83	44.71
Power cable	K6	12.00m	1.80	25.20	13.20	5.76	44.16
Three-floor linked smoke alarm system	K7	1nr	1.00	14.00	24.50	5.78	44.28
Carried to summary			8.30	116.20	129.58	36.87	282.65
PART L HEATING WORK							
15mm copper pipe	L1	9.00m	1.98	24.75	16.92	6.25	47.92
Elbow	L2	5nr	1.40	17.50	5.50	3.45	26.45
Tee	L3	1nr	0.34	4.25	1.95	0.93	7.13
Radiator, double convector size 1400 x 520mm	L4	2nr	2.60	32.50	231.36	39.58	303.44
Break into existing pipe, insert tee	L5	1nr	0.75	9.38	3.95	2.00	15.32
Carried to summary			7.07	88.38	259.68	52.21	400.26

SUMMARY

	Hours	Hours £	Mat'ls £	O & P £	Total £
PART A **PRELIMINARIES**	0.00	0.00	0.00	0.00	2,143.00
PART B **PREPARATION**	13.90	132.05	15.34	22.11	169.50
PART C **DORMER WINDOW**	192.51	1,828.85	1,914.27	561.47	4,304.58
PART D **ROOF WINDOW**	15.30	145.35	504.59	97.49	747.43
PART E **STAIRS**	54.58	518.51	624.50	171.45	1,314.46
PART F **FLOORING**	27.92	265.24	278.84	81.61	625.69
PART G **INTERNAL PARTITIONS AND DOORS**	39.12	371.64	281.15	97.92	750.71
PART H **CEILINGS AND SOFFITS**	54.63	518.99	196.29	107.29	822.57
PART J **WALL FINISHES**	12.88	122.36	51.13	26.02	199.51
PART K **ELECTRICAL WORK**	8.30	116.20	129.58	36.87	282.65
PART L **HEATING WORK**	7.07	88.38	259.68	52.21	400.26
Final total	426.21	4,107.57	4,255.37	1,254.44	11,760.36

	Ref	Qty	Hours	Hours £	Mat'ls £	O & P £	Total £
PART A PRELIMINARIES							
Small tools	A1	8 wks					280.00
Scaffolding (m2/weeks)	A3	576.00					1,296.00
Skip	A4	8 wks					800.00
Clean up	A5	8 hrs					64.00
Carried to summary							2,440.00
PART B PREPARATION							
Clear out roof space	B1	item	4.00	38.00	0.00	5.70	43.70
Remove carpet from bedroom	B2	item	2.00	19.00	0.00	2.85	21.85
Erect temporary screen in bedroom consisting of 75 x 50mm covered both sides with polythene sheeting	B3	10.00m2	6.00	57.00	11.84	10.33	79.17
Disconnect pipes from cold water storage tank, move tank to new position and reconnect pipes	B4	item	1.90	18.05	3.50	3.23	24.78
Carried to summary			13.90	132.05	15.34	22.11	169.50
PART C DORMER WINDOW							
Remove clay tiles or slates, felt and battens and remove	C1	21.70m2	16.28	154.66	0.00	23.20	177.86
Cut into rafters and purlin to form new opening size 2000 x 3000mm, trim with 150 x 50mm sawn softwood bearers	C2	0.00	0.00	0.00	0.00	0.00	0.00
Cut into rafters and purlin to form new opening size 2600 x 3000mm, trim with 150 x 50mm sawn softwood bearers	C3	1nr	16.00	152.00	38.12	28.52	218.64
Carried forward			32.28	306.66	38.12	51.72	396.50

400 Loft, size 4.5 x 6.5m

	Ref	Qty	Hours	Hours £	Mat'ls £	O & P £	Total £
Brought forward			32.28	306.66	38.12	51.72	396.50
200 x 75mm sawn softwood purlin	C4	13.60m	3.40	32.30	51.41	12.56	96.27
Cut and pin end of purlin to existing brickwork	C5	4nr	4.00	38.00	1.60	5.94	45.54
50 x 75mm sawn softwood sole plate	C6	9.40m	2.08	19.76	11.47	4.68	35.91
50 x 75mm sawn softwood head	C7	9.40m	2.08	19.76	11.47	4.68	35.91
50 x 75mm sawn softwood studs	C8	36.00m	2.82	26.79	43.92	10.61	81.32
50 x 75mm sawn joists	C9	27.00m	5.40	51.30	61.56	16.93	129.79
Mild steel bolt, M10 x 100mm	C10	18nr	2.70	25.65	22.50	7.22	55.37
18mm thick WPB grade plywood roof decking fixed to roof joists	C11	16.64m2	14.98	142.31	130.79	40.97	314.07
50mm wide sawn softwood tapered firring pieces average depth 50mm	C12	27.00m	4.86	46.17	61.02	16.08	123.27
Crown Wool insulation or similar 100mm thick fixed between joists with chicken wire and 150mm thick layer laid over joists	C13	26.64m2	23.98	227.81	290.64	77.77	596.22
19mm thick wrought softwood fascia board 200mm high	C14	18.00m	9.00	85.50	35.46	18.14	139.10
6mm thick asbestos-free insulation board soffit 150mm wide	C15	18.00m	7.20	68.40	31.50	14.99	114.89
Three-layer polyester-base mineral-surfaced roofing felt	C16	16.64m2	9.15	86.93	139.94	34.03	260.89
Turn down to edge of roof	C17	18.00m	1.80	17.10	15.12	4.83	37.05
Lead flashing code 5, 200mm girth dressing under existing tiles and over new dormer roof	C18	6.40m	3.84	36.48	42.75	11.88	91.11
Carried forward			129.57	1,230.92	989.27	333.03	2,553.21

	Ref	Qty	Hours	Hours £	Mat'ls £	O & P £	Total £
Brought forward			129.57	1,230.92	989.27	333.03	2,553.21
Lead stepped flashing code 5, 200mm girth dressing under vertical tiles and over existing roof	C19	12.80m	11.52	109.44	94.98	30.66	235.08
Marley Plain roofing tiles, hung vertically, size 278 x 165mm, 65mm lap, type 1F reinforced roofing felt and 38 x 19mm softwood battens	C20	10.00m	19.00	180.50	258.70	65.88	505.08
Raking cutting on tiling	C21	12.80m	1.92	18.24	0.00	2.74	20.98
Make good existing tiling up to new dormer	C22	16.00m	2.40	22.80	0.00	3.42	26.22
PVC-U window size 1800 x 1200mm complete	C23	0.00	0.00	0.00	0.00	0.00	0.00
PVC-U window size 2400 x 1200mm complete	C24	2nr	4.60	43.70	783.64	124.10	951.44
25 x 225mm wrought softwood window board	C25	6.40m	1.92	18.24	39.81	8.71	66.76
25 x 150mm wrought softwood lining	C26	22.40m	5.38	51.11	55.55	16.00	122.66
PVC-U gutter, 112mm half round with gutter union joints, fixed to wrought softwood fascia board with support brackets at 1m maximum centres	C27	6.40m	0.90	8.55	35.20	6.56	50.31
Extra for stop end	C28	4nr	0.56	5.32	4.44	1.46	11.22
Extra for stop end outlet	C29	2nr	0.28	2.66	2.22	0.73	5.61
PVC-U down pipe 68mm diameter, loose spigot and socket joints, plugged to faced brickwork with pipe clips at 2m centres	C30	5.00m	1.25	11.88	16.95	4.32	33.15
Extra for shoe	C31	2nr	0.60	5.70	4.18	1.48	11.36
Carried forward			160.90	1,528.55	2,026.24	533.22	4,088.01

	Ref	Qty	Hours	Hours £	Mat'ls £	O & P £	Total £
Brought forward			160.90	1,528.55	2,026.24	533.22	4,088.01
Plasterboard 9.5mm thick fixed to softwood studding, joints filled with filler and taped to receive decoration	C32	15.40m2	5.54	52.63	28.49	12.17	93.29
Plasterboard 9.5mm thick fixed to softwood joists, joints filled with filler and taped to receive decoration	C33	16.40m2	5.90	56.05	30.34	12.96	99.35
One coat skim plaster to plaster-board ceiling including scrimming joints	C34	31.64m2	15.82	150.29	38.60	28.33	217.22
Two coats emulsion paint to plaster-board walls and ceilings	C35	31.64m2	8.23	78.19	18.67	14.53	111.38
Apply one coat primer, one oil-based undercoat and one coat gloss paint on surfaces not exceeding 300mm girth	C36	22.40m	4.48	42.56	6.72	7.39	56.67
Carried to summary			200.87	1,908.27	2,149.06	608.60	4,665.92

PART D
ROOF WINDOW

	Ref	Qty	Hours	Hours £	Mat'ls £	O & P £	Total £
Remove clay tiles or slates, felt and battens and remove	D1	1.16m2	0.87	8.27	0.00	1.24	9.50
Cut into rafters and purlin to form new opening size 880 x 1080mm, trim with 150 x 50mm sawn softwood bearers	D2	2nr	6.00	57.00	30.80	13.17	100.97
Make good existing tiling up to new dormer	D3	2.16m	0.39	3.71	0.00	0.56	4.26
Velux window size 780 x 980mm complete with flashings	D4	2nr	6.50	61.75	464.00	78.86	604.61
25 x 150mm wrought softwood lining	D5	3.52m	0.84	7.98	8.73	2.51	19.22
Carried forward			14.60	138.70	503.53	96.33	738.56

	Ref	Qty	Hours	Hours £	Mat'ls £	O & P £	Total £
Brought forward			14.60	138.70	503.53	96.33	738.56
Apply one coat primer, one oil-based undercoat and one coat gloss paint on surfaces not exceeding 300mm girth	D6	3.52m	0.70	6.65	1.06	1.16	8.87
Carried to summary			15.30	145.35	504.59	97.49	747.43
PART E STAIRS							
Break into existing ceiling joists and plasterboard ceiling, trim with 150 x 50mm trimmer including temporary supports, to form new opening size 2750 x 1200mm	E1	1nr	3.00	28.50	12.00	6.08	46.58
Make good plasterboard ceiling up to new opening	E2	7.90m	1.58	15.01	3.95	2.84	21.80
Wrought softwood straight-flight staircase with 13 close treads, 2700mm going, 2600mm rise complete with 38 x 200mm strings, 66 x 66 x 1350mm newel post, 63 x 44mm handrail, balusters 848mm high and spacers	E3	1nr	32.00	304.00	565.00	130.35	999.35
25 x 150mm wrought softwood lining	E4	7.90m	1.90	18.05	19.59	5.65	43.29
Apply one coat primer, one oil-based undercoat and one coat gloss paint on surfaces exceeding 300mm girth	E5	23.00m2	16.10	152.95	23.96	26.54	203.45
Carried to summary			54.58	518.51	624.50	171.45	1,314.46
PART F FLOORING							
25mm thick tongued and grooved wrought softwood flooring	F1	31.91m2	23.61	224.30	299.95	78.64	602.88
Carried forward			23.61	224.30	299.95	78.64	602.88

	Ref	Qty	Hours	Hours £	Mat'ls £	O & P £	Total £
Brought forward			23.61	224.30	299.95	78.64	602.88
19 x 100mm wrought softwood skirting	F2	26.00m	4.42	41.99	35.62	11.64	89.25
Apply one coat primer, one oil-based undercoat and one coat gloss paint on skirting not exceeding 300mm girth	F3	26.00m	5.20	49.40	7.80	8.58	65.78
Carried to summary			33.23	315.69	343.37	98.86	757.91
PART G INTERNAL PARTITIONS AND DOORS							
50 x 75mm sawn softwood sole plate	G1	13.20m	2.91	27.65	14.78	6.36	48.79
50 x 75mm sawn softwood head	G2	14.10m	3.10	29.45	15.79	6.79	52.03
50 x 75mm sawn softwood studs	G3	33.50m	9.38	89.11	37.52	18.99	145.62
50 x 75mm sawn softwood noggings	G4	13.20m	3.70	35.15	14.78	7.49	57.42
Plasterboard 9.5mm thick fixed to softwood studding, filled joints and taped to receive decoration	G5	19.96m2	7.18	68.21	36.93	15.77	120.91
Flush door 35mm thick, size 762 x 1981mm, internal quality, half-hour fire check, veneered finish both sides	G6	2nr	2.50	23.75	110.64	20.16	154.55
38 x 150mm wrought softwood lining	G7	9.52m	2.09	19.86	8.38	4.24	32.47
13 x 38mm wrought softwood door stop	G8	9.52m	1.43	13.59	4.38	2.69	20.66
19 x 50mm wrought softwood chamfered architrave	G9	9.52m	0.15	1.43	0.46	0.28	2.17
Carried forward			32.44	308.18	243.66	82.78	634.62

	Ref	Qty	Hours	Hours £	Mat'ls £	O & P £	Total £
Brought forward			32.44	308.18	243.66	82.78	634.62
19 x 50mm wrought softwood chamfered skirting	G10	8.20m	1.62	15.39	13.04	4.26	32.69
100mm rising steel butts	G11	2 pair	0.60	5.70	7.90	2.04	15.64
Silver anodised aluminium mortice latch with lever furniture	G12	2nr	1.60	15.20	28.20	6.51	49.91
Two coats emulsion paint to plasterboard walls	G13	6.04m2	4.23	40.19	5.80	6.90	52.88
Apply one coat primer, one oil-based undercoat and one coat gloss paint to general surfaces exceeding 300mm girth	G14	m2	0.70	6.65	0.96	1.14	8.75
Carried to summary			39.57	375.92	286.52	99.37	761.80

PART H
CEILINGS AND SOFFITS

	Ref	Qty	Hours	Hours £	Mat'ls £	O & P £	Total £
Plasterboard 9.5mm thick fixed to ceiling joists, joints filled with filler and taped to receive decoration	H1	11.05m2	3.98	37.81	25.97	9.57	73.35
Plasterboard 9.5mm thick fixed to sloping ceiling softwood joists, joints filled with filler and taped to receive decoration	H2	44.20m2	17.68	167.96	103.87	40.77	312.60
One coat skim plaster to plaster-board ceiling and scrim joints	H3	55.25m2	27.63	262.49	67.41	49.48	379.38
Two coats emulsion paint to plasterboard walls and ceilings	H4	55.25m2	14.37	136.52	332.60	70.37	539.48
Carried to summary			63.66	604.77	529.85	170.19	1,304.81

PART J
WALL FINISHES

	Ref	Qty	Hours	Hours £	Mat'ls £	O & P £	Total £
Plaster, first coat 11mm bonding, and 2mm finish coat to walls	J1	19.80m2	7.73	73.44	39.44	16.93	129.81
Carried forward			7.73	73.44	39.44	16.93	129.81

	Ref	Qty	Hours	Hours £	Mat'ls £	O & P £	Total £
Brought forward			7.73	73.44	39.44	16.93	129.81
Two coats emulsion paint to plaster-board walls and ceilings	J2	19.80m2	5.15	48.93	11.69	9.09	69.71
Carried to summary			12.88	122.36	51.13	26.02	199.51
PART K ELECTRICAL WORK							
13 amp double switched socket outlet with neon	K1	3nr	1.20	16.80	24.78	6.24	47.82
Lighting point	K2	4nr	1.20	16.80	27.20	6.60	50.60
Lighting switch 1 way	K3	2nr	0.70	9.80	7.94	2.66	20.40
Lighting switch 2 way	K4	2nr	0.80	11.20	15.48	4.00	30.68
Lighting wiring	K5	18.00m	1.80	25.20	18.54	6.56	50.30
Power cable	K6	14.00m	2.10	29.40	15.40	6.72	51.52
Three-floor linked smoke alarm system	K7	1nr	8.30	116.20	24.50	21.11	161.81
Carried to summary			16.10	225.40	133.84	53.89	413.13
PART L HEATING WORK							
15mm copper pipe	L1	9.00m	1.98	24.75	16.92	6.25	47.92
Elbow	L2	5nr	1.40	17.50	5.50	3.45	26.45
Tee	L3	1nr	0.34	4.25	1.95	0.93	7.13
Radiator, double convector size 1400 x 520mm	L4	2nr	2.60	32.50	231.36	39.58	303.44
Break into existing pipe, insert tee	L5	1nr	0.75	9.38	3.95	2.00	15.32
Carried to summary			7.07	88.38	259.68	52.21	400.26

SUMMARY

	Hours	Hours £	Mat'ls £	O & P £	Total £
PART A PRELIMINARIES	0.00	0.00	0.00	0.00	2,440.00
PART B PREPARATION	13.90	132.05	15.34	22.11	169.50
PART C DORMER WINDOW	219.87	2,088.77	2,407.76	674.48	5,171.00
PART D ROOF WINDOW	15.30	145.35	504.59	97.49	747.43
PART E STAIRS	54.58	518.51	624.50	171.45	1,314.46
PART F FLOORING	33.23	315.69	343.37	98.86	757.91
PART G INTERNAL PARTITIONS AND DOORS	39.57	375.92	286.52	99.37	761.80
PART H CEILINGS AND SOFFITS	63.66	604.77	529.85	170.19	1,304.81
PART J WALL FINISHES	12.88	122.36	51.13	26.02	199.51
PART K ELECTRICAL WORK	16.10	225.40	133.84	53.89	413.13
PART L HEATING WORK	7.07	88.38	259.68	52.21	400.26
Final total	476.16	4,617.20	5,156.58	1,466.07	13,679.81

SUMMARY OF LOFT CONVERSION COSTS

	Loft conversion size					
	4.5 x 4.5m		4.5 x 5.5m		4.5 x 6.5m	
	£	%	£	%	£	%
PART A PRELIMINARIES	1,846	17	2,143	18	2,440	18
PART B PREPARATION	170	2	170	1	170	1
PART C DORMER WINDOW	4,295	38	4,304	37	5,171	37
PART D ROOF WINDOW	747	7	747	6	747	5
PART E STAIRS	1,315	12	1,315	11	1,315	10
PART F FLOORING	527	5	625	5	757	6
PART G INTERNAL PARTITIONS AND DOORS	684	6	751	6	762	7
PART H CEILINGS AND SOFFITS	665	6	823	7	1,305	10
PART J WALL FINISHES	200	2	200	2	200	1
PART K ELECTRICAL WORK	283	3	283	2	413	3
PART L HEATING WORK	400	4	400	3	400	3
Final total	11,132	100	11,761	100	13,680	101
Floor area (including dormers) m2	24.21		28.75		34.85	
Cost per square metre £	460		409		393	

Part Three

INSULATION WORK

Standard items

	Unit	Hours	Hours £	Mat'ls £	O & P £	Total £
INSULATION						
Quilt insulation						
Lightweight glasswool insulation quilt laid between joists						
60mm thick	m2	0.12	1.14	2.08	0.48	3.70
80mm thick	m2	0.12	1.14	2.74	0.58	4.46
100mm thick	m2	0.12	1.14	3.26	0.66	5.06
150mm thick	m2	0.14	1.33	5.00	0.95	7.28
200mm thick	m2	0.16	1.52	6.74	1.24	9.50
Cavity wall insulation						
Semi-rigid glass mineral wool insulation in cavity walls						
30mm thick	m2	0.12	1.14	2.15	0.49	3.78
40mm thick	m2	0.12	1.14	2.87	0.60	4.61
50mm thick	m2	0.12	1.14	3.40	0.68	5.22
60mm thick	m2	0.14	1.33	3.80	0.77	5.90
75mm thick	m2	0.14	1.33	4.51	0.88	6.72
80mm thick	m2	0.14	1.33	4.82	0.92	7.07
85mm thick	m2	0.16	1.52	5.13	1.00	7.65
100mm thick	m2	0.18	1.71	6.00	1.16	8.87
Cavity closer system 47.5mm thick in 3m lengths including stainless steel clips						
50mm wide	m	0.10	0.95	2.15	0.47	3.57
65mm thick	m	0.10	0.95	2.81	0.56	4.32
75mm thick	m	0.10	0.95	3.22	0.63	4.80
85mm thick	m	0.10	0.95	3.65	0.69	5.29
90mm thick	m	0.12	1.14	3.86	0.75	5.75
100mm thick	m	0.12	1.14	4.31	0.82	6.27
Acoustic insulation						
Semi-rigid glass mineral wool insulation in stud partitions and roofs						
80mm thick	m2	0.22	2.09	4.22	0.95	7.26
90mm thick	m2	0.24	2.28	4.71	1.05	8.04
100mm thick	m2	0.26	2.47	5.07	1.13	8.67
140mm thick	m2	0.28	2.66	7.64	1.55	11.85
150mm thick	m2	0.18	1.71	7.71	1.41	10.83

	Qty	Hours	Hours £	Mat'ls £	O & P £	Total £
Mineral wool acoustic blanket 30mm thick fixed to						
stud partitions	m2	0.14	1.33	2.74	0.61	4.68
ceilings	m2	0.18	1.71	2.74	0.67	5.12
Rigid rock wool acoustic slab 25mm thick fixed to						
timber floors	m2	0.18	1.71	5.74	1.12	8.57
Mineral wool sound-deadening quilt 25mm thick fixed to						
floors	m2	0.12	1.14	2.74	0.58	4.46
Polyethylene sound-deadening foam blanket quilt under screed						
2mm thick	m2	0.08	0.76	1.84	0.39	2.99
5mm thick	m2	0.10	0.95	1.33	0.34	2.62
Thermal insulation						
Non-combustible glass mineral wool quilt fitted between timber rafters in roof space						
60mm thick	m2	0.12	1.14	3.47	0.69	5.30
80mm thick	m2	0.14	1.33	4.60	0.89	6.82
90mm thick	m2	0.14	1.33	5.18	0.98	7.49
100mm thick	m2	0.16	1.52	5.55	1.06	8.13
140mm thick	m2	0.16	1.52	7.82	1.40	10.74
160mm thick	m2	0.18	1.71	8.93	1.60	12.24
Lagging						
Jute felt lagging in strips 100mm wide secured with galvanised steel wire to pipe diameter						
15mm	m	0.10	0.95	0.34	0.19	1.48
22mm	m	0.11	1.05	0.40	0.22	1.66
28mm	m	0.12	1.14	0.56	0.26	1.96
35mm	m	0.13	1.24	0.58	0.27	2.09
42mm	m	0.14	1.33	0.80	0.32	2.45

	Unit	Hours	Hours £	Mat'ls £	O & P £	Total £
Expanded polystryrene lagging fixed with aluminium bands to pipe diameter						
15mm	m	0.10	0.95	0.48	0.21	1.64
22mm	m	0.11	1.05	0.54	0.24	1.82
28mm	m	0.12	1.14	0.60	0.26	2.00
35mm	m	0.13	1.24	0.72	0.29	2.25
42mm	m	0.14	1.33	0.84	0.33	2.50
Expanded polystryrene lagging jacket set to bottom and sides of galvanised steel cisterns, size						
450 x 300 x 300mm	nr	0.50	4.75	6.32	1.66	12.73
650 x 500 x 450mm	nr	0.60	5.70	7.58	1.99	15.27
750 x 600 x 600mm	nr	0.80	7.60	8.61	2.43	18.64
PVC-U insulating jacket 80mm thick, filled with expanded polystyrene securing with fixing bands to hot water cylinder, size						
400 x 1050mm	nr	0.80	7.60	7.88	2.32	17.80
450 x 900mm	nr	0.80	7.60	11.57	2.88	22.05
400 x 1200mm	nr	0.80	7.60	12.69	3.04	23.33

Part Four

DAMAGE REPAIRS

Emergency measures

Fire damage

Flood damage

Gale damage

Theft damage

	Unit	Hours	Hours £	Plant £	O & P £	Total £
EMERGENCY MEASURES						
Flooding						
Instal hired pump and hoses in position	Item	1.00	9.50	0.00	1.43	10.93
Hire diaphragm pump and hoses						
50mm	hour	0.00	0.00	3.00	0.45	3.45
75mm	hour	0.00	0.00	3.25	0.49	3.74

	Unit	Hours	Hours £	Mat'ls £	O & P £	Total £
Hoardings and screens						
Temporary screens and hoardings 2m high consisting of 22mm thick exterior quality plywood fixed to 100 x 50mm posts and rails	m	2.60	24.70	29.12	8.07	61.89
Hire tarpaulin sheeting size 5 x 4m fixed in position	day	0.00	0.00	3.57	0.54	4.11
Plywood sheeting blocking up window or door opening	m2	1.00	9.50	14.57	3.61	27.68
Shoring						
Timber dead shores consisting of 200 x 200mm shores, 250 x 50mm plates and 200 x 50mm braces at centres of						
2m	m2	3.80	36.10	68.45	15.68	120.23
3m	m2	3.00	28.50	54.21	12.41	95.12
4m	m2	2.20	20.90	38.96	8.98	68.84
Timber raking and flying shores consisting of 200 x 200mm shores, 250 x 50mm plates and 200 x 50mm braces to gable end of two-storey house	Item	72.00	684.00	486.36	175.55	1345.91

	Unit	Hours	Hours £	Plant £	O & P £	Total £
Access towers						
Hire narrow width access tower size 0.85 x 1.8m, height						
7.20m	week	0.00	0.00	104.00	15.60	119.60
6.20m	week	0.00	0.00	86.00	12.90	98.90
5.20m	week	0.00	0.00	72.00	10.80	82.80
4.20m	week	0.00	0.00	58.00	8.70	66.70
Hire double width access tower size 1.45 x 2.5m, height						
7.20m	week	0.00	0.00	120.00	18.00	138.00
6.20m	week	0.00	0.00	112.00	16.80	128.80
5.20m	week	0.00	0.00	100.00	15.00	115.00
4.20m	week	0.00	0.00	84.00	12.60	96.60

	Unit	Hours	Hours £	Plant £	O & P £	Total £
FIRE DAMAGE						
Window replacement						
Take out existing window, prepare jambs and cill to receive new						
PVC-U						
size 600 x 900mm	nr	2.00	19.00	110.58	19.44	149.02
size 600 x 1200mm	nr	2.00	19.00	146.39	24.81	190.20
size 1200 x 1200mm	nr	2.50	23.75	168.94	28.90	221.59
size 1800 x 1200mm	nr	3.00	28.50	178.56	31.06	238.12
softwood painted						
size 630 x 900mm	nr	2.00	19.00	78.63	14.64	112.27
size 915 x 900mm	nr	2.00	19.00	84.34	15.50	118.84
size 915 x 1200mm	nr	2.50	23.75	89.37	16.97	130.09
size 1200 x 1200mm	nr	3.00	28.50	105.47	20.10	154.07
hardwood stained						
size 915 x 1050mm	nr	2.00	19.00	173.28	28.84	221.12
size 915 x 1500mm	nr	2.00	19.00	192.34	31.70	243.04
size 1200 x 1500mm	nr	2.50	23.75	205.84	34.44	264.03
size 1770 x 1200mm	nr	3.00	28.50	246.39	41.23	316.12
Take out existing bay window, prepare jambs and cill to receive new						
PVC-U						
size 1800 x 900mm	nr	3.00	28.50	201.48	34.50	264.48
size 1800 x 1200mm	nr	3.00	28.50	225.76	38.14	292.40
size 2400 x 900mm	nr	3.50	33.25	387.16	63.06	483.47
size 2400 x 1200mm	nr	3.50	33.25	413.67	67.04	513.96
softwood painted						
size 1800 x 900mm	nr	3.00	28.50	143.57	25.81	197.88
size 1800 x 1200mm	nr	3.00	28.50	165.87	29.16	223.53
size 2400 x 900mm	nr	3.50	33.25	247.56	42.12	322.93
size 2400 x 1200mm	nr	3.50	33.25	269.67	45.44	348.36
hardwood stained						
size 1800 x 900mm	nr	3.00	28.50	239.57	40.21	308.28
size 1800 x 1200mm	nr	3.00	28.50	268.31	44.52	341.33
size 2400 x 900mm	nr	3.50	33.25	305.85	50.87	389.97
size 2400 x 1200mm	nr	3.50	33.25	335.64	55.33	424.22

	Unit	Hours	Hours £	Plant £	O & P £	Total £
Window repairs						
Take off and replace defective ironmongery to softwood windows						
casement fastener	nr	0.20	1.90	3.65	0.83	6.38
casement stay	nr	0.20	1.90	5.55	1.12	8.57
hinges	pair	0.25	2.38	3.31	0.85	6.54
cockspur fastener	nr	0.20	1.90	9.21	1.67	12.78
sash fastener	nr	0.25	2.38	4.25	0.99	7.62
Take off and replace defective ironmongery to hardwood windows						
casement fastener	nr	0.30	2.85	3.65	0.98	7.48
casement stay	nr	0.30	2.85	5.55	1.26	9.66
hinges	pair	0.35	3.33	3.31	1.00	7.63
cockspur fastener	nr	0.30	2.85	9.21	1.81	13.87
sash fastener	nr	0.35	3.33	4.25	1.14	8.71
Take off and replace defective ironmongery to PVC-U windows						
casement fastener	nr	0.20	1.90	3.65	0.83	6.38
casement stay	nr	0.20	1.90	5.55	1.12	8.57
hinges	pair	0.25	2.38	3.31	0.85	6.54
cockspur fastener	nr	0.20	1.90	9.21	1.67	12.78
sash fastener	nr	0.25	2.38	4.25	0.99	7.62
Take out existing window cill and replace						
softwood	m	0.55	5.23	7.55	1.92	14.69
hardwood	m	0.65	6.18	22.36	4.28	32.82
Take out existing window board and replace						
softwood	m	0.45	4.28	5.78	1.51	11.56
hardwood	m	0.60	5.70	18.67	3.66	28.03
Door replacement						
Take off existing external door and frame and replace with new flush door						
softwood	nr	2.20	20.90	87.38	16.24	124.52
hardwood	nr	2.50	23.75	314.27	50.70	388.72
PVC-U	nr	2.20	20.90	289.31	46.53	356.74

	Unit	Hours	Hours £	Plant £	O & P £	Total £
Take off existing external door and frame and replace (cont'd)						
panelled door						
softwood	nr	2.20	20.90	110.12	19.65	150.67
hardwood	nr	2.50	23.75	344.56	55.25	423.56
PVC-U	nr	2.20	20.90	323.34	51.64	395.88
half glazed door						
softwood	nr	2.20	20.90	110.12	19.65	150.67
hardwood	nr	2.50	23.75	344.56	55.25	423.56
PVC-U	nr	2.20	20.90	323.34	51.64	395.88
fully glazed door						
softwood	nr	2.80	26.60	88.34	17.24	132.18
hardwood	nr	3.20	30.40	325.64	53.41	409.45
PVC-U	nr	2.80	26.60	300.51	49.07	376.18
fully glazed patio doors						
softwood	pair	3.80	36.10	198.24	35.15	269.49
hardwood	pair	4.30	40.85	457.39	74.74	572.98
PVC-U	pair	3.80	36.10	422.14	68.74	526.98
galvanised steel up-and-over garage doors						
2135 x 1980mm	nr	6.00	57.00	254.24	46.69	357.93
3965 x 2135mm	nr	7.50	71.25	889.15	144.06	1104.46
Door repairs						
Take off and replace defective ironmongery to softwood doors						
bolts						
barrel	nr	0.20	1.90	5.54	1.12	8.56
flush	nr	0.20	1.90	7.22	1.37	10.49
tower	nr	0.30	2.85	6.63	1.42	10.90
butts						
light	pair	0.20	1.90	3.32	0.78	6.00
medium	pair	0.25	2.38	3.38	0.86	6.62
heavy	pair	0.30	2.85	4.25	1.07	8.17
locks						
cupboard	nr	0.30	2.85	6.54	1.41	10.80
mortice dead lock	nr	0.85	8.08	11.24	2.90	22.21

	Unit	Hours	Hours £	Plant £	O & P £	Total £
Take off and replace defective ironmongery (cont'd)						
rim lock	nr	0.45	4.28	5.11	1.41	10.79
cylinder	nr	1.10	10.45	22.33	4.92	37.70
Take off and replace defective ironmongery to hardwood doors						
bolts						
barrel	nr	0.30	2.85	5.54	1.26	9.65
flush	nr	0.30	2.85	7.22	1.51	11.58
indicating	nr	0.30	2.85	7.28	1.52	11.65
tower	nr	0.40	3.80	6.63	1.56	11.99
butts						
light	pair	0.40	3.80	3.32	1.07	8.19
medium	pair	0.35	3.33	3.38	1.01	7.71
heavy	pair	0.40	3.80	4.25	1.21	9.26
locks						
cupboard	nr	0.30	2.85	6.54	1.41	10.80
mortice dead lock	nr	0.85	8.08	11.24	2.90	22.21
rim lock	nr	0.45	4.28	5.11	1.41	10.79
cylinder	nr	1.10	10.45	22.33	4.92	37.70
Partitions, walls and ceilings						
Pull down existing damaged partitions and walls and rebuild						
stud partition plasterboard both sides	m2	1.50	14.25	8.18	3.36	25.79
brickwork 112mm thick plastered both sides	m2	4.00	38.00	14.19	7.83	60.02
blockwork 75mm thick plastered both sides	m2	3.10	29.45	9.00	5.77	44.22
blockwork 100mm thick plastered both sides	m2	3.30	31.35	18.50	7.48	57.33
Hack off scorched plaster to walls and renew	m2	1.50	14.25	1.70	2.39	18.34
Pull down plasterboard and skim ceilings and renew	m2	1.56	14.82	3.07	2.68	20.57
Hack off damaged wall tiles and renew	m2	1.45	13.78	21.20	5.25	40.22

	Unit	Hours	Hours £	Plant £	O & P £	Total £
Take off damaged skirting and replace						
19 x 75mm softwood	m	0.27	2.57	1.87	0.67	5.10
19 x 100mm softwood	m	0.30	2.85	2.17	0.75	5.77
19 x 100mm hardwood	m	0.32	3.04	4.92	1.19	9.15
25 x 150mm hardwood	m	0.38	3.61	9.24	1.93	14.78
Take up damaged flooring and renew						
19mm tongued and grooved softwood flooring	m2	1.14	10.83	9.40	3.03	23.26
25mm tongued and grooved softwood flooring	m2	1.15	10.93	11.05	3.30	25.27
12mm tongued and grooved chipboard flooring	m2	0.80	7.60	4.88	1.87	14.35
18mm tongued and grooved chipboard flooring	m2	1.00	9.50	5.26	2.21	16.97
2mm vinyl sheeting	m2	0.50	4.75	8.94	2.05	15.74
2.5mm vinyl sheeting	m2	0.55	5.23	11.87	2.56	19.66
3mm vinyl sheeting	m2	0.60	5.70	12.10	2.67	20.47
2mm vinyl tiling	m2	0.40	3.80	7.84	1.75	13.39
3mm laminated wood flooring	m2	0.90	8.55	9.47	2.70	20.72
12.5mm quarry tiling	m2	1.15	10.93	28.56	5.92	45.41
Cut out and replace floor joists and roof members						
38 x 100mm sawn softwood	m	0.22	2.09	1.22	0.50	3.81
50 x 75mm sawn softwood	m	0.24	2.28	1.60	0.58	4.46
50 x 100mm sawn softwood	m	0.28	2.66	1.92	0.69	5.27
50 x 125mm sawn softwood	m	0.30	2.85	2.12	0.75	5.72
50 x 150mm sawn softwood	m	0.32	3.04	2.28	0.80	6.12
75 x 125mm sawn softwood	m	0.34	3.23	2.68	0.89	6.80
75 x 200mm sawn softwood	m	0.36	3.42	3.27	1.00	7.69
Take down damaged softwood staircase 2600mm rise and renew						
straight flight 900mm wide	nr	12.20	115.90	547.27	99.48	762.65
straight flight 900mm wide with balustrade	nr	13.30	126.35	628.87	113.28	868.50
two flights 900mm wide with landing and balustrade	nr	14.50	137.75	702.14	125.98	965.87

	Unit	Hours	Hours £	Plant £	O & P £	Total £
Take down damaged hardwood staircase 2600mm rise and renew						
straight flight 900mm wide	nr	12.40	117.80	928.31	156.92	1203.03
straight flight 900mm wide with balustrade	nr	13.90	132.05	1012.59	171.70	1316.34
two flights 900mm wide with landing and balustrade	nr	15.00	142.50	1117.23	188.96	1448.69
Furniture and fittings						
Take down damaged fittings and replace (material prices not included due to the wide variation in the quality and costs of the fittings)						
wall units 2000mm high						
300 x 600mm	nr	1.00	9.50	0.00	1.43	10.93
300 x 1000mm	nr	1.10	10.45	0.00	1.57	12.02
300 x 1200mm	nr	1.20	11.40	0.00	1.71	13.11
base units 750mm high						
600 x 900mm	nr	1.00	9.50	0.00	1.43	10.93
600 x 1000mm	nr	1.00	9.50	0.00	1.43	10.93
600 x 1200mm	nr	1.10	10.45	0.00	1.57	12.02
900 x 900mm	nr	1.10	10.45	0.00	1.57	12.02
900 x 1000mm	nr	1.20	11.40	0.00	1.71	13.11
900 x 1200mm	nr	1.20	11.40	0.00	1.71	13.11
sink units 750mm high						
600 x 900mm	nr	1.40	13.30	0.00	2.00	15.30
600 x 1000mm	nr	1.40	13.30	0.00	2.00	15.30
600 x 1200mm	nr	1.40	13.30	0.00	2.00	15.30
worktops						
600 x 900mm	nr	0.80	7.60	0.00	1.14	8.74
600 x 1000mm	nr	0.90	8.55	0.00	1.28	9.83
600 x 1200mm	nr	1.00	9.50	0.00	1.43	10.93
Remove smoke damaged furniture and place in skip						
easy chair	nr	0.15	1.43	0.00	0.21	1.64
settee	nr	0.18	1.71	0.00	0.26	1.97
single bed and mattress	nr	0.20	1.90	0.00	0.29	2.19
double bed and mattress	nr	0.22	2.09	0.00	0.31	2.40
set of six dining chairs	nr	0.12	1.14	0.00	0.17	1.31
dining table	nr	0.18	1.71	0.00	0.26	1.97
sideboard	nr	0.20	1.90	0.00	0.29	2.19

	Unit	Hours	Hours £	Plant £	O & P £	Total £
Remove smoke damaged furniture (cont'd)						
wardrobe	nr	0.15	1.43	0.00	0.21	1.64
chest of drawers	nr	0.20	1.90	0.00	0.29	2.19
carpet	nr	0.25	2.38	0.00	0.36	2.73
TV and hi-fi equipment	nr	0.25	2.38	0.00	0.36	2.73
Plumbing and heating work						
Take out damaged sanitary fittings and associated pipework and replace						
lavatory basin	nr	3.30	41.25	114.20	23.32	178.77
low level WC	nr	3.40	42.50	291.21	50.06	383.77
shower cubicle	nr	4.20	52.50	791.71	126.63	970.84
sink with single drainer	nr	2.30	28.75	160.41	28.37	217.53
sink with double drainer	nr	2.90	36.25	202.69	35.84	274.78
bath, cast iron	nr	4.40	55.00	412.19	70.08	537.27
bath, acrylic	nr	4.20	52.50	248.22	45.11	345.83
bidet	nr	3.50	43.75	341.02	57.72	442.49
Take out damaged cisterns and cylinders and associated pipework and replace						
polyethylene cold water cisterns						
68 litres	nr	2.00	25.00	78.28	15.49	118.77
86 litres	nr	2.00	25.00	87.75	16.91	129.66
191 litres	nr	2.50	31.25	117.72	22.35	171.32
327 litres	nr	3.80	47.50	144.25	28.76	220.51
copper cylinders, indirect pattern						
114 litres	nr	3.50	43.75	112.41	23.42	179.58
117 litres	nr	3.50	43.75	119.12	24.43	187.30
140 litres	nr	4.00	50.00	130.62	27.09	207.71
162 litres	nr	4.10	51.25	159.77	31.65	242.67
copper cylinders, direct pattern						
116 litres	nr	3.80	47.50	110.02	23.63	181.15
120 litres	nr	3.80	47.50	123.43	25.64	196.57
144 litres	nr	4.20	52.50	131.57	27.61	211.68
166 litres	nr	4.30	53.75	147.91	30.25	231.91

	Unit	Hours	Hours £	Plant £	O & P £	Total £
Disconnect damaged central heating boiler and associated pipe work and replace						
floor-mounted gas boiler						
30,000 Btu	nr	9.00	112.50	486.36	89.83	688.69
40,000 Btu	nr	9.00	112.50	486.36	89.83	688.69
50,000 Btu	nr	9.20	115.00	486.36	90.20	691.56
60,000 Btu	nr	9.20	115.00	486.36	90.20	691.56
wall-mounted gas boiler						
30,000 Btu	nr	9.00	112.50	486.36	89.83	688.69
40,000 Btu	nr	9.00	112.50	486.36	89.83	688.69
50,000 Btu	nr	9.20	115.00	486.36	90.20	691.56
60,000 Btu	nr	9.20	115.00	486.36	90.20	691.56
floor-mounted oil-fired boiler						
52,000 Btu	nr	9.20	115.00	1207.48	198.37	1520.85
70,000 Btu	nr	9.40	117.50	1251.82	205.40	1574.72
Disconnect damaged radiator and associated pipework and and valves and replace						
single panel, 450mm high						
length, 1000mm	nr	1.65	20.63	63.13	12.56	96.32
length, 1600mm	nr	1.85	23.13	88.37	16.72	128.22
length, 2000mm	nr	2.10	26.25	104.44	19.60	150.29
single panel, 600mm high						
length, 1000mm	nr	1.85	23.13	77.27	15.06	115.45
length, 1600mm	nr	2.05	25.63	108.79	20.16	154.58
length, 2000mm	nr	2.30	28.75	130.81	23.93	183.49
double panel, 450mm high						
length, 1000mm	nr	1.85	23.13	89.79	16.94	129.85
length, 1400mm	nr	1.35	16.88	121.54	20.76	159.18
length, 2000mm	nr	1.45	18.13	154.39	25.88	198.39
double panel, 600mm high						
length, 1000mm	nr	2.05	25.63	87.41	16.96	129.99
length, 1400mm	nr	2.25	28.13	123.63	22.76	174.52
length, 2000mm	nr	2.45	30.63	211.08	36.26	277.96

	Unit	Hours	Hours £	Plant £	O & P £	Total £
Electrics						
Disconnect power supply, remove damaged electric fittings and replace						
single power point	nr	0.60	8.40	6.14	2.18	16.72
double power point	nr	0.60	8.40	6.74	2.27	17.41
light switch	nr	0.60	8.40	5.12	2.03	15.55
light point	nr	0.60	8.40	5.72	2.12	16.24
wall light	nr	0.90	12.60	20.12	4.91	37.63
cooker control unit	nr	1.50	21.00	16.47	5.62	43.09
Painting and decorating						
Burn off damaged woodwork and leave ready to receive new paint-work						
general surfaces	m2	1.10	10.45	0.00	1.57	12.02
general surfaces up to 300mm wide	m	0.30	2.85	0.00	0.43	3.28
Burn off damaged metalwork and leave ready to receive new paint-work						
general surfaces	m2	1.10	10.45	0.00	1.57	12.02
general surfaces up to 300mm wide	m	0.30	2.85	0.00	0.43	3.28
Wash down existing plastered surfaces, stop cracks and rub down and leave ready to receive new paintwork						
walls	m2	0.14	1.33	0.00	0.20	1.53
ceilings	m2	0.05	0.48	0.00	0.07	0.55
Apply one mist coat and two coats emulsion paint to plastered surfaces						
walls	m2	0.30	2.85	0.59	0.52	3.96
ceilings	m2	0.32	3.04	0.59	0.54	4.17
Apply two undercoats and one coat gloss paint to plastered surfaces						
walls	m2	0.30	2.85	1.54	0.66	5.05
ceilings	m2	0.36	3.42	1.54	0.74	5.70

	Unit	Hours	Hours £	Plant £	O & P £	Total £
Prepare, size, apply adhesive, supply and hang paper to plastered walls						
lining paper						
£1.50 per roll	m2	0.30	2.85	0.36	0.48	3.69
£2.00 per roll	m2	0.30	2.85	0.48	0.50	3.83
£2.50 per roll	m2	0.30	2.85	0.59	0.52	3.96
washable paper						
£2.50 per roll	m2	0.30	2.85	0.59	0.52	3.96
£4.00 per roll	m2	0.30	2.85	0.94	0.57	4.36
£5.00 per roll	m2	0.30	2.85	1.18	0.60	4.63
vinyl paper						
£4.00 per roll	m2	0.30	2.85	0.94	0.57	4.36
£5.00 per roll	m2	0.30	2.85	1.18	0.60	4.63
£6.00 per roll	m2	0.30	2.85	1.41	0.64	4.90
washable paper						
£5.00 per roll	m2	0.30	2.85	1.18	0.60	4.63
£6.00 per roll	m2	0.30	2.85	1.41	0.64	4.90
£7.00 per roll	m2	0.30	2.85	1.64	0.67	5.16
hessian paper						
£7.00 per m2	m2	0.50	4.75	7.70	1.87	14.32
£8.00 per m2	m2	0.50	4.75	8.80	2.03	15.58
£9.00 per m2	m2	0.50	4.75	9.90	2.20	16.85
suede paper						
£9.00 per m2	m2	0.30	2.85	9.90	1.91	14.66
£10.00 per m2	m2	0.30	2.85	11.00	2.08	15.93
£11.00 per m2	m2	0.30	2.85	12.10	2.24	17.19
Prepare, size, apply adhesive, supply and hang paper to plastered ceilings and columns, butt jointed						
lining paper						
£1.50 per roll	m2	0.35	3.33	0.36	0.55	4.24
£2.00 per roll	m2	0.35	3.33	0.48	0.57	4.38
£2.50 per roll	m2	0.35	3.33	0.59	0.59	4.50
washable paper						
£2.50 per roll	m2	0.35	3.33	0.59	0.59	4.50
£4.00 per roll	m2	0.35	3.33	0.94	0.64	4.90
£5.00 per roll	m2	0.35	3.33	1.18	0.68	5.18

	Unit	Hours	Hours £	Plant £	O & P £	Total £
Prepare, size, apply adhesive, supply and hang paper (cont'd)						
vinyl paper						
£4.00 per roll	m2	0.35	3.33	0.94	0.64	4.90
£5.00 per roll	m2	0.35	3.33	1.18	0.68	5.18
£6.00 per roll	m2	0.35	3.33	1.41	0.71	5.45
washable paper						
£5.00 per roll	m2	0.35	3.33	1.18	0.68	5.18
£6.00 per roll	m2	0.35	3.33	1.41	0.71	5.45
£7.00 per roll	m2	0.35	3.33	1.64	0.74	5.71

	Unit	Hours	Hours £	Plant £	O & P £	Total £
FLOOD DAMAGE						
Pumping						
Instal pump and hoses in position	Item	1.00	9.50	0.00	1.43	10.93
Hire diaphragm pump and hoses						
50mm	hour	0.00	0.00	3.00	0.45	3.45
75mm	hour	0.00	0.00	3.25	0.49	3.74
Hire portable dryer, 350W	hour	0.00	0.00	1.90	0.29	2.19

	Unit	Hours	Hours £	Materials £	O & P £	Total £
Clean up and remove debris from basement and ground floor	m2	0.20	1.90	0.00	0.29	2.19
Hack off defective wall plaster and replace	m2	0.20	1.90	0.00	0.29	2.19
Pull down plasterboard and skim ceilings and renew	m2	1.56	14.82	3.07	2.68	20.57
Hack off damaged wall tiles and renew	m2	1.45	13.78	21.20	5.25	40.22
Take off damaged skirting and replace						
19 x 75mm softwood	m	0.27	2.57	1.87	0.67	5.10
19 x 100mm softwood	m	0.30	2.85	2.17	0.75	5.77
19 x 100mm hardwood	m	0.32	3.04	4.92	1.19	9.15
25 x 150mm hardwood	m	0.38	3.61	9.24	1.93	14.78
Take up damaged flooring and renew						
19mm tongued and grooved softwood flooring	m2	1.14	10.83	9.40	3.03	23.26
25mm tongued and grooved softwood flooring	m2	1.15	10.93	11.05	3.30	25.27
12mm tongued and grooved chipboard flooring	m2	0.80	7.60	4.88	1.87	14.35
18mm tongued and grooved chipboard flooring	m2	1.00	9.50	5.26	2.21	16.97

	Unit	Hours	Hours £	Materials £	O & P £	Total £
Electrics						
Disconnect power supply, remove damaged electric fittings and replace						
single power point	nr	0.60	8.40	6.14	2.18	16.72
double power point	nr	0.60	8.40	6.74	2.27	17.41
light switch	nr	0.60	8.40	5.12	2.03	15.55
cooker control unit	nr	1.50	21.00	16.47	5.62	43.09
Painting and decorating						
Burn off damaged woodwork and leave ready to receive new paint-work						
general surfaces	m2	1.10	10.45	0.00	1.57	12.02
general surfaces up to 300mm wide	m	0.30	2.85	0.00	0.43	3.28
Burn off damaged metalwork and leave ready to receive new paint-work						
general surfaces	m2	1.10	10.45	0.00	1.57	12.02
general surfaces up to 300mm wide	m	0.30	2.85	0.00	0.43	3.28
Wash down existing plastered surfaces, stop cracks and rub down and leave ready to receive new paintwork						
walls	m2	0.05	0.48	0.00	0.07	0.55
ceilings	m2	72.00	684.00	0.00	102.60	786.60
Apply one mist coat and two coats emulsion paint to plastered surfaces						
walls	m2	0.30	2.85	0.59	0.52	3.96
ceilings	m2	0.32	3.04	0.59	0.54	4.17
Apply two undercoats and one coat gloss paint to plastered surfaces						
walls	m2	0.30	2.85	1.54	0.66	5.05
ceilings	m2	0.36	3.42	1.54	0.74	5.70

	Unit	Hours	Hours £	Plant £	O & P £	Total £
GALE DAMAGE						
Hoardings and screens						
Temporary screens and hoardings 2m high consisting of 22mm thick exterior quality plywood fixed to 100 x 50mm posts and rails	m	2.60	24.70	29.12	8.07	61.89
Tarpaulin sheeting size 5 x 4m fixed in position	day	0.00	0.00	3.57	0.54	4.11
Plywood sheeting blocking up window or door opening	m2	1.00	9.50	14.57	3.61	27.68
Shoring						
Timber dead shores consisting of 200 x 200mm shores, 200 x 50mm plates and 200 x 50mm braces at centres of						
2m	m2	3.80	36.10	68.45	15.68	120.23
3m	m2	3.00	28.50	54.21	12.41	95.12
4m	m2	2.20	20.90	38.96	8.98	68.84
Timber raking and flying shores consisting of 200 x 200mm shores, 250 x 50mm plates and 200 x 50mm braces to gable end of two-storey house	Item	72.00	684.00	486.36	175.55	1345.91
Roof repairs						
Take up roof coverings from pitched roof						
tiles	m2	0.80	7.60	0.00	1.14	8.74
slates	m2	0.80	7.60	0.00	1.14	8.74
timber boarding	m2	1.00	9.50	0.00	1.43	10.93
metal sheeting	m2	0.20	1.90	0.00	0.29	2.19
flat sheeting	m2	0.30	2.85	0.00	0.43	3.28
corrugated sheeting	m2	0.30	2.85	0.00	0.43	3.28
underfelt	m2	0.10	0.95	0.00	0.14	1.09

	Unit	Hours	Hours £	Materials £	O & P £	Total £
Take up roof coverings from flat roof						
bituminous felt	m2	0.25	2.38	0.00	0.36	2.73
metal sheeting	m2	0.30	2.85	0.00	0.43	3.28
woodwool slabs	m2	0.50	4.75	0.00	0.71	5.46
firrings	m2	0.20	1.90	0.00	0.29	2.19
Take up roof coverings from pitched roof, carefully lay aside for reuse						
tiles	m2	1.10	10.45	0.00	1.57	12.02
slates	m2	1.10	10.45	0.00	1.57	12.02
metal sheeting	m2	0.50	4.75	0.00	0.71	5.46
flat sheeting	m2	0.60	5.70	0.00	0.86	6.56
corrugated sheeting	m2	0.60	5.70	0.00	0.86	6.56
Take up roof coverings from flat roof, carefully lay aside for reuse						
metal sheeting	m2	0.60	5.70	0.00	0.86	6.56
woodwool slabs	m2	0.80	7.60	0.00	1.14	8.74
Inspect roof battens, refix loose and replace with new, size 38 x 25mm						
25% of area						
250mm centres	m2	0.14	1.33	0.40	0.26	1.99
450mm centres	m2	0.12	1.14	0.35	0.22	1.71
600mm centres	m2	0.10	0.95	0.30	0.19	1.44
50% of area						
250mm centres	m2	0.26	2.47	0.70	0.48	3.65
450mm centres	m2	0.16	1.52	0.60	0.32	2.44
600mm centres	m2	0.18	1.71	0.50	0.33	2.54
75% of area						
250mm centres	m2	0.36	3.42	1.00	0.66	5.08
450mm centres	m2	0.26	2.47	0.85	0.50	3.82
600mm centres	m2	0.22	2.09	0.70	0.42	3.21
100% of area						
250mm centres	m2	0.44	4.18	1.30	0.82	6.30
450mm centres	m2	0.32	3.04	1.10	0.62	4.76
600mm centres	m2	0.38	3.61	0.90	0.68	5.19

	Unit	Hours	Hours £	Plant £	O & P £	Total £
Remove single slipped slate and refix	nr	1.00	9.50	0.90	1.56	11.96
Remove single broken slate, renew with new Welsh blue slate						
405 x 255mm	nr	1.20	11.40	1.30	1.91	14.61
510 x 255mm	nr	1.20	11.40	2.50	2.09	15.99
610 x 305mm	nr	1.20	11.40	5.12	2.48	19.00
Remove slates in area approximately 1m2 and replace with Welsh blue slates previously laid aside	nr	1.60	15.20	0.00	2.28	17.48
Remove slates in area approximately 1m2 and replace with Welsh blue slates previously laid aside						
405 x 255mm	nr	1.80	17.10	54.36	10.72	82.18
510 x 255mm	nr	1.70	16.15	57.89	11.11	85.15
610 x 305mm	nr	1.60	15.20	60.24	11.32	86.76
Remove double course at eaves and fix new Welsh blue slates						
405 x 255mm	m	0.70	6.65	24.56	4.68	35.89
510 x 255mm	m	0.70	6.65	30.14	5.52	42.31
610 x 305mm	m	0.70	6.65	34.87	6.23	47.75
Remove single verge undercloak course and renew						
405 x 255mm	m	0.90	8.55	19.66	4.23	32.44
510 x 255mm	m	0.90	8.55	20.36	4.34	33.25
610 x 305mm	m	0.90	8.55	22.01	4.58	35.14
Remove single slipped tile and refix	nr	0.30	2.85	0.00	0.43	3.28
Remove single broken tile and renew						
Marley plain tile	nr	0.30	2.85	0.66	0.53	4.04
Marley Ludlow Plus tile	nr	0.30	2.85	0.84	0.55	4.24
Marley Modern tile	nr	0.30	2.85	1.35	0.63	4.83
Redland Renown tile	nr	0.30	2.85	1.35	0.63	4.83
Redland Norfolk tile	nr	0.30	2.85	1.14	0.60	4.59
Remove tiles in area approximately 1m2 and replace with tiles previously laid aside						
Marley plain tile	nr	1.80	17.10	27.89	6.75	51.74
Marley Ludlow Plus tile	nr	1.20	11.40	13.56	3.74	28.70
Marley Modern tile	nr	1.10	10.45	14.27	3.71	28.43

	Unit	Hours	Hours £	Materials £	O & P £	Total £
Remove tiles in area approximately 1m2 (cont'd)						
Redland Renown tile	nr	1.10	10.45	13.65	3.62	27.72
Redland Norfolk tile	nr	1.15	10.93	16.88	4.17	31.98
Take off defective ridge capping and refix including pointing in mortar	m	1.10	10.45	1.45	1.79	13.69
Chimney stack						
Erect and take down chimney scaffold	item	4.00	38.00	0.00	5.70	43.70
Hire chimney scaffold and platform	week	0.00	0.00	94.00	14.10	108.10
Take down existing chimney stack to below roof level and remove debris						
single stack	m	2.50	23.75	1.32	3.76	28.83
double stack	m	3.50	33.25	1.32	5.19	39.76
Chimney stack in facing brick £450.00 per thousand in gauged mortar						
single stack	m	3.80	36.10	42.57	11.80	90.47
double stack	m	5.20	49.40	88.53	20.69	158.62
Terra cotta chimney pot bedded and flaunched in gauged mortar						
185mm diameter x 300mm high	nr	1.25	11.88	24.22	5.41	41.51
185mm diameter x 600mm high	nr	1.80	17.10	41.61	8.81	67.52
External work						
Remove blown-down trees, trunk girth 1m above ground level						
600 to 1500mm	nr	22.00	209.00	0.00	31.35	240.35
1500 to 3000mm	nr	38.00	361.00	0.00	54.15	415.15

	Unit	Hours	Hours £	Plant £	O & P £	Total £
THEFT DAMAGE						
Window and door repairs						
Hack out glass and remove	m2	0.45	4.28	0.00	0.64	4.92
Clean rebates, remove sprigs or clips and prepare for reglazing	m	0.20	1.90	0.00	0.29	2.19
Reglaze existing softwood windows in clear float glass with putty and sprigs						
under 0.15m2, thickness						
3mm	m2	0.90	8.55	24.57	4.97	38.09
4mm	m2	0.90	8.55	26.31	5.23	40.09
5mm	m2	0.90	8.55	29.84	5.76	44.15
6mm	m2	1.00	9.50	31.28	6.12	46.90
10mm	m2	1.05	9.98	54.61	9.69	74.27
over 0.15m2, thickness						
3mm	m2	0.60	5.70	23.48	4.38	33.56
4mm	m2	0.60	5.70	25.08	4.62	35.40
5mm	m2	0.60	5.70	28.87	5.19	39.76
6mm	m2	0.65	6.18	30.55	5.51	42.23
10mm	m2	0.70	6.65	52.37	8.85	67.87
Reglaze existing softwood windows in clear float glass with pinned beads						
under 0.15m2, thickness						
3mm	m2	1.10	10.45	24.57	5.25	40.27
4mm	m2	1.10	10.45	26.31	5.51	42.27
5mm	m2	1.10	10.45	29.84	6.04	46.33
6mm	m2	1.20	11.40	31.28	6.40	49.08
10mm	m2	1.25	11.88	54.61	9.97	76.46
over 0.15m2, thickness						
3mm	m2	0.80	7.60	23.48	4.66	35.74
4mm	m2	0.80	7.60	25.08	4.90	37.58
5mm	m2	0.80	7.60	28.87	5.47	41.94
6mm	m2	0.90	8.55	30.55	5.87	44.97
10mm	m2	0.95	9.03	52.37	9.21	70.60

	Unit	Hours	Hours £	Materials £	O & P £	Total £
Reglaze existing metal windows in clear float glass with clips and putty						
under 0.15m2, thickness						
3mm	m2	0.95	9.03	27.35	5.46	41.83
4mm	m2	0.95	9.03	29.61	5.80	44.43
5mm	m2	0.95	9.03	32.46	6.22	47.71
6mm	m2	1.05	9.98	34.91	6.73	51.62
10mm	m2	1.10	10.45	56.37	10.02	76.84
over 0.15m2, thickness						
3mm	m2	0.65	6.18	26.54	4.91	37.62
4mm	m2	0.65	6.18	28.61	5.22	40.00
5mm	m2	0.65	6.18	31.27	5.62	43.06
6mm	m2	0.70	6.65	33.62	6.04	46.31
10mm	m2	0.75	7.13	55.31	9.37	71.80
Reglaze existing metal windows in clear float glass with screwed metal beads						
under 0.15m2, thickness						
3mm	m2	1.30	12.35	32.55	6.74	51.64
4mm	m2	1.30	12.35	33.51	6.88	52.74
5mm	m2	1.30	12.35	36.19	7.28	55.82
6mm	m2	1.40	13.30	38.94	7.84	60.08
10mm	m2	1.45	13.78	61.33	11.27	86.37
over 0.15m2, thickness						
3mm	m2	1.00	9.50	31.25	6.11	46.86
4mm	m2	1.00	9.50	32.67	6.33	48.50
5mm	m2	1.00	9.50	35.63	6.77	51.90
6mm	m2	1.05	9.98	37.41	7.11	54.49
10mm	m2	1.10	10.45	60.74	10.68	81.87
Reglaze existing softwood windows in white patterned glass with putty and sprigs						
under 0.15m2, thickness						
4mm	m2	0.90	8.55	29.34	5.68	43.57
6mm	m2	1.00	9.50	34.41	6.59	50.50
over 0.15m2, thickness						
4mm	m2	0.60	5.70	29.34	5.26	40.30
6mm	m2	0.65	6.18	34.41	6.09	46.67

	Unit	Hours	Hours £	Plant £	O & P £	Total £
Reglaze existing softwood windows in white patterned glass with pinned beads						
under 0.15m2, thickness						
4mm	m2	1.10	10.45	29.34	5.97	45.76
6mm	m2	1.20	11.40	34.41	6.87	52.68
over 0.15m2, thickness						
4mm	m2	0.80	7.60	29.34	5.54	42.48
6mm	m2	0.85	8.08	34.41	6.37	48.86
Reglaze existing softwood windows in white patterned glass with screwed beads						
under 0.15m2, thickness						
4mm	m2	1.30	12.35	31.25	6.54	50.14
6mm	m2	1.40	13.30	36.37	7.45	57.12
over 0.15m2, thickness						
4mm	m2	1.00	9.50	31.25	6.11	46.86
6mm	m2	1.50	14.25	36.37	7.59	58.21
Reglaze existing metal windows in white patterned glass with putty						
under 0.15m2, thickness						
4mm	m2	0.90	8.55	29.34	5.68	43.57
6mm	m2	1.00	9.50	34.41	6.59	50.50
over 0.15m2, thickness						
4mm	m2	0.60	5.70	29.34	5.26	40.30
6mm	m2	0.65	6.18	34.41	6.09	46.67
Reglaze existing metal windows in white patterned glass with metal clips and putty						
under 0.15m2, thickness						
4mm	m2	0.95	9.03	31.25	6.04	46.32
6mm	m2	1.05	9.98	36.37	6.95	53.30
over 0.15m2, thickness						
4mm	m2	0.65	6.18	31.25	5.61	43.04
6mm	m2	0.70	6.65	36.37	6.45	49.47

Part Five

ALTERATIONS

	Unit	Hours	Hours £	Materials £	O & P £	Total £
ALTERATIONS						
Shoring						
Timber dead shores consisting of 200 x 200mm shores, 250 x 50mm plates and 200 x 50mm braces at centres of						
2m	m2	3.80	36.10	68.45	15.68	120.23
3m	m2	3.00	28.50	54.21	12.41	95.12
4m	m2	2.20	20.90	38.96	8.98	68.84
Timber raking and flying shores consisting of 200 x 200mm shores, 250 x 50mm plates and 200 x 50mm braces to gable end of two-storey house	Item	72.00	684.00	486.36	175.55	1345.91
Temporary screens						
Erect, maintain and remove temporary screens consisting of 50 x 50mm softwood framing covered one side with						
chipboard (three uses)	m2	1.15	10.93	1.38	1.85	14.15
insulation board (three uses)	m2	1.15	10.93	1.92	1.93	14.77
polythene sheeting (three uses)	m2	0.80	7.60	1.18	1.32	10.10
Underpinning						
Excavate preliminary trench by hand, maximum depth not exceeding						
1.00m	m3	3.70	35.15	0.00	5.27	40.42
1.50m	m3	4.15	39.43	0.00	5.91	45.34
2.00m	m3	4.40	41.80	0.00	6.27	48.07
2.50m	m3	6.10	57.95	0.00	8.69	66.64
3.00m	m3	7.25	68.88	0.00	10.33	79.21
Excavate trench by hand below existing foundations, maximum depth not exceeding						
1.00m	m3	4.25	40.38	0.00	6.06	46.43
1.50m	m3	4.75	45.13	0.00	6.77	51.89
2.00m	m3	5.45	51.78	0.00	7.77	59.54

	Unit	Hours	Hours £	Materials £	O & P £	Total £
Excavate trench by hand (cont'd)						
2.50m	m3	7.30	69.35	0.00	10.40	79.75
3.00m	m3	8.35	79.33	0.00	11.90	91.22
Excavate and backfill working space, maximum depth not exceeding						
1.00m	m3	5.00	47.50	0.00	7.13	54.63
1.50m	m3	5.25	49.88	0.00	7.48	57.36
2.00m	m3	5.90	56.05	0.00	8.41	64.46
2.50m	m3	8.10	76.95	0.00	11.54	88.49
3.00m	m3	9.25	87.88	0.00	13.18	101.06
Open-boarded earthwork support to sides of preliminary trenches, distance between faces not exceeding 2.0m						
maximum depth not exceeding 1.0m	m2	0.30	2.85	1.62	0.67	5.14
maximum depth not exceeding 2.0m	m2	0.40	3.80	1.62	0.81	6.23
Close-boarded earthwork support to sides of preliminary trenches, distance between faces not exceeding 2.0m						
maximum depth not exceeding 1.0m	m2	0.85	8.08	2.71	1.62	12.40
maximum depth not exceeding 2.0m	m2	1.10	10.45	2.71	1.97	15.13
Open-boarded earthwork support to sides of excavation trenches, distance between faces not exceeding 2.0m						
maximum depth not exceeding 1.0m	m2	0.35	3.33	1.62	0.74	5.69
maximum depth not exceeding 2.0m	m2	0.45	4.28	1.62	0.88	6.78
Close-boarded earthwork support to sides of excavation trenches, distance between faces not exceeding 2.0m						
maximum depth not exceeding 1.0m	m2	0.95	9.03	2.71	1.76	13.50

	Unit	Hours	Hours £	Materials £	O & P £	Total £
Close-boarded earthwork support (cont'd)						
maximum depth not exceeding 2.0m	m2	1.20	11.40	2.71	2.12	16.23
Cut away projecting concrete foundations, size						
150 x 150mm	m	0.45	4.28	0.00	0.64	4.92
150 x 225mm	m	0.55	5.23	0.00	0.78	6.01
150 x 300mm	m	0.65	6.18	0.00	0.93	7.10
Cut away projecting brickwork in footings, one brick thick						
one course	m	1.10	10.45	0.00	1.57	12.02
two courses	m	1.90	18.05	0.00	2.71	20.76
three courses	m	2.60	24.70	0.00	3.71	28.41
Prepare underside of existing foundations to receive the new						
300mm	m	0.75	7.13	0.00	1.07	8.19
500mm	m	1.15	10.93	0.00	1.64	12.56
750mm	m	1.40	13.30	0.00	2.00	15.30
1000mm	m	1.60	15.20	0.00	2.28	17.48
1200mm	m	1.80	17.10	0.00	2.57	19.67
Load surplus excavated material into barrows, wheel and deposit in temporary spoil heaps, average distance						
15m	m3	1.25	11.88	0.00	1.78	13.66
25m	m3	1.45	13.78	0.00	2.07	15.84
50m	m3	1.70	16.15	0.00	2.42	18.57
Load surplus excavated material into barrows, wheel and deposit in temporary spoil heaps, average distance						
25m	m3	1.45	13.78	0.00	2.07	15.84
50m	m3	1.70	16.15	0.00	2.42	18.57
Load surplus excavated material into barrows, wheel and deposit in skips or lorries, average distance						
25m	m3	1.40	13.30	0.00	2.00	15.30
50m	m3	1.65	15.68	0.00	2.35	18.03

	Unit	Hours	Hours £	Materials £	O & P £	Total £
Level and compact bottom of excavation	m2	0.15	1.43	0.00	0.21	1.64
Site-mixed concrete 1:3:6 40mm aggregate in foundations to under-pinning, thickness						
150 to 300mm	m3	4.35	41.33	65.26	15.99	122.57
300 to 450mm	m3	4.05	38.48	65.26	15.56	119.30
over 450mm	m3	3.65	34.68	65.26	14.99	114.93
Site-mixed concrete 1:2:4 20mm aggregate in foundations to under-pinning, thickness						
150 to 300mm	m3	4.35	41.33	72.94	17.14	131.40
300 to 450mm	m3	4.05	38.48	72.94	16.71	128.13
over 450mm	m3	3.65	34.68	72.94	16.14	123.76
Plain vertical formwork to sides of underpinned foundations, height						
over 1m	m2	2.30	21.85	7.98	4.47	34.30
not exceeding 250mm	m	0.75	7.13	2.10	1.38	10.61
250 to 500mm	m	1.25	11.88	4.24	2.42	18.53
500mm to 1m	m	1.70	16.15	7.98	3.62	27.75
Plain vertical formwork to sides of underpinned foundations, left in, height						
over 1m	m2	2.15	20.43	19.40	5.97	45.80
not exceeding 250mm	m	0.65	6.18	6.90	1.96	15.04
250 to 500mm	m	1.15	10.93	12.76	3.55	27.24
500mm to 1m	m	1.60	15.20	19.40	5.19	39.79
High yield deformed steel reinforcement bars, straight or bent						
10mm diameter	m	0.04	0.38	0.38	0.11	0.87
12mm diameter	m	0.05	0.48	0.36	0.13	0.96
16mm diameter	m	0.06	0.57	0.35	0.14	1.06
20mm diameter	m	0.07	0.67	0.34	0.15	1.16
25mm diameter	m	0.08	0.76	0.33	0.16	1.25
Common bricks basic price £200 per thousand in cement mortar in underpinning						
one brick thick	m2	5.20	49.40	26.38	11.37	87.15
one and a half brick thick	m2	7.40	70.30	39.57	16.48	126.35
two brick thick	m2	9.35	88.83	52.76	21.24	162.82

	Unit	Hours	Hours £	Materials £	O & P £	Total £
Class A engineering bricks in basic price £350 per thousand in cement mortar in underpinning						
one brick thick	m2	5.40	51.30	48.96	15.04	115.30
one and a half brick thick	m2	7.60	72.20	73.44	21.85	167.49
two brick thick	m2	9.50	90.25	97.92	28.23	216.40
Hessian-based bitumen damp proof course, bedded in cement mortar, horizontal						
over 225mm wide	m2	0.35	3.33	8.92	1.84	14.08
112mm wide	m	0.05	0.48	1.02	0.22	1.72
Two courses of slates bedded in cement mortar, horizontal						
over 225mm wide	m2	2.90	27.55	31.21	8.81	67.57
225mm wide	m	0.85	8.08	7.82	2.38	18.28
Wedge and pin new work to soffit of existing with slates in cement mortar, width						
one brick thick	m	1.90	18.05	7.82	3.88	29.75
one and a half brick thick	m	2.40	22.80	11.73	5.18	39.71
two brick thick	m	2.80	26.60	15.64	6.34	48.58
Forming openings						
Form opening for windows or doors in existing walls in cement mortar and make good to sides of openings						
75mm blockwork	m2	2.00	19.00	3.20	3.33	25.53
100mm blockwork	m2	2.22	21.09	3.20	3.64	27.93
140mm blockwork	m2	2.65	25.18	3.40	4.29	32.86
215mm blockwork	m2	2.80	26.60	3.00	4.44	34.04
half brick wall	m2	2.90	27.55	3.20	4.61	35.36
one brick wall	m2	3.78	35.91	3.75	5.95	45.61
one and a half brick wall	m2	4.80	45.60	4.92	7.58	58.10
two brick wall	m2	7.20	68.40	5.87	11.14	85.41
Form opening for windows or doors in existing walls in lime mortar and make good to sides of openings						
75mm blockwork	m2	1.80	17.10	3.20	3.05	23.35
100mm blockwork	m2	2.00	19.00	3.20	3.33	25.53

	Unit	Hours	Hours £	Materials £	O & P £	Total £
Form opening for windows or doors in existing walls (cont'd)						
140mm blockwork	m2	2.40	22.80	3.40	3.93	30.13
215mm blockwork	m2	2.55	24.23	3.00	4.08	31.31
half brick wall	m2	2.60	24.70	3.20	4.19	32.09
one brick wall	m2	3.40	32.30	3.75	5.41	41.46
one and a half brick wall	m2	4.80	45.60	4.92	7.58	58.10
two brick wall	m2	6.50	61.75	5.87	10.14	77.76
Form opening for lintels above openings in existing walls in cement mortar and make good to sides of openings						
75mm blockwork	m2	2.98	28.31	3.20	4.73	36.24
100mm blockwork	m2	3.36	31.92	3.20	5.27	40.39
140mm blockwork	m2	3.87	36.77	3.40	6.02	46.19
215mm blockwork	m2	4.23	40.19	3.00	6.48	49.66
half brick wall	m2	6.30	59.85	3.20	9.46	72.51
one brick wall	m2	10.50	99.75	3.75	15.53	119.03
one and a half brick wall	m2	12.60	119.70	4.92	18.69	143.31
two brick wall	m2	16.98	161.31	5.87	25.08	192.26
Form opening for lintels above openings in existing walls in lime mortar and make good to sides of openings						
75mm blockwork	m2	2.75	26.13	2.02	4.22	32.37
100mm blockwork	m2	3.00	28.50	2.02	4.58	35.10
140mm blockwork	m2	3.45	32.78	2.48	5.29	40.54
215mm blockwork	m2	3.86	36.67	2.97	5.95	45.59
half brick wall	m2	5.75	54.63	2.02	8.50	65.14
one brick wall	m2	9.45	89.78	2.59	13.85	106.22
one and a half brick wall	m2	11.40	108.30	3.21	16.73	128.24
two brick wall	m2	15.30	145.35	4.22	22.44	172.01
Form opening in reinforced concrete floor slab and make good to edges, slab thickness						
100mm	m2	5.65	53.68	2.50	8.43	64.60
150mm	m2	7.86	74.67	2.75	11.61	89.03
200mm	m2	9.50	90.25	3.00	13.99	107.24
250mm	m2	11.57	109.92	3.25	16.97	130.14
300mm	m2	14.55	138.23	3.50	21.26	162.98

	Unit	Hours	Hours £	Materials £	O & P £	Total £
Form opening in reinforced concrete walls and make good to edges, wall thickness						
100mm	m2	6.50	61.75	2.50	9.64	73.89
150mm	m2	7.20	68.40	2.75	10.67	81.82
200mm	m2	10.33	98.14	3.00	15.17	116.31
250mm	m2	12.10	114.95	3.25	17.73	135.93
300mm	m2	15.32	145.54	3.50	22.36	171.40
Filling openings						
Take out existing door and frame make good to reveals and piece up skirting to match existing						
single internal door	nr	6.00	57.00	10.56	10.13	77.69
single external door	nr	6.20	58.90	10.56	10.42	79.88
double internal door	nr	6.20	58.90	11.44	10.55	80.89
double external door	nr	6.40	60.80	11.44	10.84	83.08
Take out existing single door and frame complete, fill in opening, fix new skirting to match existing and plaster both sides						
100mm blockwork	m2	10.45	99.28	44.37	21.55	165.19
215mm blockwork	m2	12.47	118.47	70.82	28.39	217.68
half brick wall	m2	12.40	117.80	56.61	26.16	200.57
one brick wall	m2	14.60	138.70	80.04	32.81	251.55
one and a half brick wall	m2	16.37	155.52	110.64	39.92	306.08
Take out existing double doors and frame complete, fill in opening, fix new skirting to match existing and plaster both sides						
100mm blockwork	m2	14.59	138.61	88.73	34.10	261.44
215mm blockwork	m2	16.36	155.42	140.97	44.46	340.85
half brick wall	m2	15.54	147.63	112.61	39.04	299.28
one brick wall	m2	20.34	193.23	160.75	53.10	407.08
one and a half brick wall	m2	22.20	210.90	220.64	64.73	496.27

	Unit	Hours	Hours £	Materials £	O & P £	Total £
Brickwork, blockwork and masonry						
Take out existing fireplace, fill opening with 100mm thick blockwork plastered one side, fix new skirting to match existing, make good flooring where hearth removed						
medium size fireplace	nr	12.00	114.00	49.56	24.53	188.09
large size fireplace	nr	14.00	133.00	78.94	31.79	243.73
Cut out existing projecting chimney breast, size 1350 x 350mm, including making good flooring and ceiling to match existing and plastering wall where breast removed						
one storey	nr	36.00	342.00	39.47	57.22	438.69
Take down chimney stack to 100mm below roof level, seal flue with slates in cement mortar, make good roof timbers						
slated roof	nr	46.00	437.00	92.41	79.41	608.82
tiled roof	nr	46.00	437.00	64.28	75.19	576.47
Cut out single brick in half brick wall and replace in gauged mortar						
commons	nr	0.25	2.38	0.25	0.39	3.02
facings	nr	0.30	2.85	0.40	0.49	3.74
Cut out decayed brickwork in half brick wall in areas 0.5 to 1m2 and replace in gauged mortar						
commons	nr	5.20	49.40	12.46	9.28	71.14
facings	nr	6.00	57.00	24.12	12.17	93.29
Cut out decayed brickwork in one brick wall in areas 0.5 to 1m2 and replace in gauged mortar						
commons	nr	7.60	72.20	24.96	14.57	111.73
facings	nr	8.80	83.60	48.24	19.78	151.62

	Unit	Hours	Hours £	Materials £	O & P £	Total £
Cut out vertical, horizontal or stepped cracks in half brick wall, replace average 350mm wide with new bricks in gauged mortar						
commons	m	2.75	26.13	3.00	4.37	33.49
facings	m	3.10	29.45	4.80	5.14	39.39
Cut out vertical, horizontal or stepped cracks in one brick wall, replace average 350mm wide with new bricks in gauged mortar						
commons	m	5.20	49.40	6.00	8.31	63.71
facings	m	5.80	55.10	9.60	9.71	74.41
Cut out defective brick-on-end soldier arch to half brick wall, and replace with new bricks in gauged mortar						
commons	m	3.00	28.50	3.90	4.86	37.26
facings	m	3.40	32.30	5.20	5.63	43.13
Cut out defective brick-on-end soldier arch to one brick wall, and replace with new bricks in gauged mortar						
commons	m	3.80	36.10	7.80	6.59	50.49
facings	m	4.20	39.90	10.40	7.55	57.85
Cut out defective terracotta air brick and replace						
215 x 65mm	nr	0.35	3.33	2.05	0.81	6.18
215 x 140mm	nr	0.50	4.75	2.90	1.15	8.80
215 x 215mm	nr	0.65	6.18	7.38	2.03	15.59
Rake out joints in gauged mortar, refix loose flashing and point up on completion						
horizontal	m	0.45	4.28	0.48	0.71	5.47
stepped	m	0.65	6.18	0.64	1.02	7.84
Rake out joints and point up in gauged mortar	m2	0.65	6.18	0.64	1.02	7.84

	Unit	Hours	Hours £	Materials £	O & P £	Total £
Cut out one course of half brick wall, insert hessian-based damp course 112mm wide and replace with new bricks in gauged mortar						
commons	m	2.00	19.00	1.22	3.03	23.25
facings	m	2.00	19.00	2.82	3.27	25.09
Cut out one course of one brick wall, insert hessian-based damp course 112mm wide and replace with new bricks in gauged mortar						
commons	m	3.25	30.88	4.50	5.31	40.68
facings	m	3.25	30.88	5.72	5.49	42.08
Cut out single block, replace with new in gauged mortar, thickness						
100mm	nr	0.30	2.85	2.10	0.74	5.69
140mm	nr	0.40	3.80	3.31	1.07	8.18
190mm	nr	0.50	4.75	4.32	1.36	10.43
255mm	nr	0.60	5.70	5.61	1.70	13.01
Rake out joints of random rubble walling and point up in gauged mortar						
flush pointing	m2	0.50	4.75	0.55	0.80	6.10
weather pointing	m2	0.55	5.23	0.55	0.87	6.64
Roofing						
Take up roof coverings from pitched roof						
tiles	m2	0.80	7.60	0.00	1.14	8.74
slates	m2	0.80	7.60	0.00	1.14	8.74
timber boarding	m2	1.00	9.50	0.00	1.43	10.93
metal sheeting	m2	0.20	1.90	0.00	0.29	2.19
flat sheeting	m2	0.30	2.85	0.00	0.43	3.28
corrugated sheeting	m2	0.30	2.85	0.00	0.43	3.28
underfelt	m2	0.10	0.95	0.00	0.14	1.09
Take up roof coverings from flat roof						
bituminous felt	m2	0.25	2.38	0.00	0.36	2.73
metal sheeting	m2	0.30	2.85	0.00	0.43	3.28
woodwool slabs	m2	0.50	4.75	0.00	0.71	5.46
firrings	m2	0.20	1.90	0.00	0.29	2.19

	Unit	Hours	Hours £	Materials £	O & P £	Total £
Take up roof coverings from pitched roof, carefully lay aside for reuse						
tiles	m2	1.10	10.45	0.00	1.57	12.02
slates	m2	1.10	10.45	0.00	1.57	12.02
metal sheeting	m2	0.50	4.75	0.00	0.71	5.46
flat sheeting	m2	0.60	5.70	0.00	0.86	6.56
corrugated sheeting	m2	0.60	5.70	0.00	0.86	6.56
Take up roof coverings from flat roof, carefully lay aside for reuse						
metal sheeting	m2	0.60	5.70	0.00	0.86	6.56
woodwool slabs	m2	0.80	7.60	0.00	1.14	8.74
Inspect roof battens, refix loose and replace with new, size 38 x 25mm						
25% of area						
250mm centres	m2	0.14	1.33	0.40	0.26	1.99
450mm centres	m2	0.12	1.14	0.30	0.22	1.66
600mm centres	m2	0.10	0.95	0.25	0.18	1.38
50% of area						
250mm centres	m2	0.26	2.47	0.80	0.49	3.76
450mm centres	m2	0.16	1.52	0.60	0.32	2.44
600mm centres	m2	0.18	1.71	0.50	0.33	2.54
75% of area						
250mm centres	m2	0.36	3.42	1.20	0.69	5.31
450mm centres	m2	0.26	2.47	0.90	0.51	3.88
600mm centres	m2	0.22	2.09	0.75	0.43	3.27
100% of area						
250mm centres	m2	0.44	4.18	1.60	0.87	6.65
450mm centres	m2	0.32	3.04	1.20	0.64	4.88
600mm centres	m2	0.38	3.61	1.00	0.69	5.30
Take off single slipped slate and refix	nr	1.00	9.50	0.00	1.43	10.93
Remove single broken slate, renew with new Welsh blue slate						
405 x 255mm	nr	1.20	11.40	1.88	1.99	15.27
510 x 255mm	nr	1.20	11.40	2.35	2.06	15.81
610 x 305mm	nr	1.20	11.40	4.26	2.35	18.01

	Unit	Hours	Hours £	Materials £	O & P £	Total £
Remove slates in area approximately 1m2 and replace with Welsh blue slates previously laid aside	nr	1.60	15.20	0.00	2.28	17.48
Remove slates in area approximately 1m2 and replace with Welsh blue slates previously laid aside						
405 x 255mm	nr	1.80	17.10	0.00	2.57	19.67
510 x 255mm	nr	1.70	16.15	49.21	9.80	75.16
610 x 305mm	nr	1.60	15.20	52.39	10.14	77.73
Remove double course at eaves and fix new Welsh blue slates						
405 x 255mm	m	0.70	6.65	25.62	4.84	37.11
510 x 255mm	m	0.70	6.65	29.14	5.37	41.16
610 x 305mm	m	0.70	6.65	34.26	6.14	47.05
Remove single verge undercloak course and renew						
405 x 255mm	m	0.90	8.55	18.91	4.12	31.58
510 x 255mm	m	0.90	8.55	21.41	4.49	34.45
610 x 305mm	m	0.90	8.55	22.04	4.59	35.18
Remove single slipped tile and refix	nr	0.30	2.85	0.00	0.43	3.28
Remove single broken tile and renew						
Marley plain tile	nr	0.30	2.85	0.56	0.51	3.92
Marley Ludlow Plus tile	nr	0.30	2.85	0.79	0.55	4.19
Marley Modern tile	nr	0.30	2.85	1.22	0.61	4.68
Redland Renown tile	nr	0.30	2.85	1.22	0.61	4.68
Redland Norfolk tile	nr	0.30	2.85	1.09	0.59	4.53
Remove tiles in area approximately 1m2 and replace with tiles previously laid aside						
Marley plain tile	nr	1.80	17.10	28.12	6.78	52.00
Marley Ludlow Plus tile	nr	1.20	11.40	11.72	3.47	26.59
Marley Modern tile	nr	1.10	10.45	12.41	3.43	26.29
Redland Renown tile	nr	1.10	10.45	12.07	3.38	25.90
Redland Norfolk tile	nr	1.15	10.93	14.92	3.88	29.72
Take off defective ridge capping and refix including pointing in mortar	m	1.10	10.45	1.81	1.84	14.10

	Unit	Hours	Hours £	Materials £	O & P £	Total £
Carpentry and joinery						
Take down, cut out or demolish stuctural timbers and load into skips						
structural timbers						
50 x 100mm	m	0.10	0.95	0.00	0.14	1.09
50 x 150mm	m	0.12	1.14	0.00	0.17	1.31
75 x 100mm	m	0.14	1.33	0.00	0.20	1.53
75 x 150mm	m	0.16	1.52	0.00	0.23	1.75
100 x 150mm	m	0.18	1.71	0.00	0.26	1.97
100 x 200mm	m	0.20	1.90	0.00	0.29	2.19
Take down, cut out or demolish non-stuctural items and load into skips						
roof boarding	m2	0.28	2.66	0.00	0.40	3.06
floor boarding	m2	0.22	2.09	0.00	0.31	2.40
stud partition plasterboard both sides	m2	0.75	7.13	0.00	1.07	8.19
skirtings and grounds						
100mm high	m	0.08	0.76	0.00	0.11	0.87
150mm high	m	0.09	0.86	0.00	0.13	0.98
200mm high	m	0.10	0.95	0.00	0.14	1.09
rails						
50mm high	m	0.05	0.48	0.00	0.07	0.55
75mm high	m	0.06	0.57	0.00	0.09	0.66
100mm high	m	0.07	0.67	0.00	0.10	0.76
fittings						
wall cupboards	nr	0.25	2.38	0.00	0.36	2.73
floor units	nr	0.20	1.90	0.00	0.29	2.19
sink units	nr	0.25	2.38	0.00	0.36	2.73
staircase, 900mm wide						
straight flight	nr	4.00	38.00	0.00	5.70	43.70
landing	nr	1.50	14.25	0.00	2.14	16.39
doors, frames and linings						
single, internal	nr	0.40	3.80	0.00	0.57	4.37
single, external	nr	0.60	5.70	0.00	0.86	6.56
double, internal	nr	0.60	5.70	0.00	0.86	6.56
double, external	nr	0.80	7.60	0.00	1.14	8.74
windows						
casement, 1200 x 900mm	nr	0.50	4.75	0.00	0.71	5.46
casement, 1800 x 900mm	nr	0.60	5.70	0.00	0.86	6.56
sash, 900 x 1500mm	nr	0.80	7.60	0.00	1.14	8.74
sash, 1800 x 900mm	nr	0.90	8.55	0.00	1.28	9.83

	Unit	Hours	Hours £	Materials £	O & P £	Total £
Take down, cut out or demolish non-stuctural items (cont'd)						
ironmongery						
bolt	nr	0.20	1.90	0.00	0.29	2.19
deadlock	nr	0.25	2.38	0.00	0.36	2.73
mortice lock	nr	0.35	3.33	0.00	0.50	3.82
mortice latch	nr	0.35	3.33	0.00	0.50	3.82
cylinder lock	nr	0.25	2.38	0.00	0.36	2.73
door closer	nr	0.35	3.33	0.00	0.50	3.82
casement stay	nr	0.15	1.43	0.00	0.21	1.64
casement fastener	nr	0.15	1.43	0.00	0.21	1.64
toilet roll holder	nr	0.15	1.43	0.00	0.21	1.64
shelf bracket	nr	0.15	1.43	0.00	0.21	1.64
Cut out defective joists or rafters and replace with new						
50 x 75mm	nr	0.24	2.28	1.60	0.58	4.46
50 x 100mm	nr	0.28	2.66	1.92	0.69	5.27
50 x 150mm	nr	0.32	3.04	2.28	0.80	6.12
75 x 100mm	nr	0.33	3.14	2.60	0.86	6.60
75 x 150mm	nr	0.34	3.23	2.92	0.92	7.07
Cut out defective skirting and renew						
softwood, 75mm high	m	0.27	2.57	1.87	0.67	5.10
softwood, 100mm high	m	0.30	2.85	2.17	0.75	5.77
hardwood, 150mm high	m	0.38	3.61	9.24	1.93	14.78
Ease, adjust and oil						
door	nr	0.60	5.70	0.10	0.87	6.67
casement window	nr	0.40	3.80	0.10	0.59	4.49
sash window including renewing cords	nr	1.50	14.25	5.78	3.00	23.03
Take up defective flooring and replace with 25mm thick plain edged boarding						
areas less than 1m2	m2	1.15	10.93	11.05	3.30	25.27
areas more than 1m2	m2	1.10	10.45	11.05	3.23	24.73
Refix loose floorboards including punching in protruding nails	m2	0.20	1.90	0.00	0.29	2.19

	Unit	Hours	Hours £	Materials £	O & P £	Total £
Take down existing door, lay aside, piece up frame or lining where butts removed, rehang door to opposite hand on existing butts and hardware						
single, internal	nr	1.50	14.25	2.10	2.45	18.80
single, external	nr	1.70	16.15	2.10	2.74	20.99
Finishings						
Take down plasterboard sheeting from						
studded walls	m2	0.35	3.33	0.00	0.50	3.82
ceilings	m2	0.40	3.80	0.00	0.57	4.37
Hack off plaster from						
walls	m2	0.25	2.38	0.00	0.36	2.73
ceiling	m2	0.30	2.85	0.00	0.43	3.28
Make good plaster to walls where wall removed						
100mm wide	m2	0.95	9.03	1.90	1.64	12.56
150mm wide	m2	1.05	9.98	2.00	1.80	13.77
200mm wide	m2	1.15	10.93	2.10	1.95	14.98
Make good plaster to ceilings where wall removed						
100mm wide	m2	1.10	10.45	1.90	1.85	14.20
150mm wide	m2	1.15	10.93	2.00	1.94	14.86
200mm wide	m2	1.20	11.40	2.10	2.03	15.53
Cut out cracks in plasterwork and make good						
walls	m2	0.35	3.33	0.40	0.56	4.28
ceiling	m2	0.40	3.80	0.40	0.63	4.83
Hack off wall tiling and make good surface	m2	0.75	7.13	3.72	1.63	12.47
Plumbing						
Remove sanitary fittings and supports, seal off supply and waste pipes and prepare to receive new fittings						
bath	nr	3.50	43.75	10.00	8.06	61.81
sink	nr	3.50	43.75	10.00	8.06	61.81

	Unit	Hours	Hours £	Materials £	O & P £	Total £
Remove sanitary fittings and supports (cont'd)						
lavatory basin	nr	3.25	40.63	6.00	6.99	53.62
WC	nr	3.75	46.88	6.00	7.93	60.81
shower cubicle	nr	4.00	50.00	8.00	8.70	66.70
bidet	nr	3.25	40.63	6.00	6.99	53.62
Take down length of existing gutter, prepare ends and install new length of gutter to existing brackets						
PVC-U						
76mm	nr	0.60	7.50	9.75	2.59	19.84
112mm	nr	0.65	8.13	11.00	2.87	21.99
cast iron						
100mm	nr	1.12	14.00	22.68	5.50	42.18
Take down existing brackets from fascias and replace with galvanised steel repair brackets at 1m centres	m	0.30	3.75	2.81	0.98	7.54
Take down length of existing pipe, prepare ends and install new length of pipe						
PVC-U						
68mm diameter	nr	0.75	9.38	8.60	2.70	20.67
68mm square	nr	0.75	9.38	9.46	2.83	21.66
cast iron						
75mm	nr	0.90	11.25	34.56	6.87	52.68
100mm	nr	1.05	13.13	48.96	9.31	71.40
Take down fittings and install new						
PVC-U						
68mm diameter bend	nr	0.50	6.25	3.02	1.39	10.66
68mm offset	nr	0.50	6.25	7.10	2.00	15.35
68mm branch	nr	0.50	6.25	7.22	2.02	15.49
68mm shoe	nr	0.75	9.38	6.56	2.39	18.33
cast iron						
75mm diameter bend	nr	0.60	7.50	6.53	2.10	16.13
75mm offset	nr	0.60	7.50	15.29	3.42	26.21
75mm branch	nr	0.60	7.50	7.15	2.20	16.85
75mm shoe	nr	0.60	7.50	15.28	3.42	26.20

	Unit	Hours	Hours £	Materials £	O & P £	Total £
Take down fittings (cont'd)						
cast iron						
100mm diameter bend	nr	0.70	8.75	21.12	4.48	34.35
100mm offset	nr	0.70	8.75	29.91	5.80	44.46
100mm branch	nr	0.70	8.75	8.74	2.62	20.11
100mm shoe	nr	0.70	8.75	17.88	3.99	30.62
Cut out 500mm length of copper pipe, install new pipe with compression fittings at each end						
15mm	nr	0.80	10.00	4.81	2.22	17.03
22mm	nr	0.90	11.25	7.26	2.78	21.29
Take off existing radiator valve and replace with new	nr	0.90	11.25	9.80	3.16	24.21
Take out existing galvanised steel water storage tank and install new plastic tank complete with ball valve, lid and insulation including cutting holes, make up pipework and connectors to existing pipework						
68 litres	nr	3.70	46.25	78.28	18.68	143.21
114 litres	nr	3.80	47.50	68.20	17.36	133.06
182 litres	nr	4.90	61.25	81.07	21.35	163.67
227 litres	nr	5.00	62.50	119.97	27.37	209.84
Glazing						
Hack out glass and remove	m2	0.45	4.28	0.00	0.64	4.92
Clean rebates, remove sprigs or clips and prepare for reglazing	m	0.20	1.90	0.00	0.29	2.19
Painting						
Prepare, wash down painted surfaces, rub down to receive new paintwork						
brickwork	m2	0.14	1.33	0.00	0.20	1.53
blockwork	m2	0.15	1.43	0.00	0.21	1.64
plasterwork	m2	0.12	1.14	0.00	0.17	1.31

	Unit	Hours	Hours £	Materials £	O & P £	Total £
Prepare, wash down previously painted wood surfaces, rub down to receive new paintwork surfaces						
over 300mm girth	m2	0.28	2.66	0.00	0.40	3.06
isolated surfaces not exceeding 300mm girth	m	0.10	0.95	0.00	0.14	1.09
isolated surfaces not exceeding 0.5m2	nr	0.12	1.14	0.00	0.17	1.31
Prepare, wash down previously painted metal surfaces, rub down to receive new paintwork surfaces						
over 300mm girth	m2	0.28	2.66	0.00	0.40	3.06
isolated surfaces not exceeding 300mm girth	m	0.10	0.95	0.00	0.14	1.09
isolated surfaces not exceeding 0.5m2	nr	0.12	1.14	0.00	0.17	1.31
Wallpapering						
Strip off one layer of existing paper, stop cracks and rub down to receive new paper						
woodchip						
walls	m2	0.18	1.71	0.00	0.26	1.97
walls in staircase areas	m2	0.20	1.90	0.00	0.29	2.19
ceilings	m2	0.22	2.09	0.00	0.31	2.40
ceilings in staircase areas	m2	0.24	2.28	0.00	0.34	2.62
vinyl						
walls	m2	0.23	2.19	0.00	0.33	2.51
walls in staircase areas	m2	0.25	2.38	0.00	0.36	2.73
ceilings	m2	0.27	2.57	0.00	0.38	2.95
ceilings in staircase areas	m2	0.29	2.76	0.00	0.41	3.17
standard patterned						
walls	m2	0.21	2.00	0.00	0.30	2.29
walls in staircase areas	m2	0.23	2.19	0.00	0.33	2.51
ceilings	m2	0.25	2.38	0.00	0.36	2.73
ceilings in staircase areas	m2	0.27	2.57	0.00	0.38	2.95

Part Six

PLANT AND TOOL HIRE

	24 hours	Additional 24 hours	Week
	£	£	£
TOOL AND EQUIPMENT HIRE			
These selected rates are based on average hire charges made by hire firms in UK. Check your local dealer for more information. These prices exclude VAT.			
CONCRETE AND CUTTING EQUIPMENT			
Concrete mixers			
Petrol, with stand	12.00	6.00	25.00
Electric, with stand	11.00	5.00	22.00
Bulk mixer	26.00	13.00	52.00
Vibrating pokers			
Pokers			
petrol	40.00	22.00	80.00
electric	30.00	16.00	65.00
air poker, 50mm	26.00	13.00	50.00
air poker, 75mm	28.00	14.00	52.00
Power floats			
Floats			
power float, petrol	35.00	18.00	70.00
Vibrating screeds			
Screed units			
with 5m beam	50.00	18.00	94.00
with 3.25 - 5.20 beam	54.00	20.00	98.00
roller screed unit	100.00	40.00	200.00
Floor preparation units			
Floor saw, petrol			
350mm	50.00	25.00	100.00
450mm	65.00	32.00	125.00

	24 hours	Additional 24 hours	Week
	£	£	£
Scabbler, hand held	20.00	10.00	40.00
Diamond concrete planer	48.00	18.00	90.00
Air needle gun	20.00	10.00	38.00
Disc cutters			
Cutters			
electric, 300mm	20.00	8.00	40.00
two stroke, 300mm	22.00	9.00	44.00
two stroke, 350mm	25.00	12.00	55.00
electric wall chasers	34.00	15.00	60.00
Block and slab splitters			
Splitters			
clay	48.00	24.00	95.00
block	22.00	10.00	45.00
slab	35.00	18.00	70.00
Saws			
Electric, 150mm	34.00	18.00	65.00
Electric, 300mm	38.00	22.00	70.00
ACCESS AND SITE EQUIPMENT			
Ladders			
Double ladder, alloy			
4m	12.00	6.00	24.00
6m	14.00	7.00	28.00
9m	16.00	8.00	32.00
Triple ladder, alloy			
9m	14.00	7.00	28.00
Roof ladder			
5m	14.00	7.00	30.00
Rope operated			
11m	32.00	16.00	64.00
13m	38.00	20.00	74.00
16m	46.00	23.00	88.00

	24 hours	Additional 24 hours	Week
	£	£	£
Props			
Shoring props			
type 0	0.00	0.00	3.50
type 1	0.00	0.00	3.50
type 2	0.00	0.00	3.50
type 3	0.00	0.00	3.50
type 4	0.00	0.00	3.50
Rubbish chutes			
Chutes			
1m section	5.00	3.00	10.00
funnel	5.00	3.00	10.00
frame	5.00	3.00	10.00
Trestles and staging			
Staging			
2.4m	12.00	6.00	24.00
3.6m	18.00	9.00	36.00
4.8m	20.00	10.00	40.00
Painters' trestle			
1.8m	8.00	4.00	18.00
Alloy towers			
Single width, height			
2.30m	30.00	15.00	60.00
3.20m	36.00	18.00	72.00
3.73m	38.00	19.00	76.00
4.20m	42.00	21.00	84.00
5.20m	56.00	28.00	112.00
6.52m	60.00	30.00	120.00
7.45m	68.00	34.00	136.00
7.91m	74.00	37.00	144.00
8.20m	80.00	40.00	160.00
9.30m	86.00	43.00	172.00
10.20m	102.00	50.00	200.00
Full width, height			
2.34m	36.00	18.00	72.00
3.27m	40.00	20.00	80.00
3.73m	42.00	21.00	84.00
4.66m	50.00	25.00	100.00
5.59m	54.00	27.00	108.00

	24 hours	Additional 24 hours	Week
	£	£	£
Full width, height (cont'd)			
6.52m	58.00	29.00	116.00
7.45m	64.00	32.00	128.00
7.91m	66.00	33.00	132.00
8.84m	74.00	37.00	148.00
10.23m	82.00	41.00	164.00
11.16m	90.00	45.00	18.00
12.55m	104.00	52.00	208.00
13.94m	116.00	58.00	232.00
14.87m	122.00	61.00	244.00
15.80m	126.00	63.00	252.00
LIFTING AND MOVING			
Sack truck	7.00	3.50	14.00
Pallet truck	26.00	13.00	52.00
Plasterboard jąck lift	15.00	7.50	30.00
Stair climber	12.00	6.00	24.00
Rubble truck	18.00	9.00	36.00
COMPACTION			
Plate compactors			
Compactors			
350 x 550mm, petrol	50.00	25.00	100.00
700 x 320mm, diesel	56.00	28.00	112.00
Vibrating rollers			
Mini			
Large	50.00	25.00	100.00
	60.00	30.00	120.00
BREAKING AND DEMOLITION			
Breakers			
Hydraulic			
diesel	68.00	34.00	136.00
petrol	74.00	37.00	128.00
medium duty, electric	20.00	10.00	40.00
heavy duty, electric	42.00	20.00	80.00
Air breaker, medium	30.00	15.00	56.00
Air breaker, heavy	32.00	16.00	60.00

	24 hours	Additional 24 hours	Week
	£	£	£
POWER TOOLS			
Drills			
Cordless drill	15.00	8.00	30.00
Cordless impact	18.00	10.00	36.00
Two speed impact	8.00	4.00	16.00
Rotary drill, 16mm	22.00	11.00	44.00
Rotary drill, 20mm	24.00	12.00	48.00
Right angle drill	18.00	10.00	36.00
Combi hammer			
light duty	14.00	7.00	28.00
meduim duty	20.00	10.00	40.00
heavy duty	22.00	11.00	44.00
Grinders			
Angle grinder			
100mm	12.00	6.00	24.00
125mm	12.00	6.00	24.00
230mm	13.00	7.00	26.00
300mm	24.00	12.00	28.00
Saws			
Reciprocating saw			
standard	20.00	10.00	40.00
heavy duty	22.00	11.00	44.00
Circular saw			
150mm	14.00	5.60	28.00
230mm	16.00	6.40	32.00
Door trimmer	29.00	11.60	58.00
Woodworking			
Plane, 3.25in	15.00	8.00	30.00
Router	15.00	8.00	30.00
Worktop jig	12.00	6.00	24.00

	24 hours £	Additional 24 hours £	Week £
Fixing equipment			
Cordless nailing gun	22.00	8.00	40.00
Air nail gun with compressor	50.00	20.00	100.00
Cartridge hammer	18.70	6.80	34.00
Electric screwdriver	10.50	4.20	21.00
Sanders			
Floor	31.00	12.40	62.00
Floor edger	21.00	8.40	42.00
Orbital	10.50	4.20	21.00
WELDING AND GENERATORS			
Generating			
Generators			
petrol, 2KVA	21.00	8.40	42.00
petrol, 3KVA	23.50	9.40	47.00
silenced, 2KVA	31.50	12.60	63.00
silenced, 15KVA	189.00	63.00	315.00
PUMPING EQUIPMENT			
Pumps			
Submersible			
2.00in	21.00	8.40	42.00
Puddle pump, electric	26.50	10.60	53.00
Water pump, petrol	26.00	10.40	52.00
PAINTING AND DECORATING			
Surface preparation			
Wallpaper stripper, electric	11.00	4.40	22.00
Damp proofing			
Injection machine	32.00	12.80	64.00
Woodworn spray attachment	3.00	1.20	6.00